The Philippines and the International Monetary Fund Negotiations on Petroleum and Imports

The Philippines and the International Monetary Fund Negotiations on Petroleum and Imports

Toward a Theory of Negotiation

Kenneth Faulve-Montojo

LEXINGTON BOOKS

Lanham • Boulder • New York • London

Published by Lexington Books
An imprint of The Rowman & Littlefield Publishing Group, Inc.
4501 Forbes Boulevard, Suite 200, Lanham, Maryland 20706
www.rowman.com

Unit A, Whitacre Mews, 26-34 Stannary Street, London SE11 4AB

British Library Cataloguing in Publication Information Available

Library of Congress Cataloging-in-Publication Data

Names: Faulve-Montojo, Kenneth, author.
Title: The Philippines and the International Monetary Fund negotiations on petroleum and imports : toward a theory of negotiation / Kenneth Faulve-Montojo.
Description: Lanham, Maryland : Lexington Books, [2018] | Includes bibliographical references and index.
Identifiers: LCCN 2018029953 (print) | LCCN 2018031955 (ebook) | ISBN 9781498532709 (Electronic) | ISBN 9781498532693 (cloth: alk. paper)
Subjects: LCSH: International Monetary Fund—Philippines. | Petroleum products—Prices—Philippines. | Free trade—Philippines. | Philippines—Economic policy. | Negotiation—Philippines—Methodology.
Classification: LCC HG3881.5.I58 (ebook) | LCC HG3881.5.I58 F38 2018 (print) | DDC 338.2/32709599—dc23
LC record available at https://lccn.loc.gov/2018029953

∞™ The paper used in this publication meets the minimum requirements of American National Standard for Information Sciences—Permanence of Paper for Printed Library Materials, ANSI/NISO Z39.48-1992.

Printed in the United States of America

Contents

Introduction

Since the early 1960s, Bretton Woods twins and the Philippines have collaborated in formulating economic policies. The International Monetary Fund (IMF) and the World Bank made their reputation addressing stabilization and development issues; to address these issues, the twins designing programs that emphasized the market, which typically included monetary and fiscal prudence, structural investments, and reform. The Philippines, up to the early 1980s, collaborated with these International Financial Institutions (IFIs) to design programs that addressed their needs as well as receive financial and technical assistance to put these designs into action. These conditionalities were often one sided, the Philippines took the money, but often failed to implement many of the painful reforms and often politically sensitive policies. When the international debt crisis hit in the early 1980s, IMF, the primary focus of this work between the twins, watched its leverage increase as the Philippines desperately needed financing to resolve its macroeconomic imbalances. But this leverage did not come without struggles.

For counties like the Philippines that face mounting economic challenges and troubles, not to mention what seems like infinite political demands as the country democratizes, IMF packages typically address both stabilization and development issues. While the two are typically viewed as separate issues, they overlap as development problems can create macroeconomic imbalances, and reversing the process, balance of payments issues can derail development programs. A recurring issue that the IMF and the Government of the Philippines (GOP) consistently address is petroleum product pricing. Since the mid-1960s, the government has manipulated these prices to achieve its multiple goals of economic growth, efficient allocation of resources, revenue generation, price stability and equity, and demand management. By keeping domestic prices fixed in an environment of typically fluctuating international crude oil costs, the GOP has often found itself attempting to address that constant question in political economy when economic imbalances occur: who bears the costs of adjustment? In IMF-GOP stabilization packages negotiations, the IFI insists that private actors, not public ones, should bear the costs. For the GOP to receive access to financing needed to pay for its other obligations, it must typically accept this provision. But acceptance of policy is easy; implementation is hard. Petroleum price increases have a reverberation effect on other commodities, labor costs, transportation fares and fees, and manufacturing costs. Indeed, the GOP finds that adjusting the domestic price of petroleum prices brings on a new set of problems, especially in the context of democratization under the Corazon Aquino period. In some cases, the GOP succeeds in increasing and sustaining the prices. In other instance, the GOP found itself having to reverse the increases in whole or in part. What determined the outcome of the IMF-GOP negotiations on petroleum prices? It was the process of bargaining with various public and private veto players.

The bargaining process also has an impact on trade liberalization policies. Since the post–World War II years, the Philippines had an exchange rate regime that vacillated between tight and loose. In the end, the economy resembled the dual track economy found in many Latin American countries where a small fraction of the economy consists of an import substituting industrialization sector while the major remains mired in the production of agriculture and other products. These import-substituting industries were typically shielded from international competition by tariffs and nontariff barriers to trade, producing an inefficient sector that may or may not serve to generate the economic growth needed in a country. When IMF takes a seat at the negotiation table, it often addresses World Bank development issues as a way to generate foreign exchange to restore or fill the gap left by macroeconomic imbalances. While the exchange rate system plays a critical component in development, the GOP must address import liberalization. Much like petroleum pricing veto players, many interests are aligned with maintaining the status quo. And like the petroleum pricing issue, to announce import liberalization policies is relatively easy; to implement and sustain the policies is hard. The key toward determining the outcome is conditioned on the negotiation process.

Relying on the long history of negotiations between the two parties, this book examines five economic programs crafted by the GOP and the IMF in 1984, 1986, 1989, 1991, and 1994, programs that were formally and informally ratified by internal or domestic constituencies. Each case is disaggregated into two issues: petroleum price and import reform policies. The 1989 economic program had three petroleum episodes, as a result, I break it down to three observations (1989a, 1989b, and 1989c). Similarly, the program had two import episodes that I break down to two observations (1989a and

1989b). This approach enlarges the number of observations from ten to thirteen, allowing for the comparison of conditions both between cases and within cases. In the thirteen cases, Philippine and IMF negotiators regularly used particular strategies to achieve favorable terms to an agreement. By relying on such strategies, GOP and IMF negotiators achieved much of what they wanted in shaping the terms of import policies. In contrast, in their negotiations over the terms of petroleum pricing policy, both players came away with little of what they wanted. Given the consistent use of particular strategies in both issues, why did they work in some instances, but fail to work in others? It is in this area where this inquiry seeks to make its contribution, advancing the theoretical foundations of international political economy by addressing the following questions. In negotiations between the IMF and developing countries, what accounts for the success or failure of strategies used to secure a favorable international agreement? Which tactic is more effective, with what type of players and when? Do sequencing tactic patterns spur policy change better than simultaneous patterns? I argue that sequencing patterns successes and failures are determined by the leverage of and signals provided by international veto players and the interaction between formal and informal domestic institutions.

This study has two overarching and two intermediate goals.[1] The first broad goal is to provide a valid and descriptive account of "causal inferences" of economic negotiations and their bargains between the IMF and emerging market countries.[2] The second overarching goal is the refinement of theory, to help establish the important variables and processes in the descriptive accounts of economic negotiations.[3] Harry H. Eckstein refers to this end as "foreknowledge," the understanding of where rules do and do not hold.[4] We use the concept of causal inferences where we make a causal connection based on the data observed.[5] The study searches for the regularity of rules or "rulefulness," as termed by Eckstein, which is "the discovery of rules that phenomena observe in the concrete world, as players do in games or logicians in logic." These rules "may describe sequences like causal paths."[6] Thought of in a similar way, I look for recurring patterns of behavior, in a general sense, between the IMF and a recipient country. More narrowly, I examine the general recurring patterns regarding the domestic implementation of stabilization policies (i.e., petroleum prices) and structural adjustment (import reforms). These causal paths, as I use them, shy away from a general theory such as realism that seeks to explain both conflict and cooperation, conflict studies, or in general international economic negotiations. Bargaining on international trade, global warming, human trafficking, and IMF stabilization programs operate around different landscapes, institutions, and preferences. Furthermore, such bargaining may produce different types of outcomes. As a result, it allows us to engage in typological theorizing, producing a subclass of IMF-GOP outcomes. If these subclasses are deemed to be theoretically significant, empirically interesting, and policy practical, then future researchers will explore the instances of the subtype.[7] As a whole, these studies will help build the foundation of a midrange theory that can explore and develop IMF-recipient country conditionality negotiations.

The first intermediate goal is to refine international economic negotiation theory to better understand the process and players and how such it could be affected by causal mechanisms that can alter outcomes. The second is to provide more explanatory range and applicability to the Two-Level game by explaining these vagaries through theory development, taking the Two-Level game and providing it with more precise concepts, giving it more explanatory, descriptive, and predictive power.[8]

This research is critical on numerous points. First, it provides decision makers with policy relevant-information about instruments at their disposal and the conditions under which they are effective and marginalized.[9] Second, this research examines how decision makers can reform economically inefficient systems, serving as a critical and necessary, but not sufficient, step toward poverty reduction and ultimately the promotion of human development.[10] These two critical issues can be extended not only to the relationship, but could also be applied across a broader spectrum of IMF-recipient country negotiations and among other emerging market economies. Finally, I provide more information about the causal mechanisms regarding international economic negotiations and bargaining. As defined by Alexander L. George and Andrew Bennett, causal mechanisms are the "unobservable physical, social, or psychological process through which agents with causal capacities operate, but only in specific contexts or conditions, to transfer energy, information, or matter other entities. In doing so, the causal agent changes the affected entity's characteristics, capacities, or propensities in ways that persist until subsequent causal mechanisms act upon it."[11]

THE IMF AND GOP: JUSTIFICATION AND FRAMEWORKS

Justification

The key to this study lies in the consistent pattern inherent in the IMF-GOP relationship. From 1962 to 2000, IMF and Philippine negotiators have produced twenty-three economic programs that span over twenty-five years. This relationship prompted the IMF executive board to designate the Philippines as one of the most "extreme cases" as "a prolonged user" of IMF resources.[12] From 1962 to 1983, the IMF had limited bargaining leverage against the GOP. It could strongly recommend the government adopt austerity policies, but it lacked the "capacity" to ensure government implementation. By 1984, the IMF took a more forceful approach toward negotiations and applied various strategies to pressure the GOP to adopt and sustain reforms, especially regarding petroleum price adjustments and import reform policies.[13]

These areas have both certain similarities on economic and political dimensions and variations on the political determinants of constraints and opportunities. The IMF requires prior action, changes made first before it submits to its Executive Board the Letter of Intent (LOI).[14] It ensures upfront implementation where the IMF may have doubts and tests the government commitment toward implementation.[15] Petroleum price adjustments are instrumental toward reducing government deficits, and import reform is critical toward resource efficiency for export production.[16] Furthermore, these issues generate considerable interest from the IMF principals. In the 1980s, the World Bank pushed for import reform as part of its Structural Adjustment Loan (SAL) program while industrialized countries pushed for more open markets. Key states that financed the Philippines have preferences for petroleum price adjustments to ensure repayment of commercial loans that originate from their countries. Domestically, both areas also insulate particular groups from foreign competition and the fluctuation of international petroleum prices. Both are also placed in the context of distributive (imports) and regulatory systems (petroleum).[17] In reality, both imports and petroleum have distributive and regulatory elements. Similarly, both issues are addressed by the same Chief Negotiator (CNs), technocrats (the IMF staff and GOP negotiators), and the same Chief of Governments (COGs). These variables, in many cases, are constant across the issues, and in other cases, over time, permitting this work to examine the negotiation processes, not the outcomes, as they produced over time different ordinal results.

These results cannot be explained by international cooperation studies. Hegemony, under the realism theory, is the predominant framework that emphasizes the application of power and geography at the international system level of analysis where under the world of anarchy, states pursue power and its forms to enhance their security. In the IMF context, we can hypothesize that US allies, especially during the Cold War, would receive more IMF loans and with minimal conditionality policies or less reform.[18] Using the deterministic model, we test the theory. This outcome should be observed in the 1984 program when the anti-communist Reagan Administration supported the Ferdinand Marcos authoritarian regime to maintain US presence with the naval and air force base, of geographic significance as they served as the launching off point for US defense interests in the Asian frontier. This outcome, however, did not materialize as the IMF staff demanded tight conditionalities with prior actions. The GOP successfully implemented petroleum price increases and reversed limited success on import reforms. We expected a different outcome as the independent variable was present; as a result, we can rule out sufficient causation. The same pattern also exists with the other observations. We also expect limited import reform in 1986 and petroleum prices reform in 1989b and 1991 under the nascent Aquino regime, but these outcomes failed to materialize. The deductive model lacks the precision to capture the outcomes. Even after the end of the Cold War, the realist-oriented argument predicts that the IMF should have presented terms amendable to the GOP. However, in 1994, the IMF staff continued to require significant changes to the status quo, the creation of a new central bank, petroleum price adjustments, and import reforms. The realist framework, while parsimonious, fails to explain sufficiently the outcomes. We can then conclude that power and geography as variables are overdetermined.

Another international cooperation theory emphasizes the formal structure of principal-agent relations in which the US participates within the institutions.[19] Under this framework, principals shape the nature of IMF policies thought the use of the IMF backdoor. International actors would ensure that the IMF would secure their interests. In the context of import reforms, using this framework, the hypothesis is unsubstantiated. Industrialized countries and the World Bank pushed for trade reforms, a prior action supported by the IMF, but the GOP had no reforms in two of the six cases (see Table I.1).

The study finds the same uneven pattern in the case of petroleum reform as the GOP successfully pushed through reforms in only three of seven attempts (see Table I.1). Both system level approaches fail to examine how the IMF interacts with domestic politics.

The limitation with these deductive works stems from their orientation and their purposes. Kenneth N. Waltz articulated the levels of analysis typology where variables used to explain state behavior are categorized according to individual, domestic, and international or structural types.[20] At a rudimentary explanatory level, variables at each category could be explained as the source

Table I.1 Petroleum and Imports IMF-GOP Negotiation Outcomes and Compliance

IMF Program Date (Government)	*Petroleum Result*	*Import Results*
December 1984 (Marcos)	Reform	Reform
October 1986 (Aquino)	No reform	Reform
March 1989a (Aquino)	No reform	Reform
November 1989b (Aquino)	Reform	No reform
November 1989c (Aquino)*	Reform	N/A
February 1991 (Aquino)	Reform	Reform
June 1994 (Ramos)	No reform	Reform

*The date for petroleum price increase was September 1990.

of state behavior. For instance, individual level variables determine state behavior why states go to war. This causal argument, however, is typically overdetermined and reductionist—the part explains the whole.[21] Waltz concluded that individuals and domestic elements provide critical links toward determining state policy, but without the international landscape, scholars could not "assess their importance or predict their results."[22] In addition to their explanatory capabilities, variables within these categories also provide a different picture of the outcome. Studies from a structural level of analysis provide a more general portrait of international relations patterns, broad sweeping reciprocating patterns of interaction between the international structure and the units.[23] In contrast, individual and domestic level analysis provides more depth and detail.[24] By design, geography and power and institutional deductive theories are incapable of capturing and explaining the outcomes this book seeks to explain; they cannot fulfill this function. Ideally, a clearer and more comprehensive picture of state behavior would entail a sequenced level of analysis approach, from system to domestic to individual, or reverse; however, even this remedy maintains its bias as it would place emphasis at one level and leave the others as residuals.[25] This limitation would be accounted for by the Two-Level game model/metaphor articulated by Robert Putnam.[26]

Labeling the Two-Level game as a theory would be an overstatement. It started out as a metaphor, a simplification of real life international bargaining and cooperation. From there, it became a model, more sophisticated than the metaphor, but still less complex, a "stripped down" version of real life,[27] as scholars began to test the boundaries by which they could apply, measure, and specify the features of the model. Because of its "fuzziness" in some areas, it has yet to be deemed a theory, a more specified and defined version of the model that can be subject to rigorous testing.[28] Through inductive driven research and modeling, the Two-Level game model can shift from model to theory.[29]

Derived from the literature on international bargaining, the metaphor clarifies much of the dynamics involved in international negotiations.[30] The model delineates between international and domestic variables and illustrates how interaction between the two levels leads to (un)successful international agreements. In this respect, the Putnam model has been found to provide a more satisfactory approach toward describing and explaining international cooperation than analyses that rely solely on systemic or domestic level factors, or a sequential combination of the two. Eugénia da Conceição-Heldt concludes that the Two-Level framework can accomplish several goals. First, it can qualitatively and through a cross-country study examine institutional arrangements that explain either cooperation or rejection of international deals. Second, research should examine the interaction of levels of analysis, the argument previously made by Putnam. Third, research needs to explore the conditions under which strategies, tactics, and preferences are effective. Finally, research needs to develop midrange theories. Toward these pursuits, this book uses the behavioral-oriented Two-Level-game framework and integrates it with the formal-informal typology and the delegation/principal-agent institutional framework.[31]

Two-Level Game

The Putnam model describes international negotiations as a Two-Level game involving a negotiator who must deal simultaneously with counterparts from other nations, referred to as international Level I, and domestic constituency within his nation, referred to as domestic Level II. At Level I, the negotiator may pursue terms on what he sees to be the national interest, seeking the best possible agreement acceptable to his foreign counterpart. Put another way, does the negotiator view the deal better than the alternative, which is no deal. In bargaining parlance, this is referred to as Best Alternative to Negotiated Agreement (BANTA).[32] Given that a BANTA is unacceptable, a negotiator must have his principals or constituents ratify the deal where it falls within the "win-set" of the domestic arena, that is, any decision-process at Level II that is required to endorse or implement a Level I agreement, whether formally or informally, whatever the system of domestic decision-making. A win-set is defined as the set of all possible Level I agreements that would "win"—that is, gain the necessary majority among the Level II constituents—"when simply voted up or down."[33] According to the Putnam model, the larger the Level II win-set, the greater the possibility that an international agreement can receive domestic ratification, as it may overlap with the other country's domestic Level II.[34] In short, an international agreement requires domestic win-sets that overlap. If both negotiator win-sets are small and fail to overlap, then an agreement is impossible. If one negotiator has a small win-set and her counterpart's is larger, then she bargains from a position of strength as she argues that her hands are tied, a circumstance commonly referred to as the hands-tied strategy or "Schelling paradox."[35] The Two-Level game is broken down into Level I and Level II, but Level II serves as a catch all for all actors and institutions that fall outside the realm of the negotiations; and as a result, it cannot delineate where the state ends and society begins. To resolve this fuzziness, I disaggregate Level II into two categories, Level IIA and Level IIB where the former refers to public or cabinet principals and the latter refers to private and social veto players. At both levels, these players must ratify the Level I negotiator's international agreement. Without ratification, the terms of the agreement can be easily broken, or in game theory terms, the negotiator can defect, voluntarily or involuntarily, from the agreement.

Indeed, ratification is key as it provides "the critical theoretical link" between domestic and international politics.[36] Three elements are critical to flesh out the boundaries of ratification: the preferences of the COGs, the nature of the "win-sets," and the mechanisms of the bargaining outcomes."[37] The COG has an acceptability-set that is determined by her interests and preferences and the institutions under which she operates. Interest is "what actors want to achieve through political action,"[38] and preferences are the ranking of choices given particular interaction outcomes.[39] We can account for preferences through three ways, by assumption, observation, or deduction.[40] Using the rational-choice approach, we assume that actors are utility maximizers; in economic terms, we assume that individuals are trying to get the most for their money, the most profit for their work. In an international relations perspective, we assume that states seek to maximize their security or the national welfare of their citizens. The problem with this approach, however, is that the agent has no constraints on his decision to make complicated choices, a complete abstraction from real life.[41] An alternative assumption posits, which this work accepts, is that agents are bounded by their cognitive limitations and have malleable preferences that can be influenced by external factors.[42] We can also deduce an actor's preferences by examining her behavior, tracing backward the behavior from their starting point. Finally, we can also deduce preferences by subscribing to a particular theory. For example, simple business models assume that firms want to maximize their profits. Agent preferences are a critical component of the ratification process, on par with the "win-set."

These acceptable negotiating terms have been relatively unspecified. One work, however, began the research of attempting to provide a more rigorous application of the concept. By integrating the principal-agent analysis in the rubric of the Two-Level game, Kai Oppermann deftly delineates the boundaries of the win-set and applies his framework convincingly to the Tony Blair government in the UK case of its European policy.[43] To accomplish this constraint, Oppermann applies salience (media and opposition issues) and sanction (elections and referenda) mechanisms that define the edge of what is acceptable in foreign policy UK politics. Oppermann's aim was to make the Two-Level game more rigorous and precise, moving it from "metaphor to theory," an integration and aim that this work shares. By introducing and defining the mechanisms, he accomplished these goals, setting the building blocks for a more robust model; however, he also fell short by missing a critical feature of the game, the interaction of international and domestic arenas. It is this interaction that gives the model its appeal. Still, Oppermann led scholars down the right path; more research that incorporates the Two-Level game with the delegation framework is where, according to Eugénia Conceição-Heldt and Patrick Mello, scholars need to go.[44]

The limitation of the Two-Level framework, thus far, is its lack of precision in its causal mechanisms. Most international relations theories bypass the driving instruments that transfer the causal chain from the independent into the dependent variable, prompting a call from Andrew Bennett that researchers should tease out causal mechanism.[45] This limitation is especially significant when scholars examine how international organizations affect the domestic policies of states.[46] Some researchers have traveled down this path, shedding light on how interests and preferences are transferred from one agent to the other.[47] John Odell skillfully examines the conditions under when negotiators shift strategies between distributive and integrative in the context of world trade bargaining. While Odell examines the timing of strategy change, Conceição-Heldt expands the categories and develops a useful typology that identifies and examines the application of hard and soft strategies by rising economic powers to influence WTO rules. These tactics and strategies are instruments at the disposal of policy-makers, typically determined by the institutional rules that they have access. These rules may be formal or informal.

Formal and Informal Rules

Putnam emphasizes that formal and informal rules are used to ratify international agreements. To my point, the interplay between formal and informal institutions are instrumental to deriving particular outcomes. In this vein, Gretchen Helmke and Steven Levitsky focus on informal institutions and lay out a research agenda for their inclusion and develop a typology.[48] Formal institutions are "rules and procedures . . . that structure social interaction by constraining and enabling actors' behavior."[49] Furthermore, these constraints and processes are "created, communicated, and enforced through channels widely accepted as official."[50] The distinction between the formal and informal arrangement lies in the acceptance by the government. Informal arrangements are "*socially shared, usually unwritten, that are created, communicated, and enforced outside of officially sanctioned channels*" (emphasis in original). Helmke and Levitsky frame the typology according to how informal rules and procedures affected formal rules. These effects are exhibited in two areas: outcomes and effectiveness. If the informal institutions produce incentives to produce efficient outcomes and the formal rules are enforced and followed, then the interplay is deemed "complimentary." For instance, in the United States, the US Supreme Court "Rule of Four" facilitates the work of the justices. At the extreme end, if the two arrangements produce divergent outcomes and are in competition, then the interaction is labeled "competing." The informal arrangement produces "incentives that are incompatible with the formal rules: to follow one rule, actors must violate another."[51]

Patrimonial politics is an excellent example. But the gap between the two typologies is broad; as a result, Helmke and Levitsky bridge this gap by introducing two other subclassifications. Next to complimentary is "accommodating," where the outcomes of the informal rules and processes converge with formal ones and where those who dislike the outcome, abide by the rules regardless, but contradict the spirit.[52] Helmke and Levitsky provide as an example the consociational system of the Netherlands where "these cartel-like arrangements violated the democratic spirit of the Dutch constitution (by limiting the power of the vote), they reduced class and religious conflict, thereby enhancing democratic stability."[53] Finally, Helmke and Levitsky lay out the final category where formal rules are unable to provide incentives to produce a desired outcome, but informal rules can shape incentives to steer the processes toward the desired outcome. This genre is referred to as "substitutive." Helmke and Levitsky point out that Mexico's National Action party resolved postelection disputes through informal "gentleman's agreements," or *concertacesiones*. This arrangement served as a "pause" on conflict until the candidates accepted the credibility of election results.[54]

They also lay out how the interaction of formal and informal produces typology changes.[55] Changes in the formal institutions can produce this effect. Changes in design can reshape the costs and benefits associated with the constraints and incentives associated with informal institutions. For instance, in the US House of Representatives, changes to formal committee rules trumped informal committee structures, weakening the hold of Southern Democrats.[56] These changes have a more significant effect on complimentary and accommodation informal arrangements. Changes in state capacity can also lead to changes to informal institutions. First, increases in state strength can increase the cost and benefits of adhering to "competitive" informal institutions. Second, increases in state efficiency can provide more credibility or benefits to individuals wielded to competing or substitutive institutions. These changes have a greater impact on substitutive and competing designs. Informal institutional arrangements may also be affected from changes to the external environment. These forces may alter power and resource allocations within a grouping, "weakening those actors who benefit from a particular informal institution and strengthening those who seek to change it."[57] Changes may also occur when enough actors are convinced that an alternative is viable and if there are significant mechanisms that can coordinate expectations.[58] Clearly, change between the types of institutions is critical, just as the identification of informal institutions. This identification hinges on three points. First, do actors share expectations about the rules? One set of actors may while another may not. Two, what is the community to which informal rules apply? These cases vary and lines between groupings can be blurred. Finally, how are informal rules enforced? Are there mechanisms that identify and punish informal rule violations?[59] This typology, when fused with the Two-Level game and the delegation framework makes a more robust model.

Delegation Framework

The research on institutions has morphed over the decades. Traditionally, political scientist had relied on historical, legal, and philosophical underpinnings to examine institutions. They asked three primary questions. First, they inquired how institutional rules shaped individual behavior. Second, scholars asked how the rules were derived. Finally, they begged the normative question, how the rules worked or failed to work on behalf of the common welfare.[60] Eventually, their mooring to the traditional foundations slipped as they began to explore other disciplines for novel questions, and more importantly, different methodological techniques for answering these questions. They found themselves wading in the currents of anthropology, economics, sociology, and eventually cognitive psychology, the latter of which, ironically, anchored the discipline in the 1960s to the harbor of behaviorism. Its contributions were a sea change. It found individuals divorced from the institutional rules from which they were purported to operate, demonstrating that the effect could not be attributed to the apparent cause. Furthermore, it also laid the pathway for new techniques on conceptualization, experimentation, and measurement. Eventually, opinion-based research and its sophisticated variations eventually "had run their course," and many political scientists eventually returned to familiar waters of institutionalism.[61]

The institutionalism of our predecessors has different undertones from the institutionalism of today. In the mid-1980s, new institutionalism nudged its way back into the mainstream, focusing on the interaction between institution and individual to explain and predict individual behavior or action. Unlike its unified traditional predecessor, however, it became a collection of diverse orientations and methods.[62] Indeed, even institutionalist oriented scholars cannot reach a consensus. Some narrowly define institutions as rules, design, and structures.[63] Others expand the definition to include symbols, scripts, traditions, values, cultures, ideas, and values.[64] These contrasting views have produced at least four broad categories of institutions: historical, rational-choice, sociological, and constructivist.[65] Just the definition is broad in scope so are their functions and overlay in life. R. A. W. Rhoads, Sarah Binder, and Bert Rockman write:

> There is no doubt that institutions are said to do quite a lot. For example, they may be thought to embed history and political thought and to reflect, therefore, a set of traditions and practices, whether written or unwritten. Institutions can thus be interpreted as reflecting habits and norms, more likely to be evolved than

> to be created. But institutions also may be seen as architecture and as rules that determine opportunities and incentives for behavior, inclusion and exclusion of political players, and structuring the relative ease or difficulty of inducing change, and the mechanism through which change may be facilitated or denied.[66]

Just the definition of institutions lacks consensus, the concept of agency also lacks uniformity. Scholars typically use as a jumping-off point rational individuals that maximize their utilities, but this spring-board has met considerable criticism as it is both indeterminate (i.e., "fails to yield unique predictions") and inadequate (i.e., "the predictions fail").[67] At the heart of criticism is the assumption that agents are unbounded; they have perfect information, have no cognitive constraints, and can make complex calculations regarding their circumstances. In short, they know virtually everything. While this assumption may serve as an excellent starting point for formal modeling, not to mention the drawing board, its capacity to capture reality seems "implausible."[68] Grounding the assumption to some elements of reality requires relaxing the assumption to limits, a bounding. Rationality, as a singular framework, had to be balanced or bind by "psychological limits," as written by Herbert Simon.[69] Consequently, individuals make choices in nonoptimal situations by "aspiration, rules of thumb, standing decisions, stopping rules and satisficing." Decision makers can "satisfice either by finding optimum solutions for a simplified world, or by finding satisfactory solutions for a more realistic world. Neither approach, in general, dominates the other, and both have continued to co-exist in the world of management science."[70]Simon's position is taken by this work. The disagreement continues on to another methodological assumption—whether or not institutions should be treated as endogenous or exogenous.[71] In short, institutionalism has many strands, one of which this work integrates with the Two-Level framework.

This work adopts the rational-choice *equilibrium institutionalism* variant, championed by Kenneth A. Shepsle.[72] Whereas the historical, sociological, and constructivist variations focus on the institutions, the rational-choice variant places its emphasis on the strategic interaction of individuals. This variant has its underpinnings in the *institutions as game forms* modeled by Douglass C. North, who defines institutions as "a set of rules, compliance procedures, and moral and ethical behavioral norms designed to constrain the behavior individuals in the interests of maximizing the wealth or utility of principals."[73] This definition assumes institutions to be a *game form*, where formal and informal rules determine how individuals make choices, time their moves, react to the effect of exogenous factors, and calculate the possible alternative moves that others make. By adding player preferences, according to Shepsle, "the game form becomes a *game* (emphasis in original)."[74] Within this framework, individuals are rational, have stable and fixed preferences, seek to maximize their utilities, and have a high probability of knowing what choices others will make.[75] As a result, these individuals act within the stage of institution. The rules of the stage are fixed and stable, and consequently, institutions are treated as exogenous to the model.

The institutions as game form literature, as Shepsle points out, has three limitations. First, they place little emphasis on the origins of institutions. Second, they have little preoccupation with the durability and sustainability of rules or the institutions. Finally, they fail to provide applications about deviating from rules, accepting the notion that neither strategic defection from a game is not choice nor that an enforcement mechanism is necessary.[76] From these blind spots, the equilibrium institutional genre over time attracted more supporters (Andrew A. Schotter, Randall Calvert, Shepsle, and Avner Greif). Schotter and Calvert also neglected the first critique—Greif addressed it—they investigated the second and third points with more pronouncement. They accepted the North emphasis of political outcomes as determined by the rules, but they also departed from the game form as they examined how the rules were selected and maintained, where playing by, changing the rules, or defecting from the game are part of the equilibrium; when perturbations come into play, the institutions must be able to deter such behavior through credible punishment mechanisms, making the action of defection possible. In short, they made institutions endogenous.

This model segues into the agency framework.[77] Constructed as a principal-agent relationship, the agency model examines the interaction between the principal (constituent) and its need to resolve collective action problems to which it lacks the expertise, time, or resources to resolve. As a consequence, the principal hires, via elections or some other mechanism, and delegates to an agent the powers and responsibilities to act on her behalf. Simply, this relationship can be viewed in an attorney-client or buyer real estate agent context. Michael Jensen and William Meckling write it succinctly, the principals "engage another person (the agent) to perform some service on their behalf which involves delegating some decision making authority to an agent."[78] For our purposes, we examine the political relationship between the constituent and policy-maker. The principal also provides the agent with a degree of latitude, or slack, to pursue her responsibilities. Slack applies to the degree of latitude that domestic institutions require for ratification.[79] A narrow or tight degree of latitude constraints the agent discretion, requiring the agent to provide information constantly regarding her delegated function. A broad or loose degree of slack gives the agent room to maneuver to pursue the interests of the constituency. This line between principal and agent serves to provide a series of benefits to the former.

Why delegate? Principals delegate to derive numerous benefits. First, they benefit from specialization as the principal typically lacks the expertise, time, resources, or ability to enact a particular policy outcome. This type of delegation is likely to be founded on issues that are repetitive and frequent such as on issues of transportation or the mail. Second, principals delegate when policy externalities have the potential to be large and severe to a coordinating agent that can provide alternatives and provide solutions. Third, delegation to another party can resolve disputes. Fourth, principals delegate when they want to signal credibility to other actors. Fifth, delegation also produces policy bias or "locking in" rules.

Extracting the equilibrium institutional model with the delegation framework, principals have rules around which they operate. They may reward their agent with a job or reelection if their preferences are met or they may remove someone from office via the ballot box if they are displeased. Similarly, they may join political parties or by participating in civil or official forums to have their voices heard or amplified. They also may informally express their support or displeasure via access to decision makers. This process works both ways, from the bottom-up and from the top-down, where policy-makers may use formal rules to shape principal preferences and ratify government policies. Indeed, this is the area where this work concentrates, on the process and use of rules, formal or otherwise, norms to secure ratification for IMF supported economic policies. Within this context of changing policies, the concept of veto players is instructive.[80] Veto players, institutional or partisan, operate within a set of rules where their consent is necessary to change the status quo.[81] The greater the veto player number, the more ideological distance they have, and the more internal cohesion, the more difficult it is to alter the status quo.[82] The rules by which they operate, enable, or constrain social and government interaction and are couched by the particular institutions at play. Veto players may be the COG, bureaucratic chiefs, legislatures, or social players.

Putting It All Together

This work combines the Two-Level framework with the formal-informal typology and the delegation equilibrium as institutionalism. The extension is far from novel or imaginative. Putnam emphasizes the importance of the formal and informal ratification arrangements and procedures of international agreements and the boundaries that constrain the range of acceptable policies, the win-sets.[83] These domestic constraints are bracketed according to their society, state, or public opinion orientations.[84] Society centered research puts the primary emphasis and the antecedent for foreign policy outcomes from the seat of social conflict. Business and labor coalitions, class conflicts, trade-oriented groups, and ideology drive, shape and constrain the win-sets of principals.[85] State oriented research focus on the institutions and policy-makers, where veto players, both partisan and institutional, can provide constraints and opportunities for international agreements. This is the area within which this work is lodged. Similarly, Oppermann makes inroads on how to delineate the boundaries of the win-set, a work that is key. Finally, domestic political opinion constrains the maneuvering capacity of the international negotiator. Regardless of the orientation, all of these studies emphasize the importance of domestic political institutions in the ratification of international agreements.[86]

The discussion of the Two-Level game shows that an interaction takes place among the levels of analysis.[87] In this context, actor preferences and constraints must be specified. First, the nature of the CN's preferences must be clear. Second, within this context, the government and social constraints and institutional arrangements must also be specified. Finally, the variables that drive interstate outcomes must also be highlighted. These three specifications correspond to the three levels of analysis.

Level IIA and IIB ratification has been so far portrayed as fixed and unmalleable, but the image of a Two-Level game suggests that these constraints are not immutable and that international bargaining may profitably be directed toward loosening them. In the Agent-Principal context, the agent tries to secure more slack from his principal to secure an international bargain. Toward this end, CNs apply strategies, "a set of behaviors that are observable in principle and associated with a plan to achieve some objective through bargaining."[88] These strategies can be applied to enlarge the win-set of their opponents to secure a more favorable agreement. Toward manipulating win-set, CNs can act as agenda setters, setting up issues for consideration but also choosing among a set of alternatives they prefer. Agenda setters should know the identity and preferences of their principals, to which they have a set of bargaining range enlargement strategies. The rules also determine the sequencing of veto player moves where agenda setters make proposals acceptable to others, but also choosing among a set of alternatives they prefer.[89] In this context, CNs of the Philippines, the finance minister and the central bank governor, are the agents and the agenda setters. They initiate the process with the IMF and within their own government.

Both the IMF and GOP CNs have distributive strategies that are categorized as either "balance of power" or "ratification."[90] CNs try to restructure the domestic balance of power through side-payments, linkage to other issues, enforcement of party discipline, or the control of information about an agreement. The most interesting strategies examine the interaction between both international and domestic type of variables, typically referred to as "reverberation" and "synergy," respectively. Reverberation refers to the international pressures that tip the domestic balance of power by building coalitions.[91] Synergy refers to the strategy of interacting international and domestic level variables, creating a policy alternative absent in a purely domestic context.[92] Under the ratification procedural category, negotiators attempt to change the bargaining range by changing voting rules, placing the status of the agreement under executive management, or controlling the agenda. For the latter, synergy can also apply as domestic agenda issues can be transformed into international issues to reshape veto player bargaining range or reorder their preferences. The model, however, fails to specify the "particular actions" or tactics of reverberation or synergy strategies.[93]

To apply reverberation or synergy, CNs can choose to tactics that build coalitions simultaneously or in sequence. When CNs apply the simultaneous tactic, they first seek a consensus with their Level IIA constituents or bureaucratic counterparts then face off against their Level I adversary. When they use the latter, they typically need to overcome domestic obstacles. James K. Sebenius deduces that CNs examine the first player secured in terms of her outcome value, value to the ultimate coalition, and path effects, the success probability in building the winning coalition. Sebenius reasons that negotiators may pursue a number of patterns in seeking to ask "whom to talk to first." Two patterns are of interests.

The first pattern may follow a deference-sequencing path, "Can we find the guy who can deliver the guy?"[94] If player A defers to player B, then the negotiator should first secure the support of player B given the path effect it has on A regardless of which one is easier to secure. Securing Player A first does little to get Player B on board if B does not differ to A. Sebenius shows that the higher the level of deference that A has to B, the higher the probability that A will join the coalition. Institutions, formal or informal, can also replicate this sequenced pattern as B must sign off on a proposal first before it is submitted to A.[95] For example, in US politics revenue generating bills must originate from the lower chamber and subsequently submitted to the upper chamber.

Sebenius provides prescriptions on how to map out the deference coalition building tactic."[96] CNs should stake out the significant actors relevant to ultimate coalition. Second, CNs should also assess the preferences of these actors and the alternatives they might have to joining the coalition. Third, they should analyze and assess the potential patterns of deference, a key consideration in this process. This deference may be personally or institutionally based. Finally, the CNs should "map backward" the ideal coalition and utilize the deference patterns.[97]

The second pattern examines "external-internal" or "internal-external" sequencing.[98] Conventional wisdom dictates that getting your house in order and negotiating from a unified front then negotiating with your counterparts is a sound strategy. Sebenius points out, however, that this strategy may be doomed from the start as "internal" players may have too much leverage and be unwilling to compromise. As a result, CNs may have to negotiate internationally first then domestically second, an "external-internal" coalition building pattern. This process raises the costs for veto players at the end of the sequence to join the coalition. Sebenius lays out the pattern as a "small inside, then outside, then larger inside."[99] Within this context, the IMF and the domestic CNs typically build technocratic alliances and link issues.

While this book focuses on the sequenced pattern laid out by Sebenius, it also applies the technocratic alliance as an alternative explanation for the causal mechanism of the Two-Level framework.[100] The IMF uses a coalition building tactic referred to as a "technocratic alliance" with a recipient country negotiating team.[101] Usually educated in Western universities and versed on technical aspects of economy management, technocrats seek to run an economy based on efficiency, rationality, consistency, clear choices among alternatives, attention to prices and material incentives, and away from a system driven by political based incentives. Technocrats permeate the ranks of the IMF and its negotiation teams that conduct bargaining sessions with recipient countries. During its negotiation process, IMF technocrats attempt to build "winning coalitions"[102] of domestic support behind its preferred policy choices for curing macroeconomic ills.[103] After these connections are completed, the domestic portion of the coalition, which usually includes the CN, attempts to alter COG and other key policy-maker subjective utility functions by changing recipient country views of the value of its own demands or by changing recipient country perception of the cost of IMF proposals, real or perceived.[104] These shifts can be conducted through social learning and dialogue processes, as pointed out by Miles Kahler.[105] Inundated by technocrats and persuaded by their compelling logic and/or economic models, decision makers change their perceptions of costs and agree to terms sought by the IMF. Phrased in Putnam Two-Level game terms, the technocratic alliance efforts to build a "winning coalition" can be viewed as interaction between international Level I and domestic Level II variables. The causal pattern of a technocratic link implies that an "external-internal" pattern applies, but the causal flow may also travel the other direction. When dealing with IFIs, recipient countries form negotiating blocs that provide the IFIs with capable staffers. These staffers are typically high on the recipient country bureaucratic chain, giving them a career path that moves them into the international arena.

In theory, the technocratic alliance strategy can be an effective maneuver to shift the domestic win-set; in practice, the strategy can run into a daunting complication.[50] Kahler points out that this strategy becomes ineffective to change the perception of costs when technocrat power is circumvented by domestic factors such as institutional factors or the presence of other bureaucratic players.[106] In effect, the power inequalities present within a cabinet may be too much for the technocratic alliance to overcome. Put in a rational-choice perspective, the alliance may fail to consider the political incentives that drive a politician's behavior.

Synergy patterns can also be produced by linking international and domestic issues, a pattern envisioned with the "external-internal" sequencing strategy. Using the "lighting rod" effect, domestic negotiators absolve themselves of any preferences in the matter, and claim the IMF is the impetus behind the reform. As a result of external pressures, negotiators protest that they have little choice but to implement such reforms. In Two-Level terms parlance, domestic "hands are tied." Taken in this context, the linkage strategy has merits, but it also produces another effect. G. John Ikenberry has shown that recasting domestic issues into larger international issues provides negotiators with an opportunity to distance themselves from the grip of social actors. The strategy provides negotiators with an opportunity to advance a set of goals larger than those embraced by any particular group and potentially flood the policy space with reform support.[107] This linkage tactic goes against the "delinkage" approach now endorsed by the IFIs. Branded with a reputation as an "interfering agency"[108] and an arm of imperialist nations, the IMF is now reluctant to play the role of the "bad guy."[109] Along a similar line of contention, the World Bank has suggested that developing countries can achieve and sustain economic development programs through increasing its levels of commitment by viewing such programs as its own, to develop a sense of ownership apart from the bank.[110] Graham Bird finds that there is little reason to believe that the IMF case should be any different, that delinking the IMF from microeconomic reform would enhance the capacity of the government to implement and sustain reform.[111]

As a metaphor, the game lacked specification, definition, restriction, and the preferences of the major actors and their constraints.[112] These levels of precision have yet to be reached by subsequent works, but researchers have been moving in their direction as they build upon the works of their predecessors. At each stage, the research has moved forward, applying inductive approaches toward theory building. At this point, a formal deductive theory has yet to be constructed. This book does not aspire to such a lofty goal; rather, it sets out to produce a midrange theory on IMF-recipient country conditionality negotiations by using the Two-Level game, formal-informal interplay, the principal-agent framework, and integrating the sequencing patterns enunciated by Sebenius.

Midrange theories of political science differ from their bigger cousins, the general deductive theory. The latter have a set of assumptions and propositions that attempt to describe, explain, and predict outcomes. For example, realism attempts to explain why states balance power, bandwagon, or buck-pass, not to mention why they go to war. Furthermore, realism also attempts to explain why states cooperate for only a transitory period. In contrast to a general theory, a midrange theory has a lower aim. It attempts to explain certain subtypes of outcomes because various categories operate under different actor preferences, domestic constraints, and different international negotiating environments.[113] For instance, trade negotiations processes and institutions differ significantly from IMF-recipient country negotiations. As a result, a midrange theory is limited in scope, but it addresses important questions, puzzles, and problems peculiar to the context.

Where We Have Been

International economic bargaining and negotiations research can be categorized by its observational method.[114] A typology developed by John S. Odell has four genres.[115] International cooperation studies encompass research that focus on hegemony, international institutions, and transaction costs.[116] This genre, however, pays little attention to negotiators and their preferences and is limited to multilateral institutional building. Relatively recent research has extended this focus with the extension of domestic constraints, either institutions or veto players, which can be treated either together or separately.[117] Formal models provide another area of international bargaining where deductive game theory is applied to country interaction.[118] This research focuses on outcomes and how they are driven by structures.[119] Atheoretical case studies provide a third category, but as the genre states, it lacks theoretical rigor.[120] Verbal negotiation theory and case studies complete the genres by applying an interdisciplinary perspective.[121] This genre is also consistent with the new institutionalism framework especially in the delegation (principal-agent), formal and informal institutional, and veto player contexts.[122] This study fits in the last genre and increases our collective ability to understand international economic negotiations.[123]

Framed within the verbal negotiation context, the Putnam Two-Level game framework has shown to be an invaluable heuristic tool toward sorting out complexities, nuances, and puzzles that surround international cooperation. While it has highlighted the dynamic and reciprocal interaction present in international bargaining, the approach has also helped generate a multitude of hypotheses and postulates to enlarge perspectives, enrich insights, and open up new areas of research. The *Double-Edged Diplomacy* collaborative volume attempted to extend the rigor and precision of the model by using the framework to uncover various interactive complex patterns that exist in economic and security issues, coercive as well as cooperative bargaining, and a full range of country interactions—West-West, East-West, and North-South.[124]

In essence, *Double-Edged Diplomacy* confirms the notion that international bargaining is more than just relations between states, as it also requires an extensive look at the dynamics of domestic politics.[125] For the most part, the project concludes that negotiated outcomes can be determined through an examination of particular domestic variables: relationships between the domestic costs and benefits of international agreements, institutional configurations and the enfranchisement of domestic actors, and degrees of insulation held by the COG vis-à-vis his domestic constituency and domestic interest.[126] Few, if any, would object to the idea that these variables play a significant role in determining the outcome of international bargaining. The problem with this type of analysis is that it understates the analytical and empirical significance that strategies and tactics can have on enlarging win-sets and reducing the probability of involuntary defection. Other studies, however, examine this corollary with more precision and scrutiny.

Toward this end, several works make strong contributions. By using a negotiation analytic framework to analyze the consequences of international negotiations, Frederick Mayer argues that the outcome of international bargaining depends on the configuration of international interest, the nature of the domestic political process, and on characteristics of international bargaining.[127] He draws two conclusions that are critical to pushing forward our understanding of various international agreement outcomes. One conclusion focuses on the relationship between the context of international bargaining and internal factional interests. In cases where it is engaged in distributive international bargaining, a nation may achieve favorable terms to an agreement when it has domestic factional constraints. In contrast, in cases where it is engaged in integrative international bargaining, a nation may achieve less favorable terms to an agreement when it has domestic factional constraints.[128] The second hypothesis examines the use of side-payments on domestic factional constraints to explain international bargaining outcomes. On the one hand, side-payments can have a neutral effect on international negotiations; that is, payments that redistribute some value among internal factions without affecting national welfare. When they are used in integrative bargaining situations, side-payments used toward domestic factional constraints are more likely to increase the joint gains available in international relations. On the other hand, side-payments used under distributive international bargaining tend to be less desirable.[129] Mayer rightly points out, however, that side-payments are rarely purely neutral. Rather, they may have either a positive or negative effect, defined as adding to or subtracting from the national welfare, on the negotiated international outcome as well as to redistribute value among internal factions.[130] Mayer's work is critical to extending the theoretical rigor of the Putnam model.

Another work that attempts to add to the richness of the Putnam model is by H. Richard Friman.[131] In his analysis of the Anglo-American Financial Agreement of 1946, he suggests that policy-makers select a particular strategy based on the strength of domestic resistance and the external security threat. When they perceive domestic resistance to policy changes as relatively weak, policy-makers are likely to use side-payments to enlarge win-sets. Similarly, policy-makers are likely to play the security card by attempting to redefine economic issues as national security concerns when there is great external threat to the country. While Friman focuses his efforts on domestic bargaining tactics, Leonard J. Schoppa concentrates his research on the effectiveness of particular linkage type strategies within the context of US-Japan trade relations.[132] Schoppa finds that US negotiators can reshape their counterpart's domestic politics in two ways. First, they can transform the decision-making process in ways that expand elite participation and bring the weight to bear on policies that were previously dictated by small groups of privileged actors. Second, they can influence the way policy alternatives are considered in the decision-making process.

Robert Paarlberg examines the relationship between multilateral bargaining and microeconomic reform.[133] He poses the question: to what extent did domestic agricultural reforms in the United States and the European Union (EU) derive from the 1986–1993 Uruguay Round of GATT (General Agreement on Tariffs and Trade) negotiations? Paarlberg examines how synergistic linkage can create greater political space to pursue domestic reform. The results were somewhat surprising. The Uruguay Round added little to the pace of EU reform and it exacerbated distributive conflicts in the United States. As a corollary to this finding, Paarlberg finds that domestic concerns dramatically affected the completion of the Uruguay Round, as US-EU differences over agriculture dominated the final agenda of the negotiations.

The cited works use empirical and behavioral grounded evidence to support contentions regarding strategies and international agreement outcomes. In contrast, Keisuke Iida attempts to strengthen the Putnam model through formal modeling.[134] By examining ratification rules that can shape the possibility of international cooperation, Iida developed a model that targets the median voter's evaluation of the status quo, and negotiator's use of information and particular institutional features to push through ratification. Indeed, the model has sufficient power and precision to capture the interplay between institutions and social actors in industrialized countries, a key toward explaining involuntary defection. Iida's model, however, loses explanatory power to explain negotiations within the context of IMF-recipient country bargaining. Three reasons account for this poor fit. First, many developing countries rely on informal ratification procedures. Second, the model relies on the median voter's preferences, and it assumes that all voters voice their opinion at the same volume, where the notion of one-person, one-vote is the norm. The model fails to consider variances of political efficaciousness under various political institutions. Finally, the model could provide strong explanatory power in narrating Philippine's involuntary defection on the American military bases negotiations, as the Philippine Senate rejected ratification of the agreement, but it is too crude to handle the nuances and complexities of IMF-recipient country negotiations and microeconomic reform. Given these limitations, formal modeling requires severe modifications if it is to have sufficient explanatory powers in an IMF-developing country dialogue.

What separates this work from the others? This research builds on these works and uses them as a point of departure. It is similar to the others in that it examines how particular conditions affect strategies. Save for the Schoppa work, this research is different from the others in that it attempts to tease out the conditions that hinder or promote the effectiveness of particular strategies. It also deviates from the path paved by the previous scholars in that it examines the complexities, nuances, and peculiarities inherent in North-South relations, especially as one party attempts to consolidate its political support. In this respect, this work is distinguished from the work of Paarlberg. I examine IMF-recipient country negotiations, bargaining that has different dynamics from multilateral negotiations. Furthermore, a technocratic alliance, a strategy I use as an alternative explanation, is peculiar to IMF-recipient country bargaining.[135] The two others are common to most types of negotiations depicted in political science literature.

This book seeks to address these issues in the area of IMF-recipient country studies. Studies on the IMF focus on two broad categories, the organization itself and its relationship to a recipient country. James R. Vreeland breaks down the typology further.[136] These subtypes are institution itself, its connections with the major executive board members, primarily the United States, and other creditors and lenders.[137] Recently, these links have been framed in the delegation context.[138] These studies are empirically rigorous and cut across a range of states and time periods. Similarly, Vreeland breaks down the interaction into participation, effects, and compliance. Like the IMF studies genre, these studies also for the most part follow the same quantitative approach. Participation is broken down into macroeconomic imbalances, recidivism and sovereignty, and domestic desire for conditions.[139] The literature suggests that balance of payments problems are an inconclusive indicator of why countries turn toward the IMF.[140] Recidivism, past participation makes future participation more likely, appears to be statistically valid.[141] The desire for conditions has three rationales, the blame game, signaling, and leverage, each has been empirically supported with the latter found to be the most compelling. The leverage argument was perhaps initially articulated first by Putnam.[142] Andrew Kiondo later makes this the case for Tanzania; Vreeland supports this argument in the cases of Nigeria and Uruguay.[143] Both Kiondo and Vreeland apply case studies to derive their conclusions. Teri Caraway, Stephanie Rickard, and Mark Anner examine the role of the IMF on labor markets.[144] Using a statistical model from 1980 to 2000, they find evidence that governments with strong domestic labor receive less pressure from IMF conditionality. In fact, they find that the IMF is likely to bend to labors' demands.[145]

The outcomes of IMF programs have generated considerable research. Before and after studies produced by Cheryl Payer and Tony Killick produced such earlier works. Recent studies examine the IMF/World Bank programs effect on crises and whether or not they induce military coups.[146] Compliance is the final genre, but its study is far messier than the others due to measurement problems.

How do we know when an IMF program goes offline? Until the 1990s, IMF programs were typically unavailable to the public; as a result, researchers applied various types of proxies. Since the Asian Financial Crisis in the mid-1990s, the IMF promoted transparency in its programs. While data became easier to acquire, the programs themselves provided the problems as they tied together a combination of fiscal and monetary policies, governance issues, and structural issues. This bundling became the problem. If target is missed, does the program fail? This uncertainty notwithstanding, scholars in this genre have applied aggregated and disaggregated approaches. One aggregated strategy examines the disbursement of the financing. This assessment assumes that the recipient government always request 100 percent of the entire disbursement allocation and that the IMF halts the disbursement in the event program targets are unmet.[147] According to Vreeland, this measure rest on shaky grounds as governments may not request 100 percent of their disbursements, compliance issues may be internment but the program could be successful overall, or the United States may play a significant role in the duration of fund withholdings.[148] These problems have forced other researchers to develop more refined compliance indices that categorize them according to the types of conditionalities.[149] This new strategy starts to move down the road toward disaggregation.

According to Vreeland, the disaggregation approach that started in the 1980s began to uncover a peculiar puzzle.[150] IMF programs were found to have uneven patterns of compliances. Governments would reach fiscal policy targets in some years, but not in others. This unevenness, which the IMF literature has barely scratched the surface, has led Vreeland to speculate that politics and the institutions around which they operate may produce this variation in outcomes.

> Perhaps compliance rates vary across different types of conditions because certain governments face veto players and interest groups that have more say over some policy areas than others. . . . Because there is such variance in the constellations of conditions that are found in IMF arrangements, an alternative

> approach to studying conditionality is to be highly focused on a narrow type of conditions.[151]

Vreeland praises the pioneering work of Erica Gould as a razor sharp approach at conditionality applied compliances driven by a recipient country's foreign commercial creditors.[152] This approach generates questions about why certain issues are selected and how they perform. "Why are certain conditions imposed in some cases but not in others? Is conditionality driven by purely economic actors or do political factors—both international and domestic—play a role? Which countries comply with which types of conditions? Is compliance uniformly high across certain policy areas and low across others?"[153] The disaggregation approach mirrors this book's focus.

IMF studies are a unique breed. They focus on a number of wide-ranging issues, but approach seems provides a qualitative look that overlooks the nuances inherent in the Two-Level game framework. Miles Kahler provides a critical integration of the framework with the dynamics of IMF negotiations. He examines the IMF and its negotiations with Somalia and Jamaica, where the former failed and the latter "succeeded."[154] He asked the question of why the respective governments were reluctant to accept IMF lending given that the costs of no-agreement is worse than the conditionalities.[155] The answer lied with domestic interests, which was complicated by the domestic rules of ratification where informal rules obfuscate the ratification process and complicate attempts to influence it.[156] Ironically, the technocratic alliance and executive board (United States) pressure strategies proved fruitless. The technocratic alliance coalition building strategy is central to the next work.

Taking a similar line, Ngaire Woods qualitatively examines the IFIs negotiations with recipient countries. Showing how the reach of power countries is limited, Woods uses a constructivist-oriented approach to examine the IFI bureaucracies and their interactions with Mexico, Russia, and African countries. She relies on the role of ideas, concepts, and methodology to push forth liberal agendas that ultimately shape policy preferences. The Polak model, the Washington Consensus, and debt servicing and relief programs were constructs that filled the halls of the IMF. These ideas were delivered by technocrats to persuade governments and social actors to accept liberal conditionality policies. The interlocutors' presence was critical, but their influence varied.[157]

Numerous conditions determined when and where an IFI could be influential. Typically, interlocutors insulated from politics determined that rationalization of economic policies could be formulated. This proposition, however, was qualified by Kahler. Woods points to several conditions that may be amenable to new ideas. A crisis can serve as a catalysis toward pushing through policy changes.[158] Similarly, special agencies or boards that shield the decision-making process from politics can also assist in policy change. Inside bureaucracies, recruitment, administration processes, and innovation patterns that are receptive to new ideas may also transmit the liberal agenda to governments.[159] Instruments at the technocrat's disposal also matter such as resources and information. Since the early 1990s, the IMF has realized that it must help sustain and implement reforms. In short, the IMF is effective to push through reforms when the decision-making process is centralized and insulated but these arrangements are difficult to maintain when a government is in crisis or in domestic transition, creating a messy and fluid dynamics of democratic transitions; a trade-off.[160] Furthermore, Woods also shows how the IFI use persuasion to interest groups and others. As agenda setters, domestic interest groups would have to rely on information and a high level of certainty, but since these elements are in short supply, these actors play the role of veto players where they respond to government initiatives. Over time, these interests groups may have their preferences reflected in the government and bureaucracies that overtime capture them via kickbacks and other instruments.[161] According to Woods, "Their capacity to do this is greatest in systems that are not transparent and where formal systems of accountability do not function."[162] Another work that examines IFI-recipient country negotiations focuses on the World Bank and the Philippines.[163]

Robin Broad provides an in-depth and insightful account. She seeks to examine how the economic policy can shift from a primary commodity oriented economy to an export driven industrialized one in the late 1970s and early 1980s. Furthermore, she inquires how the IFIs were able to push the Philippines in such a direction. Using a critical theory approach, Broad demonstrates the interaction with two sets of "Triple Alliances." Derived from the work of Peter B. Evans, the first set links the transnational capital, domestic government, and domestic interests. Broad introduces the second link, the transnationalists factions of the government, industry, and finance.[164] Prior to the late 1970s, the Philippine Central Bank and its nationalist orientation blocked these alliances from building partnerships that would push through export-oriented policies. This obstacle was overwhelmed by the World Bank as it maneuvered around the central bank and linked up with bureaucracies that were receptive to the IFI's preferences, the Ministry of Finance and the Ministry of Industry. It used the technocratic alliance to mount its coalition building efforts. Broad's narration and interviews made this case study stand out in Philippine-IFI relations in richness and detail.

Almost thirty years have passed since the release of Broad's work, and during that time, much has "continued and changed."[165] In the Philippines, the IMF and GOP have continued their relationship through the 1991 and 1997 crises. The Philippines political process continues to have relatively smooth transitions among presidents, with the exception of Joseph Estrada to Gloria Macapagal-Arroyo. Regretfully, poverty continues to be a pervasive issue as its decline has been slow and uneven. From an academic side, much has changed with regard to the tools of analysis. More instruments have been developed that provides a combination of precision, rigor, and richness. Surprisingly, no works have been developed to examine the IMF-GOP relationship in a comprehensive and detailed fashion. This book seeks to fill this void.

This book also follows along the line of the Woods work, but with some critical differences. She examines the negotiation process from the perspective of the IFI versus the recipient countries. In contrast, this book uses the Philippine CNs as the starting point. They are the agenda setters to both their international counterparts and their Level IIA and IIB principals. They initiate the process with the IMF, provide it with data and information, and access to various parts of government and society. A government submits its LOI, with both the government and IMF contributions, to the IMF executive director. In this vein, CNs may request an IMF program or defer. Woods points out that during the Asian Financial Crisis South Korea initially sought to pursue its stabilization policies independent of the IMF, but it eventually did after strong US pressure.[166] Furthermore, a government can request that the IMF extend its persuasion tactics to its domestic principals.

Methodology

To meet the goals of the book, I use an embedded multiple-"within-case study" approach. I use a series of IMF-GOP negotiation observations—1984, 1986, 1989a, 1989b, 1989c (for petroleum prices only), 1991, and 1994. The 1989 program had numerous phases that require disaggregation. The parties in March 1989 concluded an IMF program, designated 1989a. The GOP subsequently provided in November 1989 an addendum, designated 1989b, and implemented policies central to the program but it was forced to backtrack. Later, the GOP implemented another reform policy (1989c), which was subsequently reversed. I use the convergence and process tracing methods to highlight the variable-orientations among the multiple observation points, which provides rigor to midrange theory's validity under similar and different international and domestic settings.[167] The 1984 observation captures the weakening soft authoritarian Ferdinand Marcos regime and the US Ronald Reagan Cold War years. The 1986 observation captures the democratizing years under Corazon Aquino still under the context of the US Reagan years. The 1989a, 1989b, 1989c, and 1991 observations contain changes to the domestic context as the Cold War began to wane. The 1994 observation, under the Fidel Ramos administration, had the United States lose the military base leases. I also breakdown the observations into two issues, petroleum price and import reform policies, a tactic justified by the bundling of IMF programs, which contain numerous topics and issues that obfuscate attempts at measurement, qualitative, or otherwise.[168] Breaking down the program into two microeconomic issues produces thirteen observations with key variables still in play but where the outcome varies. The prime hypothesis is stated as A (IMF) -> B (Reform Outcomes) with the dichotomous outcomes depicted in Table I.1.[169]

An explanatory hypothesis draws out the causal chain connection of the prime hypothesis, giving us A -> q -> r -> s -> t -> B.[170] The Two-Level game explanatory causal link is categorized into two stages. Level I negotiations entail the CNs for both the IMF and the GOP. Level II negotiations consist of the COGs—the IMF, the managing director, and the GOP president, for their respective sides—the cabinet, and other government and social support and veto players ratify Level I negotiated terms. To provide more precision, I break down Level II into subcategories—Level IIA and IIB—where the former addresses the COG and other government players and the latter encompasses the legislative and social veto actors.[171] This breakdown indicates that Levels IIA and IIB outcomes are separate and distinct, where IIA should precede IIB outcomes, a sequence that may occur in most cases but not always. Level IIA and IIB principals, who allow some discretion to their negotiating agents, providing more or less slack, the "looseness or tautness of domestic rules require consultation, accountability, and approval by the principal during the negotiation itself, prior to the ratification phase."[172]

Table I.2 depicts the variation in outcomes at Levels IIA and IIB. The outcomes on negotiated IMF program compliance are nominal. "Success" indicates that the GOP implemented the policy reforms. "Limited Success" indicates some level of reform, but not at the levels the GOP intended. For instance, it could increase petroleum price levels, but at marginal levels insufficient to resolve a particular fiscal deficit issue. "Qualified Success" indicates phased-in success where the GOP drags its feet on reform. This can be done either in incremental stages or as a "deal breaker" at the end of negotiations. "Reversed Limited Success" indicates reforms with some degree of reversal. "Reversed Failure" indicates reform policies were completely reversed or rescinded. "No attempt" indicates that authorities did not attempt reforms. See Table I.2 for the variations.

As there are variations in outcomes at both levels, there are also variations in causal patterns for the outcomes, but I focus on the causes, not the effects, as causes may not produce the expected outcome.[173] I find both multifinality (same causes, different effects) and equifinality (different causes, same effect) patterns. As a result, I treat each observation as a "deviant case," focusing on explanatory variables of general relevance[174] that traces unique patterns of events.[175] This strategy permits me to generate a midrange theory.[176] Smoke and Bennett justify the development of subtypes as having few simple independent variables to explain many dependent variables.[177]

Table I.2 Levels IIA and IIB Petroleum and Imports Outcomes

IMF Program Date (Government)	*Petroleum Reform Level IIA Outcomes*	*Petroleum Reform Level IIB Outcomes*	*Import Reforms Level IIA Outcomes*	*Import Reforms Level IIB Outcomes*
December 1984 (Marcos)	Success	Qualified success	Qualified success	Reversed limited success
October 1986 (Aquino)	Qualified success	No attempt	Qualified success	Reversed limited success
March 1989a (Aquino)	Limited success	No attempt	Success	Qualified success
November 1989b (Aquino)	Success	Success	Success	Reversed failure
November 1989c* (Aquino)	Success	Success	No attempt	No attempt
February 1991 (Aquino)	Success	Limited success	Success	Success
June 1994 (Ramos)	Success	Reversed failure	Success	Success

*The date for petroleum price increases was September 1990.

The dependent variable is the CN's behavior to change the status quo.[178] He determines if status quo changes can be made and through which means. The IMF, its principals, and domestic veto players, specifically the COG and domestic groups, are the independent variables as their behavior and constraints influences the CN. This process chain focuses on the interaction between the CN and the IMF and its principals and the cabinet, the COG, and social veto players, and his use of sequencing tactics to push for reform. Other variables such as IMF principals, the World Bank, and the Philippine legislature are implied in the above chain. I treat deference and "external-internal" tactics as clear and distinct, but in reality, they overlap. For instance, deference used in this study refers to securing actor A to get to actor B. In some circumstances, CNs may use the IMF (variable A) to get to the COG, weakening the boundary between the two tactics, but the "external-internal" tactic makes a clear demarcation between the two in coalition building. The latter is also consistent with the Two-Level tactic of synergy, where gains are created by mixing the two levels where they would be absent by domestic circumstances alone. Taking these tactics, I present the two hypotheses.

Hypothesis 1: If they can clearly map out the coalition pattern process, then CNs can successfully apply the deference tactic.

Hypothesis 2: If they can change the costs/benefits calculations of veto players, then CNs will successfully apply the "external-internal" tactic.

To test these outcomes against the theories and generate a midrange theory, I use a hybrid format.[179] I use a multiple variable-oriented convergence method to test the extreme variation on both the independent and dependent variables against variables with minor variations, strengthening the connection in our causal inference and also limiting any possibility of an external variable having any significant effect on the outcome.[180] In combination, I also use process tracing to unravel the chain of events between the independent variable and the dependent variable, providing a causal inference among the variables.[181] These formats enable me to test and develop theories and find antecedent and condition variables.[182]

Process tracing is a social inquiry technique that examines the intermediate process between cause and effect. Andrew Bennett and Jeffrey T. Checkel define the strategy as "an analysis on evidence on processes, sequences, and conjectures of events with a case for the purposes of either developing or testing hypotheses about causal mechanisms that might causally explain the case."[183] Process tracing is consistent with "within a case" events. Cases are instances of a class of events.[184] Within cases information is "evidence from within the temporal, spatial or topical domain defined as a case."[185] By examining the process within cases, we can examine the "causal process observations on context, process, or mechanism."[186] This technique relies upon "histories, archival documents, interview transcripts, and other sources."[187] These sources are used to the extent possible to piece together the narration of the observations. In some instances, interviews that express policy-makers preferences may not be available so this book pieces together the causal trace from available resources.

Focus and Limitations

Important points must be raised about for what this book omits. On the observations, I do not include IMF-GOP program under the 1997 Asian Financial Crisis, an omission made on three points. First, this study's two overarching goals are to examine the causal mechanisms of economic agreements and contribute to international political economy theories. It is to tease out and assess the mechanisms inherent in the Two-Level game and hopefully contribute to driving more research on economic negotiations. The IMF-GOP negotiations, while critical, provide useful evidence with which to ground and inductively derive the conditions under which these causal mechanisms succeed or fail. In short, we should not miss the forest for the trees—the theory is what drives this work, not the evidence. Second, at this juncture, I find no theoretical justifications for adding two more petroleum prices and import observations. The external context of the negotiations have been sufficiently captured in the existing observations, with the shortage of international credit and the withdrawal of US bases, primarily in the 1994 observation. With the exception of the financial contagion surrounding South Korea, Malaysia, and Indonesia, the variables are relatively consistent. Similarly, the 1994 observation also captures the domestic conditions in play. In short, the international and domestic variables and constraints are relatively constant across and between the observations; it is the values of the independent and dependent variables that differ. Third on more practical level, research and funding resources are finite, and while such an endeavor would be of personal interests, it would be impractical. Other scholars may take up this challenge.

Another important note must be made regarding the emphasis. The Two-Level game focuses on the "Janus-face" role of international and domestic negotiators to hammer out a deal.[188] They strike a bargain then turn around and secure their principal's acceptance of the agreement. In this study, I emphasize the role of the GOP negotiator to bargain with his principals, taking one important part of the equation, the domestic politics portion, and de-emphasizing the other, the capacity of the IMF negotiator to influence his principals. I take this position for two reasons. First, the trend in the literature has a different emphasis. Traditionally, studies of international organizations examine either the organization itself or the relationship between itself and its principals. Recently, the focus has shifted toward "how and what way did they influence politics within and between states?"[189] Using this demarcation as a jumping-off point, this study downplays the institutional dynamics of the IMF and its principals, fleshing out these only when necessary to point out the causal mechanisms, and instead focuses on the link between the IMF negotiators and their GOP counterparts and the GOP negotiators/agents and their principals. Furthermore and along the lines of the first point, the contemporary view of the IMF is that it forces stabilization policies down the throats of recipient countries. This perception is misplaced. Rather, the government and its CNs devise a stabilization plan and present it to the IMF for approval. The latter then reviews the design and recommends program design adjustments. These changes can be either in degree, an increase in taxes or interest rates, or in kind, a shift from one tactic (i.e., privatization) to another (i.e., taxation). The two sides then meet and strike a bargain. In this process, the recipient country initiates, drives, and implements the change. Finally, on a logistic note, resources limited my capacity to take the study to the IMF and its principals, a gap that other scholars can perhaps take up. While critics might point out these holes in this study, its emphasis is still solid as it examines the causal mechanisms that drive the variations in outcomes.

The book traces out the negotiation process into numerous sections. The following chapter covers the Philippine and IMF background. This portion goes over some of the nuts and bolts that serve as the foundation for both actors. For the Philippine side, geography, colonialism, and the domestic elites shaped in large part the political and economic landscape of the country. Ferdinand Marcos and the 1972 declaration of martial law not only pushed the country toward development, but it also marginalized one set of elites and replaced them with another, those beholden to Marcos. Economic development appeared to take hold, but this growth was an illusion as it could not be reconciled with the crony capitalism built up by Marcos. The chapter then examines the role that international factors have on development, primarily the arrangements made with other principals to the IFIs. The section shows that the GOP searched for avenues to make its voice louder in the international arena cacophony. The section then moves to the IMF, where it briefly covers the organization's structure, but just as importantly, it goes over the institution's preferences over fiscal policy and structural adjustment. This section is critical because the book argues that the pressure the IMF applied to the reform of petroleum prices and imports were relatively constant throughout the observations. In a similar fashion, the following section paints the Philippine CNs as having similar preferences, but with some variation in logic, for pursuing IMF programs.

The next five chapters (2–6) detail the negotiation process between the IMF and the GOP on petroleum prices. Chapter 2 sets up the Philippine Petroleum Industry and then moves into the 1984 Petroleum Price Observation. This observation is set up with a political economy precursor context, followed by an international political economy and a domestic political economy sections. Afterward, the epilogue section covers concluding information. The last section covers an assessment of the causal mechanism (deference and external-internal sequencing) and a presentation of the alternative explanation, the technocratic alliance. For good measure, a section of what the GOP learned is also incorporated. This pattern is consistent through chapters 2–6, with the exception of chapter 4, the 1989 observation. This chapter follows a similar pattern to the others, but is divided into three sections, 1989a, 1989b, and 1989c, each corresponding with the GOP dealings with petroleum prices. The following five chapters on imports (7–11) also follow the same format, with chapter 7 introducing the political economy of imports from 1946 to 1983. Like chapter 4, chapter 9 also breaks down the observation into sections.

The organization of the chapters follows the Two-Level game framework. While the approach argues for the simultaneous interaction between the international and domestic levels, for expository purposes, I categorize the narrative into the international and domestic levels. I first examine the international political economy, then the domestic political economy second. The starting point is arbitrary as the interaction between both levels produces incentives for the GOP CNs to approach the IMF. In short, I have to start somewhere. For the international dimension, the CN attempts to move address his counterpart's win-set. In some cases, they already agree on the principles, but differ on the terms. This portion also examines the international landscape and how they affect, or are affected, by the United States, the World Bank, other donating agencies, and commercial creditors. Following this, the domestic portion examines the CN attempts to move the government cabinet, in Two-Level game parlances, the Level IIA. After the policy is declared, domestic Level IIB veto players must ratify the provision, through the interplay of formal and informal institutions. In some cases, veto players accept the provisions, in others, the GOP must retract the policy.

The empirical chapters can be read on an issue or companion basis. Reading on the issue, the reader gets a taste for the petroleum price from 1984 through 1994 (chapters 2–6) or imports from 1984 through 1994 (chapters 7–11). Readers can also cover the chapters in a companion basis, where chapters 2 and 7 correspond to the 1984 IMF program with the emphasis placed by issue. As chapter 2 deals with petroleum prices, the IMF emphasis on stabilization and government deficit reduction are emphasized, and the GOP focus on pricing veto players is also addressed. In contrast, chapter 7 examines the focus of imports in the context of structural adjustment. Reading on a companion basis provides a different focus.

The final chapter uses the inductive tactic to generate the midrange theory and its typology. Based on the thirteen observations, I derive two major propositions and six minor propositions. The two major propositions are driven by the issue of leverage clarity at the international level and the type of formal and informal institutional interaction. The clarity issue can also be couched in terms of formal and informal governance processes. The typologies are broken down into Level IIA and Level IIB categories, which demonstrate an equifinality effect. For instance, even when two outcomes are similar, they may have been driven by different causes or variations in the process. I conclude that sequencing does matter at Level IIA, but that an "external-internal" tactic should precede a "deference" move. Furthermore, I also conclude that "deference" is more effective in petroleum price reform while an "external-internal" strategy is more effective in an issue like import reform. In the end, this book hopes to push the boundaries of IMF-recipient country negotiations and provide policy advice for decision makers. Despite these lessons, however, we should not lose sight of the ultimate goal—to make peoples' lives better.

In the context of IMF-recipient country negotiations, domestic, or internal forces can place considerable constraints on the terms negotiated in an IMF economic package.[190] For the IMF team, various internal departments, the managing director, and the executive board members constrain the ability of the IMF negotiators to authorize a program that is perceived as "loose" in its conditionalities. They are also constrained by public and private donors and creditors as well as the World Bank. For the developing country, constraints on an agreement that is perceived as "tight" in its conditionalities derive from a number of sources. Groups that are ideological, nationalistic, or economic in nature can form resistance to IMF programs. Further consideration of the IMF program must also be given by the existence of a technocratic core and the periodic political test (elections) that those in power must eventually clear. Finally, the COG also plays an important role where domestic interests and institutions may in fact determine the set of agreements preferred by the COG to the status quo, referred to as "acceptability-set," in negotiations with the IMF.[191] The closeness of fit between the COG's own acceptability-set and the ratification win-set in a society is determined by two key COG considerations: economic policy and political entrepreneurship. These considerations are laden with "veto players," institutional or partisan actors whose consent is required to implement changes to the status quo.[192]

NOTES

1. See Collier, Brady, and Seawright, 153–154.
2. Ibid, 154; and King, Keohane, and Verba, Chapter 3.
3. Collier, Brady, and Seawright, 154; and King, Keohane, and Verba, 19–23.
4. Eckstein, 88.
5. King, Keohane, and Verba, 8.
6. Eckstein, 88.
7. George and Bennett, 78.
8. Ibid., 87.
9. Ibid., 69.
10. Ndikumana, Boyce, and Ndiaye, 16.
11. George and Bennett, 137.
12. International Monetary Fund 2002.
13. Broughton, 625.
14. Martin, 159–160.
15. Thomas and Ramakrishnan, 3.
16. Montes, 1991, 36.
17. Lowi, 690–691.
18. Thacker points out this outcome is unsubstantiated. See Thacker.
19. See Martin 2006.
20. Waltz 1957.
21. Ibid., 1979, 60.
22. Ibid., 1957, 238.
23. Ibid., 1979, 40.
24. Singer, 90–92.
25. Moravcsik, 14–15.
26. Putnam 1988.

27. Shepsle, 7.
28. Ibid, and Waltz, 1979, 13.
29. Moravcsik, 23.
30. For seminal works in this field, see Walton and McKersie; Raffia; and Fisher and Ury.
31. da Conceição-Heldt, 12–13.
32. Odell, 27.
33. Putnam, 437.
34. Ibid., 436.
35. Evans 1993, 402–3; Schelling, 21–8; Conceição-Heldt and Mello coin the phrase "Schelling paradox." 10.
36. Moravcsik, 23.
37. Ibid.
38. Frieden, Lake, and Schultz, 42.
39. Conceição-Heldt and Mello, 16; Shepsle, 20–29; and Odell, 19.
40. Conceição-Heldt and Mello, 16.
41. Odell, 18.
42. Ibid., 19.
43. Oppermann 2008.
44. Conceição-Heldt and Mello, 21.
45. Bennett 2013.
46. Checkel, 76.
47. See Conceição-Heldt 2004, Odell 2009, Wallander 1999, and Schimmelfennig 2003.
48. Helmke and Levitsky, 2004.
49. Ibid., 727.
50. Ibid.
51. Ibid., 729.
52. Ibid.
53. Arend Liphart as cited by Helmke and Levitsky, 729.
54. Todd Eisenstadt as cited by Helmke and Levitsky 2002, 729.
55. Helmke and Levitsky 2000, 731–3.
56. North 1990, 88.
57. Helmke and Levitsky 2002, 732.
58. Schelling, 90–2.
59. Helmke and Levitsky 2002, 733.
60. Rhodes, Binder, and Rockman, xii.
61. Ibid., xiii.
62. March and Olson 1984; Hall and Taylor 1996; and Lowndes 2010.
63. Rhodes, Binder, and Rockman, xiv.
64. March and Olson 1984; Hall and Taylor, 1996, Lowndes 2010; Levi, Cook, O'Brien, and Faye, 1990, 11–15.
65. March and Olson 1984, Hall and Taylor, 1996, Lowndes 2010, and Rhodes, Binder, and Rockman Binder, and Rockman confine the broad categories to three. Hay adds the constructivists variant to the mix.

66. Ibid., xiii.
67. Elster, 19.
68. Shepsle 2017, 26.
69. Simon 1955, 101.
70. Simon 1979.
71. Hall and Taylor 1996; and Shepsle 2008.
72. Shepsle 2008; Shepsle 2017.
73. North 1990, 201–2.
74. Shepsle 2008, 1033.
75. Odell, 2000, 18; Hall and Taylor 1996, 944–5; and Shepsle, 2017, 24–6.
76. Shepsle 2008, 1034.
77. Gutner 2017, 18–19.
78. Jensen and Meckling 1976, 308.
79. Odell and Eichengreen 2000, 159.
80. Mansfield, Milner, and Pevehouse.
81. Tsebelis, 2.
82. Ibid., 2 and Chapter 1.
83. Putnam 1898, 435–441.
84. Da Conceição-Heldt and Mello 2017, 7–11.
85. Ibid., 7.
86. Ibid., 11.
87. Moravcsik, 23.
88. Odell, 31.
89. Tsebelis, 2.
90. Moravcsik, 24–27.
91. Putnam, 454.
92. Ibid., 448.
93. Odell, 31.
94. Bill Clinton strategist Bill Daley as quoted by Sebenius, 328.
95. Tsebelis, 2.
96. Sebenius, 332.
97. Ibid., 328.
98. Ibid., 335–337.
99. Ibid., 337.
100. Sebenius examines the role that external players have to reduce domestic opposition. See Sebenius 2012.
101. On this issue, I have profited from the works of Robin Broad; C. David Finch; Miles Kahler 1992.
102. I borrow the phrase from James K. Sebenius.
103. C. David Finch, 9. A technocratic alliance is similar in concept to epistemic communities. Both consist of professional groups that believe in the same cause-and-effect relationships, truth tests to accept them, and share common values. The difference lies in the matter of emphasis. The term technocrat is more specifically bound as individuals who focus on economy management. In contrast, the term epistemic community is a generic label for a group of professionals who subscribe to a common

ideological framework. In essence, technocrats are a subset of an epistemic community. For a discussion on epistemic communities see Peter Haas 1990 and 1992.

104. Adapted from Walton and McKersie, 72 and 75.
105. Kahler 1992, 89–136.
106. Ibid.
107. Ikenberry, 163.
108. Broad, 20.
109. Kahler 1992, 122.
110. World Bank 1989, 3.
111. Bird, 494.
112. Moravcsik, 23.
113. Ibid.
114. Van Evera, 27–30.
115. Odell 2000, 12–17.
116. See Koehane 1984.
117. Mansfield, Milner, and Pevehouse 2007.
118. See Powell 2015; Chaudoin, Milner, and Pang 2015; and Bapat and Kwon 2015.
119. Sebenius 1996, 326–327.
120. Walker 1992.
121. Walton and McKersie 1965; Schelling 1980; Sebenius 1996; Odell 2000; Schoppa 1993; and Sebenius and Singh 2012.
122. Hall and Taylor 1996; Helmke and Levatsky 2004; and Tebelis 2002.
123. Odell 2000, 6.
124. Evans, Jacobson, and Putnam 1993.
125. For more on this theme, see Evans, 397–430.
126. Evans, 412–23.
127. "Managing Domestic Differences in International Negotiations: The Strategic Use of Internal Side-payments."
128. Ibid., 796–804.
129. Ibid., 810.
130. Ibid., 812–15.
131. "Domestic Tactics in Economic Negotiations."
132. Schoppa 1993.
133. Paarlberg 1997.
134. Iida 1993.
135. A solid treatment of technocratic alliance can be found in Miles Kahler 1993, 377; and C. David Finch, 7–9.
136. Vreeland 2007.
137. See Lavelle 2011, Truman 2006, and Gould 2003.
138. Hawkins, et al 2006; Stone 2008.
139. Vreeland 2007, 51–67.
140. Ibid., 52 and 145n 3.
141. Ibid., 55–62.
142. Putnam, 454.

143. Kiondo and Vreeland 2003, 33–52.
144. Caraway, Rickard, and Anner 2012.
145. Ibid., 28.
146. Dreher and Gassebner 2012; and Casper 2015.
147. Killick as cited in Vreeland 2007, 98.
148. Vreeland 2007, 98–99.
149. Ibid., 101–2.
150. Ibid., 104.
151. Ibid., 105.
152. Gould 2003.
153. Vreeland 2007, 106.
154. For Jamaica, the first round of negotiations centered on devaluation and ended in a stalemate. The next round excluded devaluation and had looser conditionalities. They were accepted. Kahler 1993, 387.
155. Kahler 1993, 364.
156. Ibid., 389.
157. Woods, 71.
158. MacIntyre, Pempel, and Ravenhill 2008.
159. Woods, 76–80.
160. Ibid., 80.
161. Ibid., 81.
162. Ibid., 81.
163. The IMF-GOP relationship dates back to the early 1960s. See Payer 1974; Jose 1984; Bello, Kinley and Elinson 1982; and Thompson and Slayton 1985.
164. Broad, 13, 17, and 228.
165. Timberman 1991b.
166. Woods, 70.
167. Van Evera, 58–6.
168. Vreeland 2007 103–6.
169. Van Evera, 11–15.
170. Ibid.
171. Schoppa also breaks down the levels.
172. Odell 2000, 159.
173. Van Evera, 22–23.
174. George and Bennett, 20; George and Smoke, 514–515; Van Evera, 22–23.
175. Van Evera, 65.
176. George and Bennett, 31.
177. Ibid., 114.
178. This variable is consistent with George and Smoke's assessment (1974) of the recipient of US deterrence. This actor is constrained by deterrence and initiates moves to change the status quo. Also, it is consistent with George and Bennett's definition of causal mechanisms. See Fn. 12.
179. Van Evera, 66.
180. George and Bennett, Chapter 9.

181. George and Bennett, Chapter 10; Van Evera suggests that these investigative strategies can bleed into one another, 66.
182. Van Evera, 61–73.
183. Bennett and Checkel, 7.
184. George and Bennett, 17.
185. Bennett and Checkel, 8.
186. Henry Brady and David Collier, 12, as cited by Bennett and Checkel, 8.
187. George and Bennett, 6.
188. Moravcsik, 15.
189. Checkel, 76.
190. Kahler, 376–377.
191. Moravcsik, 30.
192. Tsebelis, 2.

Chapter 1

Background of the Philippines and the IMF

This chapter covers numerous threads. Its intent is to provide pertinent background. First, it provides a historical political economy of the Philippines. These cover the geography, culture, colonialism period of the country, under both the Spanish and the United States, and the role of elites. It then moves into the Ferdinand Macros years, which provides a sound political economy jumping-off point. The following section covers the international institution and financing context, which shows the pathway created by Marcos' Finance Minister Cesar Virata to link up with IFIs. Afterward, the chapter moves into a brief coverage of the IMF, which goes over the institution's structure, and just as importantly, its preferences for conditionality policies. This section is critical to the book as it shows the constant pressure the IMF puts on the GOP under the negotiation observations. This chapter then takes the flip side and examines the CNs' (Chief Negotiators) preferences under the auspices of IMF negotiations. It also extends the description by examining briefly the institutional arrangements and the social actors embedded under the petroleum pricing and import and tariff liberalization process.

HISTORICAL POLITICAL ECONOMY OF THE PHILIPPINES

The Philippines is an archipelago composed of approximately 7,600 islands. Of these islands, about 1,000 are inhabited, but the bulk of the inhabitants live on three principal islands: the northern island of Luzon, the southern island of Mindanao, and the central cluster of islands known as the Visayan Islands. The Philippines was a major producer and exporter of sugar, exporter of coconut products, and copper and gold. The Philippines formally engaged in the export of logs and timber, but deforestation practice and more recently, attention to conservation, has mitigated this export. The country is subject to the vagaries of the weather and natural disasters, where typhoons, droughts, earthquakes, or volcanic activity can cause major disruptions in agricultural production.

In large part, the archipelago dimension of the islands shaped the cultural context of the Philippines. Its proximity to Malay accounts for the principal descent orientation of the population, as many of the inhabitants in the southern region subscribe to Islam for their cultural orientation. The teachings of Islam, however, did not secure its reach far enough up to the central or northern islands to establish a national identity or a central culture. As a result, the hierarchical order extended only as far as the provincial administration. Although the identification and subservience of the inhabitants remained primarily local, the strength of this kinship serves as the basis for traditional and contemporary Filipino culture. In this relationship the datu, or chieftain, supervised the land and its loyal cultivators in exchange for homage and a portion of the years' crops. This type of rule continued on through the Spanish era. Foreign influence was a mainstay in the country's economic and social development. The Spanish virtually wiped out many local cultures in its attempt to proselytize Christianity. The country also received little in the way of social or economic aid as it was administered from the Spanish empire in Mexico rather than directly from Spain. British and American investment firms are given credit for propelling the Philippines in the global economy.[1] Despite its new role as a supplier of raw products (sugar, indigo, and Manila hemp [abacci]) for the world market, the Philippines remained a poor and impoverished country as Spanish and Mexican administrators and indigenous wealthy landed elites reaped the rewards.

Although they watched their gains increase with the country's move into the global economy, the elites did not receive political participation rights and pushed for political reforms from the colonial administration. The Spanish answered these demands with harsh and repressive tactics, leaving the Filipinos with little alternative but to secede. Under the stewardship of General Emilio Aguinaldo, the struggle for independence began as early as the 1880s. Hostilities between the indigenous elites and the Spanish continued for some time, eventually culminating under the Spanish-American War of 1898.

THE UNITED STATES AND THE PHILIPPINES

The portrayal of the United States as the liberator of the Philippines from the chains of Spanish rule is greatly exaggerated. Approximately eighteen years before the arrival of Commodore Matthew Perry, Philippine rebel forces consistently engaged the Spanish military in battles. By the late 1890s, the Spanish forces suffered numerous losses and were on the verge of surrender. Perry, however, managed to convince the Spanish leaders to surrender to him rather than to the indigenous forces. Shortly before the United States agreed to terms with Spain officially ending the Spanish-American War, Washington decided to annex the Philippines. This decision was not well received by the Filipino revolutionary forces. Again they took to arms, only to be suppressed by a better-trained and better-equipped American military. The insurrection took little time to quell.

Unlike its predecessor, the United States treated the Filipinos with "respect and admiration." When it took over the reins of the Philippines, the United States found problems that reached such gigantic proportions. To resolve these problems, the American administration injected massive amounts of aid toward the education system, one of the most neglected areas under the Spanish tenure. In 1903 alone, the United States assigned an additional 900 teachers and invested over $100,000 for books and supplies,[2] along with a curriculum that imparted the virtues of democracy and "Americanization." In addition to education, the United States also introduced administration reforms, industrial investment, new commercial institutions, improved infrastructure, and a wider circulation of modern ideas.[3] Although the indigenous population benefited from these reforms, the United States proved to be the true beneficiary as it maintained an environment of domestic stability,[4] as the populace would be "less liable to be led by political leaders into insurrectionary schemes."[5]

Not content to rely on social-economic reforms to maintain domestic stability, the United States also provided economic incentives to win the cooperation of indigenous elites. In 1909, the United States established the Payne-Aldrich tariff that created a free trade zone between the two parties. As a result of guaranteed access to the foreign market, trade between the two countries intensified as 63 percent of Philippine imports came from the United States and 79 percent of Philippine exports went to the United States. Intensifying trade ensured the United States that Filipino landed elites received increased windfalls.[6] In addition to providing these incentives, the United States also provided limited political participation rights and favorable fiscal and monetary policies to secure the support of elites and establish a stable domestic political climate.[7] Filipino elites received significant short-term benefits from this political-economic arrangement, but in the long run, the relationship proved damaging for the economic development of the Philippines as it had failed to provide significant incentives to increase economic efficiency, a problem that became evident in the early 1970s. In times of prosperity, the relationship benefited both parties, especially Filipino elites. In contrast, in times of economic hardship, the relationship received significant pressure for change from the American side. During the economic crisis of 1929, American agricultural interest groups called for cuts to the importation of cheaper Philippine agricultural products. By 1934, the American executive branch responded to these calls through the establishment of the Tydings-McDuffie Act, altering the status of the Philippines from a colony to a commonwealth.

The change in Philippine status proved to be of limited real consequence as the United States continued to maintain final authority over Philippine domestic and foreign policy. The United States maintained final approval of Philippine loans. The US Supreme Court could review all decisions of the Philippine courts. American citizens were to have equal rights with Filipino citizens on the Islands. The United States limited Filipino immigration to the United States to a minimum quota of fifty a year.[8] The United States also curtailed its preferential treatment of Philippine exports. Sugar exports were reduced to 850,000 long tons, coconut oil to 200,000 tons, and cordage to 3 million pounds each year. Exports in excess of these quotas were subject to pay full duty.[9] The terms of the Tydings-McDuffie Act clearly maintained the supremacy of American interest in the US-Philippine relationship, but it also had provided a timetable for formal Philippine independence. Within ten years, the American president had formal authority to grant the Philippines full independence. These plans were put in stasis with the outbreak of World War II and the Japanese invasion of the Philippines on December 1941.

After a two-year postponement, Philippine independence changed little in the Philippine-US relationship. World War II devastated the Philippine economy, which required a massive infusion of capital to rebuild. The United States offered the Philippines an opportunity to become the recipient of $620 million in financial aid, an offer laden with strings. In exchange for access to American funding, the Philippines had to agree to be a signatory to the Philippine Trade Act of 1946 (otherwise known as the Bell Act). Like the Tydings-McDuffie Act, this treaty contained several provisions that allowed the United States to maintain control over Manila's foreign and domestic policies. First, reciprocal free trade would be reestablished for eight years until 1954; afterward, each side would be subject to gradually increasing duties until 1973, at a time when full duties could be imposed. Second, the Philippines would not be allowed to impose taxes on exports to the United States. Third, absolute quotas were established on seven Philippine export products, including sugar, coconut oil, abaca, and tobacco. Fourth, the exchange rate of the Philippine peso would be pegged to the dollar at the prewar rate of P2 per dollar. The Philippines could not adjust the exchange rate or impose restrictions on the transfer of funds from the Philippines to the United States without the consent of the United States president. Fifth, the United States president was given the right to suspend all or part of the act if he found that the Philippines was "in any manner" discriminating against United States citizens or business interest. Finally, the treaty stipulated that US nationals who wished access to Philippine national resources and public utilities would be granted parity rights.[10] The Philippines accepted these conditions. In 1955, the Laurel-Langley Agreement amended the Bell Act and adjusted the scope of the relationship. The amendment dropped two clauses. One clause required that the Philippines peg the Philippine Peso to the US dollar. The other clause prohibited the Philippines from imposing an import tariff. The agreement also broadened the "parity" agreement by extending the rights of Philippine nationals to invest in American properties. The widening of the "parity" agreement also extended US investors' access to all areas of the Philippine economy, not just limited rights to areas of natural resources and public utilities.[11] Indeed, post–World War II arrangements between the United States and the Philippines made it difficult to distinguish any real difference between a commonwealth Philippines and an independent Philippines.

In the same manner the Laurel-Langley Agreement manifested the economic relationship between the Philippines and the United States; the Military Base Agreement of 1947 exemplified the same relationship in security matters. This accord provided the United States with extensive use of military facilities, most notably Clark Air Base and Subic Naval Bay for ninety-nine years.[12] The Rusk-Ramos agreement signed in 1966 amended the expiration date to expire after twenty-five years (to 1991) and thereafter to be subject to one year's notice of abrogation by either party. Under these arrangements, the United States was required to consult with the Philippine government before placing any long-range missiles in the country and before using these bases for non-mutual security purposes. In 1979, an agreement gave the United States the right to "unhampered" military operations, and the right to store nuclear weapons in the Philippines without the need for Philippine consent.[13]

The importance of these bases to the security of the United States cannot be underestimated. As the largest American air force and naval facilities outside of the United States, these installations played an important part in America's cold war containment strategy as they were used as forward bases that enabled the United States to project its military might in the far East at the height of the Vietnam War, these bases served as a critical launching pad for American incursions into insular Southeast Asia. In 1991, the United States and the Philippine government under Corazon Aquino renegotiated the treaty terms of the bases. The end of the Cold War as well as the damage sustained during the Mt. Pinatubo disaster left the United States with little incentive to hang on the bases. From the Philippine side, the negotiations presented an excellent opportunity to craft a "balanced" treaty. The two nations had agreed to new terms of a treaty, but the Philippine Senate chose not to ratify the agreement. American military forces left the Philippines in late 1993.

DOMESTIC ELITES

Just as foreign powers were instrumental in shaping the context of Philippine economic policies, various domestic forces also contributed to perpetuate these policies. These actors, primarily consisting of land holding elites, also played a significant role in determining the shape and content of Philippine economic policy. Under Spanish rule, domestic competition among private groups was relatively muted. This condition changed under American rule as domestic elites had institutionalized competition for the spoils of public policies. In the early 1930s, Filipino elites crafted a constitution that mirrored the American constitution with several exceptions, most notably the establishment of a unicameral legislation. Years later, domestic elites amended the charter to create a bicameral legislature, with a Senate chamber composed of twenty-four members elected at large body. By creating this institution, the framers built an institution that held "24 rival candidates for the presidency."[14] As a chamber that owes its allegiance to the landed elite, the new Senate served as an arena for true elite competition. After it was granted independence, the Philippines initiated a wave of economic policies that reflected the influence of private elites.

Elite capacity to influence public policy was especially evident during the 1950s and 1960s. In the immediate postwar period, the exhaustion of US foreign aid, political business cycles, high demands for consumption and capital goods, and the pegging of the Philippine peso to the dollar at the prewar exchange rate exerted considerable pressures on the country's balance of payments.[15] To resolve this problem, the Philippine Central Bank intervened in the foreign exchange market, a move that shifted economic incentives toward sectors that had favored industrialization at the expense of agricultural elite windfalls.[16]

Responding to these residuals, landed elites petitioned Congress to bypass these exchange controls through the introduction of a No-Dollar Law in 1955. Under this bill, producers circumvented exchange controls by bartering for imports outside the exchange control system.[17] In 1962, landed elites pressured the Congress and had the exchange controls gradually lifted.[18] In 1965, the Philippine government, under Ferdinand Marcos, initiated a campaign of unprecedented expansionary fiscal and monetary policies. These policies created inflationary pressures as well as stabilization problems, and required the Philippines to initiate foreign exchange controls.

THE MARCOS YEARS

By handily winning the 1965 presidential election, Ferdinand Marcos ushered in a new era of Filipino politics. During his first term in office, Marcos set new political standards and precedents in his struggle for political power. He faced a hostile Congress, controlled by the opposition Nacionalista Party, but he managed to wrestle power away from the legislature through both front and back door maneuvers. By using parliamentary procedures, Marcos usurped Congressional vested control over the expenditure of public monies.[19] By relying on surreptitious approaches, Marcos attempted to disrupt the role of legislators as power brokers. He broke this link by making himself more available to the rural sector and increasing his exposure to these areas by injecting massive levels of government funds in the form of infrastructure development programs. Marcos did not devise this strategy; rather, his predecessors had used this formula as a way of strengthening their positions vis-à-vis Congress. The crucial difference between Marcos and his predecessors, however, rested on the extent to which they used their capacity to generate rural infrastructure programs. Past presidents had funneled money to rural areas incrementally; by contrast, Marcos dispensed massive amounts of rural aid.[20] This maneuver enabled Marcos to undermine the traditional power of the Congress and channel power directly to the executive branch.

By relying on massive expenditures to maximize his political preferences, Marcos also drove the government coffers into new levels of debt. During his first term in office, government expenditure rose in real terms by approximately 43 percent. In terms of Gross National Product (GNP), government expenditures increased from 11.5 percent to 14 percent of GNP.[21] As revenues were insufficient to pay for its expenditures, the government borrowed heavily through short-term lending practices from both domestic and foreign sources. The GOP lowered the discount rate and greatly increased rediscount ceilings, making loans from domestic sources easier to obtain. Between 1965 and 1967, the adjustment of interest rates increased domestic credit by 40 percent compared to a rise in nominal GNP of 18 percent.[22] From late 1965 to 1968, the GOP increased by nearly two-thirds its foreign debt, as the external debt increased from $491 million to $828 million.[23] In 1969, the mounting government debt increased dramatically as Marcos opened government coffers to finance his presidential reelection bid.

By increasing government expenditures by over 25 percent in 1969, the Marcos administration easily had engaged in the most expensive election campaign in Philippine history.[24] The Marcos administration's continued practice of heavy borrowing from domestic and foreign sources inevitably spelled financial disaster. By the end of 1970, outstanding public and private foreign debts totaled more than $1.6 billion. Of this total, $450 million was due in 1970 and two-thirds within four years. Of the debt maturing in one year, $196 million was owned by the Central Bank, $58 million by the government, and $198 million by the private sector.[25] After the election, Marcos sought to ease the government's financial burden. In January 1970, the major foreign official creditors formed a Consultative Group for the Philippines and agreed to restructure the Philippine external debt in return for the successful negotiation of an IMF stabilization program.[26] In exchange for its third credit tranche from the IMF, the Philippine government agreed to float the Philippine peso, which had fallen from P3.9 to P6.4 per dollar by the end of 1970.[27]

While appeasing his financial creditors, Marcos also entrenched his hold on political power. The 1935 Constitution prohibits a Filipino president from serving more than two terms in office. To circumvent this legal constraint, Marcos sought to redraft the constitution in a manner that benefited his efforts to maintain his hold on power. A constitutional convention was called and it modified the constitution to remove presidential term limits and to change the system of governance from a separate legislative and executive system to a parliamentary system. In 1973, the new constitution was ratified by local village assemblies through a show of hands. The Marcos power grab was not lost on students and the intelligentsia. They became cynical over the government's inability to initiate real and meaningful political and economic reforms, a concern that forced them to take to the streets to demonstrate their dissatisfaction.[28] In addition to these demonstrations, alleged communist activities against the state drew the attention of the business community as they became concerned about instability to the business climate. Responding to these concerns, Marcos declared martial law in September 1972.

Martial law had two significant effects. Technocrats were given relatively free rein to govern the country's economic development program.[29] The Philippines experienced a 13 percent rise in the terms of trade, an expansion of exports, and a 9 percent growth of real GNP.[30] Economic growth was assisted by a swift increase in government expenditure, particularly public investment, supported by somewhat higher tax collections and an increased use of foreign funds. Martial law also provided Marcos with the means to transform the domestic political structure. By skillfully using patronage mechanisms to entrench and consolidate his power, reinforce or strengthen those of his political allies, and weaken those of his adversaries, Marcos managed to fortify himself in an incontestable political position through policies designed to capture and channel economic rents toward newly created political support groups. Monopoly and monopsony positions were created in key industries (particularly in the traditional export sector) that granted exclusive rights to particular firms or individuals who supported Marcos. Furthermore, government power was used to force the transfer of assets from one owner to another to nurture these new relations.[31] Marcos used patronage mechanisms brilliantly, but it was not new to the pages of Philippine politics. In the end, Marcos eclipsed the power of elites that had dominated Philippine politics for centuries and developed a rent-seeking and rent-distribution government that over time would sap the energy of the domestic economic and contribute significantly to the economic crisis of the 1980s. A generic definition of rent-seeking is "to describe behavior in institutional settings where individual efforts to maximize value generate social waste rather than social surplus."[32]

INTERNATIONAL FINANCE, DEVELOPMENT, THE WORLD BANK, AND THE IMF

As a developing country, the Philippines lacks the capacity to develop independently its economy. Countries require capital for infrastructure, agriculture and manufacturing projects, and ultimately education, among other things. The Philippines had the tax base to provide the domestic component of revenues, but it lacked the bureaucratic capacity to secure them due to high levels of bureaucratic corruption. It had difficulty collecting taxes on existing activities and products; the GOP also had difficulty deriving new forms of taxes. This problem stemmed from the nature of the political structure where the Congress was responsible for taxation. Frequently, heads of the financial bureaucracy would always prod the chief executive to push Congress along, but the legislature always resisted these efforts, especially near election time. The executive was also reluctant to move on revenue generating measures. While corruption fueled the country's incapacity to generate sufficient domestic revenues, it typically also led to the shoddy quality of work on such projects. For projects fully financed by domestic sources, pork barrel type relationships between governments and recipients often made for inefficient enterprises and of dubious long-term quality.[33]

Cesar Virata, Marcos' second finance minister, was charged with both reforming the government and developing the country.[34] The Congress and the president always resisted the implementation of new revenue generating measures, which required Virata to search for a measure that could achieve both. To generate revenues, he needed international assistance, but the foreign creditors and donors often uncoordinated their financial assistance so a mechanism was needed to organize and coordinate the foreign actors. With the advice of his predecessor Eduardo Romualdez, Virata inquired with the World Bank if it would lead and coordinate a Consultative Group to promote Philippine development.[35] With the World Bank in charge, it coupled financing with reforms and development projects. Japan and the United States were the two main public players of this consortium, not surprisingly the largest private commercial lenders also came from these countries.[36] The Consultative Group also provided the auspices for the Philippine Assistance Plan, sometimes referred to as the Multilateral Aid Initiative.[37] Not only could they generate the necessary capital needed for massive projects, but the World Bank could ensure that work quality would be top rate.[38] Solely domestic endeavors could lead to suboptimal projects. Corruption, inadequate or poor oversight or audit practices could account for this. International-domestic reverberation ensures, at least in theory, competitive bidding practices, vetting of contractors, high quality techniques and equipment, and more rigorous audit practices.[39]

The international benefits were critical, but the Consultative Group also provided a more significant role, break the domestic policy obstacles by forging coalitions. Philippine cabinet members are equal in rank, each guarding and promoting his or her portfolio and typically committed to maintaining the status quo. Agents that represent industry and trade, since they represent a multitude of interests, are more likely to support protectionism over export-oriented policies.[40] To soften these positions, Virata would set the agenda for cabinet meetings on development opportunities, informing them that foreign financing would be available for development projects that followed the World Bank "rationalization" path.[41] This trail was to be a collaborative effort between international and domestic players where foreign financing would account for 60 percent while domestic sources would have to make up the remainder.[42] As a result, the finance minister also has to argue for new revenue measures to complement foreign financing.[43] The World Bank also served a functional purpose. According to Virata, the World Bank also provided powerful leverage in dealing with domestic resistance.

> Well, we—well, at that time Minister [Jaime] Ongpin [Corazon Aquino's first Finance Minister] I would say was more agreeable to the reforms because we had to deal with the protected sector, we had to deal with them and they have large political power. So and of course there would be arguments: where are you going to get foreign exchange if you liberalize imports? And precisely the objective of the loan was to help us bridge this gap where there will be an apparent increase in imports so we got supported by the World Bank.[44]

Alliances relied on more than just capital; it also relied on information. Frequently, the World Bank would produce reports that addressed areas of critical need. The assumption is typically made that the World Bank and its staff originate these reports then disseminate them to interested parties. In reality, the impetus for the report begins with the GOP finance minister, who sets the agenda. The GOP report is then handed off to the World Bank, which then produced its own report based on the same information and most of the time, the same conclusion. The interaction between the GOP and the World Bank would then create a constant network of policy dialogue. At times, the World Bank required information that particular government agencies would provide. National Economic Development Agency (NEDA) eventually served as the domestic research link between the GOP and the World Bank.

NEDA sometimes provided the major input of the IFI's reports.[45] Most of the time, this exchange of information and ideas was cordial, but disagreements arose from time to time. When such instances occur, staff members from both sides articulate, refine, and deliberate on issues such as the pace or appropriateness of such reforms. In all cases, the government's position takes precedent.[46] The prime minister also relies on the World Bank for its technical experts or recruit individuals with specific development expertise. Capital, information, and expertise served as leverage for the finance minister against domestic opposition to reform.

Similarly, the finance minister also needed to build coalitions at the international level. Virata and his successors realized that influence could also be manufactured with the World Bank and the IMF. Counties with large quotas automatically have loud voices. Small countries such as the Philippines can be typically seen, but not heard. To amplify its voice, the Philippines under Virata attempted to influence the institutions decision-making process by putting Filipinos in key positions on the governing board, namely the executive director, the alternate directors, and the technical assistant. The board members for both institutions were filled by distinct regional groups of voting bloc. The key for Virata was to get the Philippines into one of these coalition-voting blocs. The Southeast Asian block would have been the logical choice with which to join, but the weight of Indonesia and Malaysia made being able to secure the coveted position virtually impossible.[47] Furthermore, some blocs would allow access to only particular positions such as the alternate director, not the executive director. Instead, it initially entered the coalition bloc of East Asian countries, Taiwan, Vietnam, and Korea that applied a rotation scheme to the appointment process. Being the last to join, the Philippines eventually assigned Ernest Leung, one of Virata's most trusted lieutenants, to the IMF technical assistant position. After some issues with Taiwan later developed, the Philippines managed to temporarily place Placido Mapa Jr. to the IMF executive director position. Over the course of Virata's tenure, the GOP "butterflied" its way to regional group after regional group to secure a plum position. With the World Bank, it ended up joining the Latin group headed by Brazil.

For the IMF, the Philippines took advantage of an opening. For the IMF, the Philippines joined the Middle East bloc headed by Saudi Arabia. The rotating scheme it used had the advantage of being a regularized system where the parties would have certainty about who would be assigned and when. But the opportunities for board placement were far and few between. In contrast, for the IMF the Australia, New Zealand, and South Africa group used a less formal system where appointments would be assigned periodically on a regular basis. When the bloc received significant international pressure to kick out South Africa for its politically untenable apartheid policies, it asked the Philippines to join. This invitation brought together the Philippines and Australia into one grouping where both would have relatively equal voting weights. The Philippines was a developing country with a large population, and Australia was a highly developed country with a relatively small population. South Korea would eventually join this bloc. Over time, the system became institutionalized so it would outlast Virata's stint as finance minister.[48] The virtue of this strategy was that it not only served as an entry the Philippines to have its interests represented within international organizations but also as a training ground to have well qualified Filipinos demonstrate their skills at the international level.[49] The regional bloc to valued board positions served the GOP well.

Another path toward board positions was through issue area and common positions. This method allows countries to influence a broader network of issues. The Philippines was eventually elected to the Group of 24 representing Southeast Asian nations, a nomination that made it eligible to have headed the Development Committee. Created in 1974, the committee advised the World Bank and IMF Board of Governors on development issues such as oil shocks, macroeconomic adjustments, and eventually trade and environmental issues. In 1976, Virata was nominated to chair the committee. As chair, he could set the agenda and had access to numerous policy-makers.[50] Under these institutional paths, Virata, his successors, and others had access, but not necessarily influence, with critical policy-makers such as Robert McNamara, Paul Volcker, H. Johannes Witteveen, and Jacques de Larosière.[51] Access, not meeting of the minds, was the intent behind these institutional trails as Mapa would clash with McNamara on population growth issues.[52] These back channels provided the Philippines with an excellent opportunity to make its voice louder in a noisy international setting.

These positions were critical for delivering credible input in the financial process. By design, members appointed to these coveted positions were skilled, technically qualified, and well versed in economics. For instance, in the case of the Asian Development Bank (ADB), where Malacañang appointed directors, occasionally a candidate would fill a position based on political largess rather than meritorious qualification, as done by Presidents Joseph Estrada and Gloria Macapagal-Arroyo, which would sully the reputation of the organization.[53] Fortunately, appointments like these were not translated to the international level. Political connections did occasionally play a role in appointments as did economic interests. When Philippine teams attend the annual IMF-World Bank meetings, delegates include former executives of foreign banks that advise the Philippine Central Bank. This included Tristram Belplat, a confidant of Marcos and a top executive of Manufactures Hanover Bank, the company that led the private lender advisory committee.

During the 1980s debt crisis, private creditors formed an arrangement to bargain with debtor countries. The Philippines had approximately 480 such creditors and negotiations with all parties would prove untenable. As a result, the creditors selected twenty banks to serve as representatives for the remaining. Typically, when negotiations between the GOP and the international creditors took place, the Philippine Central Bank took the lead with a team that comprised both members of the Central Bank and the Ministry of Finance.[54] The negotiation process was also informal and quite painstaking as the GOP team would have to address each advisory creditor and listen to unique problems, proposals, and recommendations. Once they convinced the twenty, convincing the remaining creditors was easier.[55] Finance delegates took the lead when negotiating with the Paris Club, a more structured process where their counterparts would consist of a group of negotiators typically from Japan, Germany, US development agencies.

The commercial creditors were typically difficult to deal with. According to Felix Enrico Alfiler, researcher with the Central Bank, the negotiations typically took place under terse circumstances.

> There were 480 commercial banks we had to deal with, under different regulators with different view on how to approach debt restructuring. To begin with, they were already mad at us, thinking that we screwed them.
>
> That is why the IMF becomes important. These private banks rely on the ability of the IMF to surveil and assess a country, and to tell them when to bail out a country. It's a symbiotic relationship, but there can also be some tension between the commercial banks and the IMF, as when the commercial banks rely on the IMF to know what's happening in a country, and then they find out that the IMF didn't exactly know what was happening. So, in our case, in a way, nobody really knew. But at the end of the day they still had to rely on the IMF.[56]

As a side-bar, when the Philippine delegation would negotiate with foreign public and private creditors, Leung would use a tactic to get a "psychological edge," discussing and showing the private creditors the financial limitations of the delegation. Leung would say, "We're from a deb-ridden Third-World country. It's only proper that we scrimped on lodging (in a three star hotel), unlike you guys. You can afford it."[57] These psychological tactics, in addition to the demonstration of modesty, also translated to Virata.[58] Other lead delegates such as Gregorio Licaros, Jose Fernandez, and Finance chiefs Vicente Jayme Jr. and Jaime Ongpin lived extravagantly.

THE IMF: AN INTERNATIONAL FINANCIAL NETWORK AND PREFERENCES

In 1944, the Bretton Woods Conference gave birth to twins, the IMF and the World Bank.[59] The IMF, our primary concern, was given the responsibility to manage its principals' fixed exchange rate system for the purposes of promoting free trade that serves as a means toward international economic growth. It guarded against the 1930s "beggar-thy-neighbor" policies that saw trading partners devalue their currencies to gain a competitive advantage against one other. The result of these policies, in addition to protectionist policies, was a reduction of global economic output and growth. Invariably, principals could run into short-term macroeconomic imbalances, driven by apparent short-term capital or trade deficits. To assist its principals, the IMF was also chartered to provide an opportunity to adjust these imbalances through short-term financing. Functioning like a Christmas Club, the IMF generates this liquidity through its principals' contributions. The members contribute money to the IMF based on their quota, a set amount based on the size of a country's economy, the level of international trade, and the size of its international reserves.[60] As of this writing, the United States has the largest quota at approximately 17 percent of the total. Japan has the second largest with approximately 6 percent with China a close third. These quotas also nearly reflect the weighted voting of the IMF members. The United States has a little more than 16 percent of the total votes, Japan has 6.16 percent, and China has a little more than 6 percent.[61] These loans have strings attached where governments accept the money but must make policy adjustments to correct the source of the imbalance. These loans append the term "conditionality," where financing is exchanged for policy adjustments. When the West went off the gold standard in 1973, the IMF lost one of its raison d'être, but kept the other major function—the international lender of last resort.

The IMF is organized in a hierarchical fashion. Each principal has a governor and an alternate governor seat in the Board of Governors, the controlling body of the IMF. Technically, it controls the IFI, but since it meets only once a year, it must delegate the day-to-day responsibility of control to another body, the Directors of the Executive Board. Serving two-year stints, the twenty-four directors control the IMF. It hires an executive director, typically a European, whereas the World Bank has a US citizen as its president. The executive director functions like the chief executive officer of a company and is the chair of executive board. He directs the IMF staff and its lead negotiator. The director functions as the COG (Chief of Government) in the Two-Level game framework; the lead negotiator as the CN.

The IMF bases the terms of its conditionality policies on two tenets. These concepts are lodged on the foundation of classical and neoclassical economics, the concept of equilibrium as the aim of IMF analysis, assessment, and prescription to resolve economic problems and crises.[62] In simple terms, an equilibrium occurs when supply and demand are rationalized, meaning they efficiently allocate resources so that surpluses or shortages are absent, or at the most, tolerable. A system reaches disequilbrium when "changes in demand, opportunities, or relative prices that gives an economic actor an incentive to change his or her behavior in order to increase his or her gains or decrease his or her costs."[63] According to the IMF, economies in disequilibrium or macroeconomic imbalances need to be stabilized. One reason for this imbalance, according to the IMF, derives from what it refers to as the "absorption approach."[64] This theoretical construct posits that a country in macroeconomic imbalance "absorbs" excessive imports relative to its capacity to produce.[65] Aggregate demand is increased by government expansionary fiscal and monetary policies. The absorption approach was formally modeled by then Director of Reserarch at the IMF Jacques J. Polak, who developed a formal equilibrium construct that staked out the role of excessive credit through expansionary monetary policy.[66] This extension simplified the assessment and prescription for the IMF staffers, allowing them to point to monetary policy as the culprit for short-term macroeconomic imbalances. Its simplicity was the key as it was then "relatively easy" for IMF staffers to assessment economic problems, but it also had some flexibility to take particular circumstances in to account. In short, the Polak model required a credit contraction, an intentional slowing of the economy, to allow for market forces to correct themselves. The Polak model took hold as the theoretical construct upon which 1960s IMF analysis, assessments, and prescriptions would be based.[67] The model has subsequently been refined by "financial programming" models.[68]

Despite the formalized construct with which the IMF operated, conditionalities were loosely applied, if at all. The first oil shock of the 1970s witnessed a number of industrialized counties or developing countries that approached the IMF window with little or no requirements for changes in policies.[69] Furthermore, loans given had increased after the shock, but then it strangely dropped several years after, producing the perception that the IMF was standing on the sideline.[70] This reality and perception changed after the second oil shock when IMF lending accelerated dramatically. This quicken pace was due to both the criticism the IMF received for its seeming disappearance prior to the second oil shock and the push from Managing Director Jacques de Larosière to engage and assist countries in need.[71] Subsequently, the IMF increased its loans, but with relatively loose conditionalities. The new lending pattern produced expected results. Without strict conditionality, the IMF could no longer safeguard its resources. Borrowers had difficulty repaying the lender; some countries went into arrears.[72] By mid-1981, this type of lending reversed as conditionality "retightened." The source of this reversal emanated from industrialized countries.[73]

The 1982 Debt Crisis gave the IMF more leverage to bargain with tight conditionality policies. The IMF, in addition to industrialized country governments and commercial banks, provided credit in exchange for policy adjustments. It required strict conditionalities to "reduce public sector expenditure and investment, eliminate government subsidies, increase the costs of goods supplied by the government, increase income and the sales tax, set positive real interest rates to discourage capital fight and increase savings, rationalize and stabilize the exchange rate, and reduce inflation."[74] These prescriptions would later be termed by John Williamson as the "Washington Consensus," the neoliberal guide to economic policy-making.[75] In general, these policies through deregulation and privatization extracted the state from the economy. The IMF recognized that government price controls, perhaps just as much as exchange rate policies, helped to produce macroeconomic imbalances.[76] The central point to these policies was to safeguard the international financial system, primarily its creditors, by forcing the debtor countries to tighten their belts and absorb the cost of adjustment while freeing up savings to repay the principle and interests payments on their debts. In essence, the debtor countries bore the costs of adjustment. Ngaire Woods aptly phrases the process for debtors: "The result was that debtors had to adjust hard."[77]

Embedded with the Washington Consensus was the emphasis to generate economic growth. Prior to the debt crisis, the IMF equation had included structural economic issues, which was included in the original Article IV consultations. These reforms were required to combat persistent structural forms of unemployment and stagnant economic growth.[78] Structural unemployment argues that the supply and demand for jobs are in equilibrium, but the supply of jobs may exceed the demand of jobs in one sector while the demand for jobs might exceed the supply of jobs in another. Cyclical unemployment, in contrast, is the macro-level demand for jobs that is not met by the macro-level supply of jobs due to a reduction of economic activity. By the time the Ronald Reagan administration came into full swing and pushed its agenda for "supply side economics," the IMF had already begun fleshing out its themes and propositions of structural adjustment and included them in their conditionalities. According to James Boughton, the IMF staff was consistent on the issue of petroleum prices, "almost all arrangements approved in the aftermath of the 1979–1980 increase in world oil prices required the borrowing country to price petroleum products domestically so as to cover fully the cost of importing or producing the oil."[79]

For the IMF, the preferences of applying the "Washington Consensus" was consistent from the 1980s into the early 1990s, the time duration of this study. According to Woods, these ideas found a nesting place in the halls of the IMF where they would be sustained by particular mechanisms. First, the staff derives primarily from industrialized countries, especially English speaking ones. Second, the staff shares similar graduate training in approach, application, and conceptualizations. In this context, they train in deductive logic taken from neoliberal oriented models, paying little attention to cultural and social particulars. Finally, they steer away from political ideological applications.[80] The distinction between the Washington Consensus and the structural adjustment programs was the formality. The former was driven in large part by the Polak model, but its structural adjustment provisions were not. Boughton sums it up. "In contrast to the Fund's conventional adjustment levers—monetary, fiscal, and exchange rate policies—here was no generally accepted model or paradigm linking specific structural policies either to macroeconomic performance or to external viability. The (Executive) Board in effect was authorizing the staff to experiment further in trying to develop a general approach, but structural conditions would still have to be applied case by case."[81]

In the case of the Philippines, the IMF and the World Bank worked in tandem to drive home structural adjustment via trade liberalization policies. In the early 1980s, the World Bank and the GOP negotiated a Structural Adjustment Loan (SAL) that eventually went off line. The IMF added the World Bank's preferences to the list of conditionalities. Furthermore, industrialized countries were constantly pushing for trade liberalization, a preference that the IMF had to honor.

The IMF preferences for the reduction of petroleum subsidies and trade liberalization had more to do with just balance of payments deficits adjustments; they also had to deal with debt issues. Moving toward rationalization of fiscal policies would contribute to the GOP's capacity to repay its debts. Furthermore, liberalizing imports would, ceteris paribus, rationalize the intermediate products that contribute to the added value of export products, increasing the level and value of exports to aid in balance of payments. Over the course of the observations, foreign financing became the key point in dealing with the IMF. As foreign creditor and donors, both private and public, would only deal with the GOP after it secured an economic program, the IMF saw its leverage increase. The Baker and Brady plans dealt with new ways to deal with developing country debt, although the plans did not favor indebted countries as much as they would have preferred. Still, the context of foreign financing revolved around IMF negotiations.

GOP PREFERENCES AND ISSUE CONTEXT

Just as the IMF applied relative constant pressure for petroleum and import reforms, GOP preferences, too, varied, but at the margins. For GOP CNs, primarily the finance minister and secretaries, a successfully negotiated IMF program produced a multifinality effect. It opened the spigot to private and public international donors and creditors. It also kick-started financing to an economy that stalled because it lacked foreign exchange to pay for its imports. Finally, it was used as leverage to reform sectors of the economy that were obstructed by political concerns and interests. An IMF program served numerous purposes.

For Marcos' Finance Minister Virata, the IMF served two purposes. First, it allowed him to rationalize the economy, and he used the IMF/World Bank for both leverage and higher levels of financing.[82] A technocrat who gained the favor of the IMF and World Bank, Virata used their technical expertise to provide international solutions to domestic problems and used the combined weight of the institutions to broaden the agenda.[83] The causal process also worked in the other direction. The World Bank would only increase development funds if the GOP could increase its revenue capacity.[84] Second, Virata needed the IMF in the mid-1980s to kick-start a moribund economy that lacked the financing necessary to pay for critical imports such as oil and other products. COGs typically shared the development preference but they also had patrimonialism and legitimacy preferences.[85] COGs preferred to meet all three preferences simultaneously, but in the long run, they were incompatible. While COGs had variations in preferences, none of these can be viewed as monolithic. For instance, development goals have been a "wavering constant," but even technocrats had varied approaches and paces that would split the cabinet.[86] In addition to these divisions, factions existed that supported the position of Marcos cronies and the First Lady, Imelda Marcos. Similarly, the early Aquino government years had significant cabinet factions. By the mid-Aquino years, technocrats had more influence on economic policy-making, but divisions on the pace and timing remained.

Legitimacy preferences anchored domestic petroleum prices. Starting with the Marcos years, the GOP socialized petroleum pricing by subsidizing prices through a price buffer fund, all toward the goal of social peace.[87] Invariably, this insulation from market forces required some adjustment as neither the GOP nor oil corporations could maintain constant prices in the face of rising international prices. Marcos and his NEDA often discussed prices but inaction was often the solution to "allow conditions to reverse themselves and thereby resolve the problem."[88] This "substitutive" informal process was challenged as the IMF required government deficit reduction. Once the agenda was set, an acknowledgment on the principle of price increases, an "accommodative" informal corporatist tripartite meeting with NEDA, oil companies, and the Marcos controlled Trade Union Congress of the Philippines (TUCP), which was represented by the labor department, discussed the levels.[89] Typically, labor demanded increases in wages or cost of living adjustments. Afterward, an "accommodating" institutional arrangement proceeded where the Board of Energy publicly formally ratified the decision. Then, groups outside the tripartite, typically transportation organizations, which affect commuters, followed by businesses, informally voice their preferences via either access to policy-makers or demonstrations and strikes.[90]

The tripartite process continued under the Aquino and Ramos governments.[91] The GOP invited all labor sectors to participate in a transparent and quicker process in exchange for a restraint on strikes.[92] The TUCP left the process in 1986, and the *Kilusang Mayo Uno* (KMU/May First Movement), a conglomeration of nine labor unions with more left ideological leaning, headlined the union banner. In exchange for their participation and their restraint on striking, Aquino instituted union friendly policies.[93] She later shifted her policies when domestic and foreign businesses pressed, forcing KMU and other left oriented unions to strike and demonstrate against petroleum price increases.[94] Rivalries and different expectations among the unions, however, muted the effectiveness of this mechanism to challenge government policies.[95] This process varies from imports.

Import controls, as well as exchange rate policies, were shaped by the prevailing industrial policy. In the early 1970s, the Marcos regime shifted government policies toward export-oriented industrialization, but cronyism, domestic resistance, and corruption hampered the transition. The 1980s marked a significant departure, at least in intent, when the GOP secured a World Bank Structural Adjustment Loan (SAL) program that attempted to liberalize gradually the economy from 1981 to 1985 through the Tariff Reform Program (TRP) and the Import Liberalization Plan (ILP). The former successfully reduce tariffs, and the latter ran into numerous difficulties reducing nontariff barriers.[96] These policies were pushed by government technocrats, individuals that pushed to rationalize the economy and who sometimes pursued policy projects sponsored by the IMF.[97] Despite their presence, technocrats faced a substitutive institutional framework with Marcos as the ultimate Level IIA veto player, pitting the country's development interests against "cronyism (which) ultimately determine(s) the agenda of economic policymaking."[98] Level IIB actors also had some influence in the "accommodating" decision-making process. Often, industrial associations shielded from market forces threatened to either reduce production or shut down completely. Frequently, their interests were protected by government officials with ties to these corporations and often served as their mouthpieces. To offset their reform costs, the GOP often used "revitalization" policies, giving industries cheaper credit and technical assistance or by lowering tariffs for intermediate and capital goods.[99] In October 1983, the GOP aborted the SAL program due to balance of payments problems.

Economic policy-making during the early Aquino years focused both on development and legitimacy. Constrained by the country's $28 billion foreign debt,[100] the GOP renegotiated the terms of previous loans, and restarted new financing, giving it room to pursue its legitimacy and development goals. As the country democratized, many foreign actors (the IMF, WB, and the United States) gave the Aquino administration plenty of leeway to reform the previous structures. Commercial banks were not so generous. On the domestic side, Aquino tried to balance the preferences of a broad-based coalition that swept her into office, often juggling developmentalism and legitimacy, and to some extent, patrimonial preferences. Initially, she used her unchecked powers to sacrifice Marcos cronies, but the reintroduced legislature in 1987 constrained her efforts to reform the non-crony elements of the economy.[101]

Unlike the narrow decision-making process of its predecessor, the Aquino administration emphasized consensus, especially with IMF negotiations. Under the Marcos government, the CNs often made commitments with their IMF counterparts they could not keep since they failed to consult with other agencies. To resolve this problem and provide a defensible position in their IMF dealings, in 1986 the GOP widened the scope of participation as the CNs first talked to GOP agencies, securing a common position, before they bargained with the IMF, an "inside-outside," approach.[102] Aquino remained the ultimate veto player.

The Level IIA petroleum price process shifted to a "complementary" arrangement. A cabinet subgroup determined increases, with the COG sometimes reinforcing the decision. A September 1987 military coup attempt, which correlated the two developments, produced a joint-agency cabinet committee that addressed the implication of price hikes. Later adding the military, the committee would be given veto rights over all cabinet proposals. The military's role in politics during the democratization process never wavered; furthermore, it also pressed the Aquino administration for budget increases.[103] On the mission and the budget, both the military and the government saw eye to eye. Then General Fidel Ramos succeeded in lodging the military's role in politics, at first on a de facto basis,[104] but eventually institutionally through its inclusion with the Cluster for Political and Security Affairs, later called Cluster E.[105] This standing committee had a gatekeeping function on petroleum price adjustments, among other things, and frequently frustrated numerous finance secretaries' attempts to increase prices. Furthermore, Aquino admonished cabinet members who submitted price increases to the full board without the cluster's approval.[106] To secure this consent, finance ministers would attempt to increase the military budget[107] or use the prospect of foreign loans as leverage to bypass resistance.[108]

The Ramos administration was perceived to be impervious to a coup given his background. This was not the case. His finance minister relayed, "Every time we tried to raise the issue of oil prices, someone in the (cabinet) meeting would just make a short statement that considering continuing threats of coups, it was untimely to adjust the prices. And that was it."[109] To maintain military support, Ramos used financial instruments, promises of modernization, and appointments; yet, financing was in short supply as the military base rents accounted for 90 percent of the military budget.[110] This shortfall had to be filled either via domestic resources, which were limited, or foreign financing, which gave more leverage to the CNs. Ramos could authorize a modernization deal, done in 1996, only after he secured an IMF program in June 1994.

Like the petroleum price process, the decision-making process for trade was every bit as political. During the Marcos years, trade policy was shaped in relatively closed quarters. Technocrats from the NEDA and the Ministry of Trade and Industry pushed for liberalization while the Tariff Commission (a division of NEDA) held legally mandated public hearings before any tariff adjustments. These accommodating hearings served not only as the mechanism for consultation between the ministries and Level IIB actors, but also for determining an incentive priorities list.[111] Before the SAL program, these hearings were taken lightly by the industrial associations since they produced little and that Marcos typically determined many critical rates by decree.[112] The Aquino government opened up the process considerably by allowing for consultation through the trade and industry ministry[113] and other bureaucracies. The process continued to resemble the "substitutive" Marcos process as new Trade and Industry Minister Jose Concepcion Jr., interjected his protectionist preferences.[114] In 1987, when Congress was restored, it joined the fray as it gave itself the power to delay action on tariff changes and conduct public hearings.[115]

CONCLUSION

Like many developing countries, the Philippines has a long history replete with economic and political struggles. From the arrival of the Spanish to the withdrawal of the American military presence, foreign sources of power have had a significant role in determining the content and scope of Philippine public policy. As the main beneficiary of these policies, land holding elites continued to perpetuate their hold on political power. This political domination was eventually challenged and undermined by Ferdinand Marcos, but the power shift came at a high cost as the Philippines became mired in unmanageable debt, arrears that significantly contributed to the balance of payments crisis in the early 1980s. Since the 1983–1984 crisis, the GOP has "partnered" with the IMF in what seems like countless times. The following chapters cover these pains.

NOTES

1. Hawes, 21.
2. Official Handbook of the Philippines and Catalogue of the Philippine exhibit, 215–217.

3. Ibid., 106.
4. Karnow, 200 and Owen, 105.
5. Karnow, 200.
6. Steinberg, 75.
7. Owen, 105–106.
8. Jenkins, 34–37.
9. Ibid., 37.
10. For the key provisions of the Bell Act, see Schirmer and Rosskamm, 88–90.
11. Ibid., 94–96.
12. Ibid., 96–100.
13. Ibid., 247–248.
14. Wurfel, 81.
15. Ibid.
16. These phases were originally suggested by Jagdish N. Ghagwati and Anne O. Krueger and are explained in Baldwin, 12–16.
17. Ibid.
18. The International Monetary Fund also had a hand in the dismantling of foreign exchange controls. Licbauco, 31.
19. Stauffer, 12–25.
20. Doronila, 131–132.
21. Dohner and Intal, 382.
22. Ibid.
23. Baldwin, 67.
24. Dohner and Intal, 382.
25. Baldwin, 76.
26. Dohner and Intal, 383.
27. Baldwin, 76.
28. Wurfel, 113.
29. Dohner and Intal, 389.
30. In large measure, the economic upturn can be attributed to a worldwide upswing of commodity prices in 1972–1974. Ibid.
31. See Dohner and Intal, especially Chapters 2 and 4. For an in-depth look at the sugar and coconut industries see Hawes; for an insightful look at the financial sector, see Paul D. Hutchcroft, 1993.
32. Buchanan, 4.
33. Sicat, 279.
34. Ibid., 270.
35. Ibid., 251.
36. Ibid., 492.
37. Dolan, 185.
38. Sicat, 270.
39. Ibid., 270, 279–280.
40. Ibid., 291.
41. Ibid., 271–275.
42. Ibid., 275.

43. Ibid., 272.
44. Interview with Virata, World Bank History Project, 13–14.
45. Sicat, 287.
46. Ibid., 276.
47. Ibid., 642–643.
48. Ibid., 646.
49. Ibid., 643–646.
50. Ibid., 650–652.
51. Ibid., 670, 741n. 5.
52. Interview with Virata, 12.
53. Sicat, 648.
54. Yuson, 58.
55. Ibid.
56. Ibid., 59.
57. Ibid., 61.
58. Sicat, 656.
59. This section is derived from Boughton, Woods, and Vreeland 2003 and 2007.
60. Vreeland 2007, 14.
61. IMF Website, "IMF Members' Quotas and Voting Power, and IMF Board of Governors." <https://www.imf.org/external/np/sec/memdir/members.aspx>. Site accessed on September 6, 2016.
62. Gilpin, 55–57.
63. Ibid., 55.
64. Boughton, 559.
65. Woods, 40.
66. Boughton, 559.
67. Ibid., 559.
68. Woods, 41; Polak 1997; and Sau 1992.
69. Boughton, 560.
70. Ibid., 560.
71. Ibid., 562–563.
72. Ibid., 564.
73. Ibid., 564–566; Woods 42–43.
74. Woods, 48.
75. Williamson, 1990.
76. Boughton, 589.
77. Woods, 53.
78. Boughton, 90.
79. Ibid., 589.
80. Woods, 55.
81. Boughton, 590.
82. Broad 1998, 74–75; and Montes 1991, 30–31.
83. Tadem 2012, 27; Tadem 2014, 354.
84. Sciat 2014, 271–273.
85. Montojo 1999, 66–70.

86. Tadem 2012, 27–29; and Tadem 2014, 354–357.
87. Sicat 2014, 529.
88. Ibid., 53.
89. Ibid., 529; Kuruvilla 1995, 131; and Hince and Lynch 1997, 155.
90. Boudreau 1996, 68.
91. West 1997, 108–124.
92. Ibid., 109.
93. Ibid., 109; Lindsey 1993, 170–171.
94. Lindsey 1993, 170.
95. West 1997, 110.
96. Dohner and Intal 1989, 446; Alburo 1991, 160; and Baustista 1989, 21.
97. Montes 1991, 31.
98. Ibid., 33.
99. Bautista 1989, 21 and 39.
100. Timberman 1991, 335.
101. Haggard 1990, 250.
102. Interview with Senior Budget and Management official, anonymity requrested, May 23, 1995.
103. McCoy 1999, 261.
104. Timberman 1991, 259.
105. Chambers 2014, 108.
106. Interview with Senior Budget and Management official, anonymity requested, May 23, 1995.
107. Timberman 1991, 259.
108. Velasco 1991, 191; Chambers 2014, 105.
109. Author interview with Ramon del Rosario, Secretary of Finance, Department of Finance, Republic of the Philippines; Metro-Manila, Philippines, May 1995.
110. Chambers 2014, 115.
111. Alburo 1991, 161; Montes 1991, 33.
112. Baustista 1989, 39.
113. Alburo 1991, 161.
114. Author interview with former Senior Trade and Industry official, anonymity requested, August 7, 1995.
115. Alburo 1991, 163.

Chapter 2

1984 GOP-IMF Petroleum Price Observation

This chapter focuses on GOP-IMF bargaining process as the parties negotiated the terms of an economic program and applied particular bargaining tactics to derive a favorable agreement and change perceptions of costs. This chapter is divided into six sections. Following this introduction, the next section paints a brief overview of the government controlled Philippine petroleum environment. The following sections narrate the story behind the politics of petroleum in the Philippines, with an introduction to the context, the international and domestic political economy dynamics, and an epilogue to the program. Squeezed between the domestic political economy narrative and the epilogue is a section that assesses the causal mechanisms at work.

PHILIPPINE PETROLEUM INDUSTRY

To comprehend fully the nature of IMF-GOP negotiations regarding the link between the GOP's fiscal problems and the oil sector, we should lay out the nuances of the Philippine petroleum industry. In particular, this section briefly narrates the price buildup of petroleum products, government policies and procedures, and the price buffer scheme. In the mid-1960s, the GOP began its modest involvement in the domestic petroleum market, an affair that intensified over the decades.[1] During this period, the GOP involved itself in myriad of petroleum activities: transportation, exploration, refining, and the process of petroleum pricing.[2] The petroleum-pricing process proved to be a politically sensitive and highly controversial issue for the GOP. The heart of the controversy surrounded the price build-up schema, a formula that changed in value to meet GOP policy preferences. The price build-up formula of local petroleum products was as follows:

1. DOCT + Tax + price buffer impost = WPP
2. WPP + hauler margin + dealer margin = pump price
   *DOCT = direct oil company take
   *Tax = ad valorem and/or specific
   *price buffer impost = Republic Act 6173 Special Fund or Consumer Price Equalization Fund (CPEF) or Oil Price Stabilization Fund (OPSF)
   *WPP = Wholesale Purchasing Price

The GOP policy preferences revolved around a myriad of goals: attainment of economic growth, efficient allocation of resources, revenue generation, price stability, price equity, and demand management.[3] Since these objectives can hardly be met simultaneously, trade-offs were often inevitable. Based on its policy preferences, the GOP judicially adjusted the tax component to meet its goals. Over time, the GOP discovered the effectiveness of taxing petroleum products. In the early 1970s, the GOP adjusted the tax component on local petroleum products to shape consumer behavior, curtail public consumption, and minimize the country's dependence on foreign oil.[4] Simultaneously, the GOP used taxation measures to promote one sector over another as it highly taxed gasoline, a commodity used primarily by the upper- and middle-income sectors, and marginally taxed diesel and fuel oil, commodities used primarily by the low-income sector and the country's manufacturers. Later, scarce foreign and domestic financing forced the GOP to develop other methods of generating revenues. The GOP found that the taxation of petroleum products helped restore the government's financial position. As a result, the tax component value reached staggering proportions. In 1973 the total taxes of premium gasoline consisted of 22 percent of the pump price. By comparison, the government's share of the pump price reached an unprecedented 52 percent in 1986.[5] Levels of DOCT increased as well during this period, but the rate did not match the pace of tax increases.

Whatever the policy preferences of the GOP, domestic oil companies had to receive an adequate and constant DOCT level. The GOP fixed the DOCT according to a complex formula that involved the following components: the rate of return, the cost of freight and insurance, and the landed cost of crude. Of these components, the landed cost of crude was more likely to be subject to extreme fluctuations. The GOP set a reference rate for the landed cost of crude based on the cost of Arab Light crude and the foreign exchange rate between the Philippine peso and the US dollar. Changes to these variables ultimately altered the cost of crude for domestic oil companies, either providing them with more revenues or less revenues; however, since the GOP determined the DOCT value, oil companies were prevented from earning excess or suffering deficient revenues. Should the landed cost of crude drop, oil companies remitted the excess profits into a special price buffer fund and withdrew funds when the opposite occurred. The buffer fund became the instrument that addressed a multitude of interests.

The price buffer funds were the catchall mechanisms, functioning for both principals and agents. The 1973 Republic Act 6173 created a special reserve funded by earmarked imposts on petroleum products. First, the funds generated revenue from imposts on petroleum products transactions by both oil companies and consumer activities, a Philippine "petro-peso" if you will. In a closely related function, they also financed energy related projects, a reserve that was frequently tapped by the Ministry of Energy (MOE).[6] Third, they also kept at arm's length the distributive conflict between consumers and producers of petroleum by insulating the former from domestic changes to the costs of petroleum products. This protection was not restricted to domestic cost as the funds also protected oil companies from changes to the landed costs of crude, especially from the frequent depreciations of the Philippine peso. Oil companies were permitted to request reimbursement funds when unfavorable changes in the landed costs exceeded the GOP set reference rate. The funds served their purposes when receipts were in surplus; when they were in deficit that was another matter. In several cases the Special Fund and its successor buffer mechanism were depleted of funds or in deficit, and the oil companies had little choice but to petition the Board of Energy (BOE), an arm of the MOE, to restore or reset the value of DOCT by adjusting petroleum prices. Despite the GOP's best intentions to keep the Special Fund and one of its successor price mechanisms, the CPEF, financially solvent, both schemes ran into liquidity problems and had to be dismantled.[7] In October 1984, the GOP reintroduced the price buffer fund by establishing the OPSF.

Unlike the previous pricing mechanisms, the OPSF insulated both domestic and international transactions from vacillating market forces. Upon its inception, the OPSF subsumed the Philippine Central Bank responsibility of maintaining the controversial forward cover practice.[8] Oil companies normally engage in oil procurement contracts with payment terms of thirty days. Due to the foreign exchange crisis of 1983, the GOP sought to limit the shrinking supply of foreign exchange by prompting domestic oil companies to enter into contracts that extended the terms of payment to one year. In exchange, the GOP guaranteed to reimburse oil companies for costly deprecations to the Philippine peso. GOP fiscal losses in this agreement were partially offset by attaching a small financing cost on the extended credit period. For the oil companies, a grace period of one year on their oil payments permitted them the rare opportunity to "double dip." First, they profited from their domestic sales without paying for their oil cost. Second, financing earmarked for petroleum imports could be invested into other projects before the procurement contracts were due.[9]

The price buffer funds principally served to protect the financial interest of domestic oil companies, but they also produced other effects. First, it insulated the BOE from undesired public scrutiny. Should the oil companies submit price increase petitions to the BOE, a hearing must be held in public. These hearings were typically met with public protest from labor, transportation, and consumer groups. Since oil companies could request price buffer fund reimbursement, they rarely had to request formal pump price increases due to changes unfavorable to the landed cost of crude. As a result, bureaucrats were frequently shielded from making difficult economic decisions, which ultimately had political implications, regarding price increases. In essence, the price buffer funds minimized the amount of public "protest in front of his [the bureaucrat's] office."[10] Second, the price buffer funds violated the GOP principle of price equity. Under the conditions of a favorable landed cost of crude and a buffer fund surplus, the price buildup required consumers to pay more for the price of petroleum products. They pay an additional price buffer impost and derive no immediate benefit. In effect, consumers were overcharged.[11] Finally, it served as a redistribution mechanism that minimized distributive conflict between the principals, the oil companies and consumers, keeping the principals at arm's length.

Despite the price buffer mechanisms' residual effect on the consumer, the GOP attempted to maintain a price scheme designed to preserve consumer "price equity." The Marcos regime established the "socialized pricing" scheme, a setup that give top consideration to lower income and manufacturing sectors in structuring price levels and minimizing the domestic impact of a price increase. As a result, gasoline products and other so-called luxury fuels bore the cost of subsidizing other petroleum products such as diesel, kerosene, fuel oil, and liquid petroleum gas (LPG). The Marcos government maintained this policy for the better part of its years to placate the consuming public. The Aquino administration had designs to move away from this practice, but it reluctantly continued the scheme, as did the Ramos administration. The OPSF regime was effectively dismantled in 1998.

The socialized pricing scheme also extended to other sectors in an attempt to achieve "social peace."[12] Under strict market conditions, increases to the price of petroleum reverberated to other parts of the economy. Unions would demand that wages increase to shield them from the costs of adjustment. Non-unionized labor groups would simply pass on the cost to consumers. Transportation companies would request fee and fare increases. Business would also pass along the cost to consumers. In the end, the costs of recovery or adjustment would typically be fully born by the non-organized lower-income sector. This process was what Marcos tried to arrest. According to former general director of the National Economic Development Agency (NEDA) and economic planning minister under Marcos, Gerardo P. Sicat, the president was reluctant to allow market forces to determine outcomes that would create tension among the sectors.[13] To maintain stability, Marcos instituted a corporatist or tripartite arrangement among government interests, businesses, and labor.[14] This institution functioned to coordinate the preferences among the participants, getting them to "cooperate," getting them to mutually adjust their policies/demands.[15] The government would act as a buffer between the rational demands of NEDA against the Ministry of Labor, acting to shield its principals from the market erosion of benefits by granting cost of living adjustments.[16] The usual outcome would be a union wage increase. The wage adjustment did not end the process. It continued as subsequent changes would be made by transportation associations, which would petition government commissions, for fare and fee increases. Business enterprises would then bear the costs of adjustment since the government would institute price controls. Typically, the costs would be diffused among principals and the bureaucracies, not one bearing fully the costs of adjustment.

The tripartite process continued under the Aquino and Ramos governments.[17] The GOP invited all labor sectors to participate in a transparent and quicker process in exchange for a restraint on strikes.[18] The TUCP left the process in 1986, and the Kilusang Mayo Uno (KMU/May First Movement), a conglomeration of nine labor unions with more left ideological leaning, headlined the union banner. In exchange for their participation and their restraint on striking, Aquino instituted union friendly policies.[19] She later shifted her policies when domestic and foreign businesses pressed, forcing KMU and other left oriented unions to strike and demonstrate against petroleum price increases.[20] Rivalries and different expectations among the unions, however, muted the effectiveness of this mechanism to challenge government policies.[21]

The BOE and its variants, the price buffer fund, and the tripartite process formed a set of formal rules and norms that assess, implement, and ratify the price of petroleum products. The rules and norms of the process are clear. Petroleum companies and the public interact in a rational game where the price reflects the equilibrium point of supply and demand. When external shocks inflect the price, the companies draw down from the buffer fund to maintain the price, either drawing down the surplus or producing a deficit. Once the latter reaches a threshold, a line determined by the GOP, it signals the publicly owned petroleum company to petition for a price increase, the move then triggers a move from the two other companies, coordinating the petitions of the oil companies. The commission then convenes a public forum to determine the validity and documentation for the price increase. Afterward, affected bureaucracies and private veto actors look to offset their costs and pass through the tripartite or their respective agents, requesting increases in the costs of their products at levels commensurate to the price increase. After the process has been completed, chief executives maintain the prerogative to ask the BOE to reevaluate its decision, which will either reverse in the entirety or in part. The rules and norms on this process are precise and unequivocal; however, the rules and norms are also subject to a series of fuzziness.

The rules and norms have been adapted to secure an equilibrium. At the price request end, a bargain can be reached if the hearing announcement and the hearing's duration are shortened or lengthened. Furthermore, a more inclusive tripartite that accounts for the broader range of labor would mute the number of unions that would protest at the increase, reducing the intensity of strikes, protests, and demonstrations. In addition, Tripartite meetings dates and times are variable, not fixed, which usually precipitates some type of price hike. After the announcement of price increases, organizations with no immediate interests access the backdoor of bureaucracies, the legislature, and the COG. The process also works in reverse where organizations and their representatives would lobby veto players that resist the rationalization of prices.

1984 GOP-IMF PETROLEUM PRICE OBSERVATION

In the early 1980s, the Philippines exemplified the economically troubled developing country.[22] The second international oil crisis forced the Philippines to pay nearly double the value of its oil imports while importing almost 20 percent less than the quality of oil imported in 1973.[23] The international economic recession made it difficult for the Philippines to export its goods as industrialized countries reduced their levels of imports. In 1983, foreign commercial credits were completely severed. These developments magnified the already precarious Philippine economic structure. Since its inception, the Marcos government squandered billions on unproductive infrastructure, agricultural, and energy projects, leaving the country with little to show for its massive investments. In 1982, a paper magnate with several billion pesos in outstanding loans fled the country, leaving behind a battered and shaken financial community. During this time, the country experienced extensive levels of capital flight, a trend intensified by the August 1983 assassination of Marcos' political nemesis Benigno Aquino, which forced the authorities to announce a ninety-day moratorium on its principle payments to its foreign commercial creditors—the first of five consecutive declarations to be issued. Later, the GOP surreptitiously overstated the country's international reserve levels to the IMF, forcing the international financial institution (IFI) to withdraw from the December 1983 Standby Agreement negotiations. These developments contributed to the formation of a precarious Philippine economy where the country's external capital and current accounts were pushed severely out of balance and the authorities begun to lose control of the government deficit.

To resolve the external and internal imbalances, the GOP primarily focused on the need to restart trade financing, a move that would lead to the restoration of economic activity. To reach this goal, the GOP had to go through the IMF and secure an economic stabilization program and subsequently IMF "seal of good housekeeping." In the negotiation process, the GOP delicately balanced the consistent requirements of the IMF with the preferences of its domestic Level II constituents. The IMF was prepared to take a hard line, moving from a relatively "loose" distributive position to a "tight" one given the overstatement debacle. It coordinated its efforts with the World Bank to link stabilization and structural adjustment measures into an integrated package and utilized its control over access to other sources of external finance to force policy actions prior to reaching an agreement.[24] In short, the road to external financing went through the IMF. It focused its position on strengthening the country's external current and capital accounts while allowing for the elimination of arrears and other restrictions on current payments and on imports. To meet these ends, the IMF insisted that the GOP drastically reduced its domestic demand without fueling inflation.[25] Although the GOP realized that it needed to increase petroleum prices as part of the remedy, it delayed these measures due to the May 1984 parliamentary election and the existence of traditional opposition to petroleum price increases. By altering the decision-making process and issuing side-payments, the GOP managed to assuage Level II dissent and implement a critical component of its agreement with the IMF.

INTERNATIONAL POLITICAL ECONOMY OF THE 1984 PETROLEUM PRICE OBSERVATION

Negotiations between the two parties progressed intermittently, a process that normally takes three months was extended to a year. After the talks began, the GOP postponed the process and its implementation measures due to concerns regarding eroding domestic political support for the Marcos administration. After the GOP resumed its implementation measures, the IMF found these moves to be inadequate and issued a stem warning to the GOP to intensify its commitments. The GOP heeded the IMF admonition and complied, but the talks reached a crescendo as the GOP initiatives failed to keep pace with the deteriorating economic conditions, a situation that prompted the IMF to threaten to sever negotiations unless the GOP extended its commitments. The GOP again complied but the talks stalled as the IMF backed off a third time and placed additional requirements before the GOP to secure financial commitments from the country's private and public creditors. Nearly one year after the negotiations started, the GOP managed to fulfill all the IMFs requirements and secured an economic stabilization program.

In the early 1980s, both domestic and international developments staggered the Philippine economy. In late 1981, industrialist Dewey Dee fled the country and left behind approximately P635 million in unpaid debts.[26] This development left numerous large domestic financial institutions scrambling for financing to cover their loses. In August 1982, Mexico announced a moratorium on its private foreign debt payments, caused in large part by the increase in US interests rates. Strung by this announcement, large and small commercial lenders became skittish and either stopped or reduced continued lines of credit. This reaction had an obvious effect on other sovereign borrowers as they had to either follow suit or continue to pay. A year later, Benigno Aquino returned to the Philippines from his exile in the United States and was assassinated when he attempted to disembark from the airline. These three developments slowly strangled the Philippine economy, each one tightening the hold more as foreign lines of credit were near impossible to secure. It was under this backdrop that Philippine Finance Minister Cesar Virata attempted to secure a deal with the IMF.

Virata's mission was to get commercial banks to restart lending the Philippines. This goal started in October 1983 after the Mexican announcement. The international financial community concluded that Mexico had serious problems and needed to be rescued. Just as importantly, big creditors stopped lending and were soon followed by smaller banks. This perception and response dramatically impacted the Philippines, which was considered a minor issue as the Dee fallout was relatively contained, but it still contracted the economy. Virata knew that the country's reserves would not hold out and wanted to reschedule the country's debt payments, not pursue debt relief. Virata's calls to meet with financial creditors were rebuffed, save for Manufacturers Hanover Bank, the lead in the consortium of international credit lender, which agreed to provide $60 million.[27] He also shuttled to Washington and Tokyo to meet with government officials to perhaps leverage their commercial lenders but the response was the same in both instances—the governments and the banks were separate entities and influence from the government would not be forthcoming.[28] Meanwhile, the World Bank approved its SAL II application. With World Bank money on its way and the credit lines freezing, the country's economic conditions were steadily deteriorating.

After the Aquino assassination, the macroeconomic deficit nose-dived. Speculation had increased. Reduced and arrested credit lines to the government and businesses strangled economic activity. In one day, the creditors withdrew $770 million.[29] This development pushed up the demand for dollars, which sent the Philippine peso into a spiral downfall. With the greenback in short supply, credit lines closed and there was no rollover of debt. The country's debt servicing had to be paid on schedule forcing the country to dip into its reserves and cash earnings, which were soon to be exhausted.[30] The IMF and the World Bank were needed to stabilize the economy, but this required Washington to be on board. The conditions were now significantly different now as Marcos' friends had abandoned him and were skeptical of his weakening grasp on power. Indeed, Washington was demanding significant economic reforms.

Bridging finance was necessary to continue to edge the economy along, but this financing was not forthcoming from traditional sources. Marcos subsequently instructed Trade and Industry Minister Roberto Ongpin seek financial assistance from friendly Southeast Asian neighbors. Malaysian Prime Minister Mahathir Mohamad and Brunei Sultan Hassanal Bolkiah were amenable to the request. Mahathir loaned $75 million; the Sultan, a good friend to Ongpin, loaned $150 million. The result from Singapore's Lee Kwan Yew was different. Instead of a loan promise, Ongpin and Lee debated the developments in the Philippines, with the latter concluding that after watching the crowds that attended the Beningo Aquino funeral, the Philippines was "undergoing a 'catharsis.'"[31] The loans were helpful, but terribly insufficient. They served to bridge the financial gap, not restore the Philippine economy's activities.

Washington played a significant role in the GOP's negotiations. A friend of Ronald Reagan, Marcos was viewed as the stronghold for American interest in the Pacific. Weakening of the Marcos regime would be sure to invite communists. Furthermore, the lessons of South Vietnam and specter of the communist takeover just proved too daunting to ignore.[32] The 1981 change in the US administration from President Jimmy Carter to Ronald Reagan proved dramatic. The Carter Administration pivot to human rights goals chastised Marcos for both martial law and his human rights violations. The change in administrations put the Philippines under a different light as Reagan embraced Marcos as a friend rather than keep him at a distance the way Carter did. This treatment played a significant role in Marcos' decision to end martial law in January, shortly before Reagan's inauguration.[33] Reagan himself was loyal to Marcos, but some of his cabinet principals and other staffers were skeptical of his rule. By the time of the 1983 assassination, many of the principals had shifted their preferences and were seeking ways to push him out of office. Others in the administration opposed the move, but some slowly started to shift their position. By early 1984, the State Department recommended using economic leverage to force reforms. Reagan "conceded to modest pressure."[34] This included using financial statecraft to "ease or tighten the burden of his foreign-debt burden."[35] While there were cracks in the armor of support, support still prevailed. The creditors, however, had another perspective. It viewed economic troubles in the Philippines a minor issue, a blip, where the Dee fallout precipitated a slowing economy. As a result of this perception, the international financial community scattered when Virata approached them about rescheduling debt payments, not debt relief. Mexico was viewed as a serious problem, the Philippines, which suffered a recession from the Dee fallout, was not considered as dire. As a result, creditors and the international financial community paid little attention to Virata.

The GOP started negotiations with the IMF on a new program as the old one was no longer sustainable. The development that followed next pushed negotiations to a standstill. When examining Philippine Central Bank statements, the IMF found discrepancies. These incongruences can be viewed through various perspectives. One position asserts that these overstatements were purely accidental and overblown. One bank dealing with the Central Bank allegedly forgot to log in the additional payment to the bank, inflating the Central Bank foreign exchange holdings. Another interpretation presents the event as a duplicitous effort. The Central Bank apparently shuffled its dollar holdings. The department of finance, principally Deputy Finance Minister Ernest Leung, investigated and discovered the trouble. Needless to say, the Central Bank leadership became suspect from here on out. The international reserve overstatement observation had severely damaged GOP credibility in the eyes of both the IMF and the domestic financial community, a confidence that desperately needed to be restored. After the overstatement discovery in December 1983, the IMF demanded the removal of Central Bank Governor Jaime Laya.[36] Marcos complied with the IMF demand, transferring Laya to another post. Meanwhile, Finance Secretary and concurrent Prime Minister Cesar Virata, a technocrat, led the GOP's efforts to find a suitable replacement acceptable to both the IMF and the domestic banking community.[37]

After careful consideration, Virata settled on the selection of Jose "Jobo" Fernandez, who was versed in orthodox oriented economics and an established member of the banking community. Virata and Leung drew up a list of possible candidates for the position, which included the names of Fernandez, Placido Mapa Jr., president of Philippine National Bank (PNB) and a former senior official of the World Bank, and Rolando Gapud, president of the Security Bank and Trust Company. Eminently qualified for the position, Mapa received serious consideration, but the PNB role in the overstatement raised concern that he would be tacitly implicated with Laya's mismanagement of the Central Bank records.[38] Given this constraint, Virata and Leung shifted their endorsement for the nomination of Fernandez. Despite Virata's standing in government and his impeccable credibility with the foreign financial community, he did not have Marcos' ear; instead, they informed Minister of Trade and Industry Ongpin, a confidante of Marcos, of the best candidate and Ongpin relayed the selection to Marcos.[39] According to a senior official, "Bobby said that Marcos didn't have much of a choice."[40] Ongpin was technically, "the guy to get the guy," who joined the Marcos cabinet as an outsider, and eventually earned the trust of Marcos.[41] Several days later, Marcos selected Fernandez to head the Central Bank.

Fernandez quickly gained overwhelming approval of both the IMF and the banks.[42] Having removed some of the tarnish left by the international reserve observation, Virata was anxious to revive the country's credit lines that had ceased in October 1983 by concluding the IMF negotiations and thereby getting the virtually stagnant economy back on its feet.[43] To resolve the crisis, the GOP concentrated on four strategies: demand management policies, exchange reform, incomes policies, and structural adjustments.[44]

Heading into negotiations, both parties guarded their optimism as an atmosphere of uncertainty surrounded the Marcos administration. Marcos did not have a clear policy mandate from the people and there was every reason to begin seriously to doubt the ability of the Marcos government to govern or to uphold or implement any agreements—particularly those being suggested that would cause more austerity and difficult conditions in an economy already reeling from four-and-a-half years of recession and stagnation.[45] To complicate matters, even Virata himself "was not clear of Marcos' hidden agenda."[46] As the final arbiter of Philippine economic policy, Marcos put up a veil of uncertainty between him and Virata, as he always carefully calculated the effects of IMF demands on his political constituency, cronies as referred to by Stephan Haggard.[47] Virata had to cut through the shroud, desperately needing Marcos' support to reach an agreement. To shift Marcos' cost perception closer toward the IMF target point, Virata often informed Marcos that a measure needed to be implemented because the IMF wanted it, using what is commonly referred to as the "lightning rod" strategy.[48]

Marcos was meticulous for being unresponsive, but this was planned rather than a character flaw. When crisis decisions needed to be made, he typically would default to inaction. The reasons for this behavior were threefold. First, he hoped that the conditions that created a crisis would somehow reverse itself and resolve the problem. This wishful thinking only postponed the need for inevitable cost recovery. Second, when conditions were no longer sustainable, he would gallop on his white steed into the fray to save the day, reinforcing his image as a concurring hero.[49] Finally, he would allow for social actors to petition their issues to the GOP and respond to them, again, also building up his image as a hero.

Marcos' bargaining strategy notwithstanding, the ideal first choice economic solution would have been to float the Philippine peso and implement austerity measures to reduce the impact of inflation.[50] By cutting government expenditures, increasing interest rates, and allowing exporters to continue their trade, the GOP would generate foreign exchange earnings and limit expenditures for "nonessential" imports. This plan would have been in line with the World Bank SAL programs to provide a structural shift with the economy, giving the GOP a "good bargaining chip with the IMF."[51] However, the GOP chose to not move in this direction and instead implement what could be considered a second-best policy—the installment of capital controls. One instrument of control was the creation of the parallel Binondo Central Bank, an institution designed to stabilize the exchange rate by intervening in the black market currency system. This mechanism is covered further in chapter 7.

The IMF typically dislikes the use of exchange rate controls as it violates its letters of agreement. This instrument, however, does buy the GOP time to slow the downward spiral of the macroeconomy, doing it gradually, not with shock therapy. This time also helps the GOP to phase in the cost of adjustment to the social actors. Its other advantage would be to allow the IMF to provide funding to close the foreign exchange gap. It also had the advantage of being used as a bargaining chip against the IMF to remove once the IMF provided funds for a new agreement.[52] While the foreign exchange regime was critical to stabilizing the economy, government fiscal shortfalls were also on the table. With these measures, the IMF required GOP prior commitments to offset the uncertainty surrounding the Marcos regime, measures that involved tightening fiscal and implementing monetary reform and structural adjustments.[53] These were designed to reduce domestic demand while at the same time restructuring the economy. The IMF took additional steps to ensure GOP compliance as it included other departments in the negotiations, such as budget, Trade and Industry, the NEDA and even Marcos himself met with the IMF staff on two occasions.[54]

Early in the negotiations, the IMF made it clear that the GOP needed to make substantial cuts to the proposed P6.5 billion government deficit by slashing the government's Consolidated Public Sector Deficit (CPSD).[55] The IMF insisted the GOP attain this target through cuts in the investment programs of the state-owned corporations, an increase in taxes, and tax burden shifts away from international trade and toward domestic transactions.

After the first round of negotiations, the GOP initially hesitated, then begun to take small steps in the direction prescribed by the IMF. The GOP hesitated to implement wholly the Fund's prior commitments due to the May 14 election of 183 *Batasang Pambansa* (Parliament) members. In consideration of the election, the GOP felt that initiating the IMFs prior actions at such a time would prove to be damaging politically for the Marcos government's performance.[56] This concern was manifested in the manner that the GOP attempted to meet a multitude of goals: raise revenue, curtail consumer and producer import behavior, and transfer foreign exchange savings into the importation of producer goods destined for the export sector. In April, the GOP took steps to meet these goals as it increased the rate of the import duty from 5 percent to 8 percent.[57] Although this measure was targeted to raise an additional P1.1 billion over the remainder of the year, the levy proved ineffective to alter consumer demand as net domestic petroleum consumption registered 5.2 million barrels for the 1984 first quarter and increased to 5.6 million barrels for the 1984 second quarter.[58] The reason for the lack of change in petroleum consumption was simple: the GOP neglected to allow the oil companies to increase the price of petroleum products to adjust for the new levy rate.[59] In June, the GOP continued to deepen its April move by increasing the import duty on crude oil by 2 percent and the specific tax on petroleum products by P0.08 centavos per liter on gasoline and diesel and by P0.04 centavos for fuel oil.

Increases in ad valorem and specific taxes and eventual devaluations of the peso put pressure on the GOP to pass on the increase in the price of domestic petroleum products proportionally. Indeed, during the IMF-GOP negotiations, the GOP allowed the domestic oil companies to increase the price of petroleum products three times (see Table 2.1). The first observation (May 18, 1984) occurred several days after the May election, a hike that resulted in a marginal increase of P0.40 centavos per liter. According to the BOE, oil companies were allowed to increase petroleum prices to recoup their losses incurred from

Table 2.1 Historical Petroleum Product Prices (Per Date of Adjustment 1-19-1971 to 12-11-1984) (Metrol Manila) in Peso per liter

Dates	*Premium*	*Regular*	*Diesel*	*Kerosene*	*Fuel Oil (Wholesale)*
1/19/1971	0.36	0.3	0.25	0.25	0.16
1/20/1973	0.38	0.32	0.27	0.27	0.18
10/20/1973	0.54	0.48	0.4	0.34	0.2
11/27/1973	0.65	0.59	0.48	0.45	0.25
2/18/1974	1.02	0.96	0.79	0.74	0.51
4/16/1974	1.15	1.09	0.88	0.85	0.59
5/16/1975	1.28	1.22	0.95	0.92	0.65
1/16/1976	1.5	1.4	1.13	1.04	0.76
4/21/1977	1.81	1.66	1.21	1.12	0.79
3/23/1979	2.23	2.07	1.42	1.37	0.93
8/1/1979	3	2.8	1.69	1.66	1.12
2/2/1980	4.5	4.3	2.4	2.4	1.53
8/8/1980	4.95	4.75	2.81	2.81	1.79
3/22/1981	5.25	5.05	3.11	3.12	2.09
7/1/1983	5.47	5.27	3.43	3.44	2.4
11/3/1983	6.47	6.27	4.43	4.44	3.4
5/18/1984	6.89	6.69	4.89	4.89	3.66
6/7/1984	8.28	8.03	6.2	6.21	4.65
10/20/1984	9.24	8.81	7.26	7.17	5.22
12/11/1984	9.24	8.81	7.26	7.17	5.23

Source: Republic of the Philippines, Energy Regulatory Board.

the April 3 percent levy increase.[60] The second round of price hikes came at the heels of the June devaluation of the Philippine peso and the increased import duty. This time, the price increase produced a more transparent effect as the average price of petroleum products increased by Pl.27 per liter. Oil companies were the beneficiary of this price increase round as they received an average DOCT increase of Pl.19 per liter while the GOP received a marginal increase of P0.06 centavos per liter increase for its coffers.[61] The third round of increase occurred in October 1984, when talks came on the verge of breaking off. This increase will be discussed in more detail in the following section.

The second round of price increases came at the heels of the Fund's dissatisfaction with the GOP prior measures. On a critical point, the GOP continued to practice an expansionary monetary and fiscal policy despite the tight credit crunch. In effect, the GOP was stepping on the gas pedal while it was also stepping on the brake. Another issue of contention with the IMF concerned the GOP exchange rate policy. In early June, the GOP amended its 1983 policy that required banks to sell all of its foreign exchange to the Central Bank by reducing the requirement to only 80 percent of their foreign exchange holdings. The gradual "freeing up" of foreign exchange trading exerted dramatic pressure on the peso and forced the GOP to devalue the Philippine currency from P14 to P18 to the $1. This exchange rate, however, had different effects on various sectors of the economy as the GOP also initiated a stabilization tax on numerous external transactions, thus effectively creating a multiple exchange rate system.

The IMF insisted that the GOP continue its efforts to deepen the reform prerequisite measures. The GOP measures, however, steered toward the wrong direction and jeopardized the talks. Afterward, the IMF negotiation team warned the GOP of IMF Managing Director Jacques de Larosière's insistence that the exchange regime be dismantled and the reduction of Central Bank foreign exchange holdings from 80 percent to 50 percent. Should the GOP fail to comply with these demands, the IMF would "defect" from the talks.[62] In addition to these demands, the IMF wanted the GOP to introduce a P8 billion package of tax measures to the 1985 budget; P7 billion would come from increased sales and excise taxes, taxes on coconut and sugar industries, and petroleum product taxes. These moves fell in line with IMF recommendations that the GOP shift the weight of its revenue sources from taxes on foreign trade (particularly exports) to domestic activities. Furthermore, the IMF also insisted that the GOP remove tax exemption preferences from the Philippine National Oil Company (PNOC), National Food Administration (NFA), and National Power Company (NPC).[63]

Loud signals for reform also came from other powerful international sources. The IMF held a precarious position in its dealings with the GOP. Taking a distributive bargaining position, the IMF demanded that the

GOP initiate harsh stabilization and structural adjustment measures, but its bargaining position was compromised by the United States, Japan, and the World Bank. These international Level II actors continued to funnel money to Marcos and minimized any incentive to surrender his position. In fact, the Level II actors reinforced Marcos. Why should he capitulate to IMF demands for reforms when funds continued to roll in, albeit at levels lower than he desired? But eventually, the domestic landscape, especially in the United States, changed. In early August 1984, the US Congress announced that it had lost confidence in Marcos, a position that the Reagan administration eventually adopted. In early October, the United States and Britain voted against a World Bank-GOP agricultural loan because not enough had been done to dismantle the sugar and coconut monopolies. The Reagan administration was under significant pressure to free overseas markets. Approval of the loan went through, although the United States officially refused to grant any further economic assistance to the Philippines until an agreement was reached with the IMF.[64]

Backed into a corner, the GOP yielded to the IMF demands. Given the dramatic shift in international support, Virata had indirectly received support for his appeals to Marcos to initiate reform programs. His arguments on the Philippine position became reinforced.[65] These acknowledgements for change, however, could not reconcile the hemorrhaging of actual foreign obligations with the promises of future foreign aid. As a result, in October 1984, the GOP was forced to announce its fifth moratorium on the principle payments of its foreign loans. Meanwhile, it also initiated a series of policy changes designed to address stabilization. The crux of these reforms rested with dismantling the tacit multiple exchange rate system and replacing it with a system classified by the IMF as "free floating."[66] Afterward, the peso exchange rate immediately fell to Pl 9.99 to $1, a depreciation of 11 percent. The currency depreciation subsequently affected the price of domestic petroleum products. The GOP did more than merely allow oil companies to increase petroleum product pricing as it also made considerable adjustments to petroleum products price buildup. These taxes led to the third increase in the price of petroleum products by an average of P0.87 centavos per liter. Instrumental to the increase was the manner in which the increase occurred, as the IMF insisted that the distortion among the fuels be removed.[67] While simultaneously increasing the price of petroleum products, the GOP also assured the public that afterward, stable prices would be in place through the reintroduction of the petroleum price buffer fund in the form of the OPSF.[68]

The reestablishment of the petroleum price buffer scheme did not develop from a vacuum. Throughout the negotiations the IMF pushed the GOP to initiate a rational petroleum product-pricing scheme.[69] The GOP repeatedly acknowledged the value of a market oriented pricing structure in terms of oil company benefits and a reduction of government subsidies, but it also realized the political value of maintaining stable domestic prices in the face of constant changes to the landed cost of crude and insisted that it be allowed to establish another price buffer fund. Both sides compromised on the creation of the OPSF, but under the condition that it be phased out, the duration of which was never discussed.[70] In conjunction with the creation of the OPSF, the GOP agreed to remove the forward cover credit mechanism from the auspices of the Central Bank and transfer it to the OPSF. Under the new scheme, the MOE became responsible for administering the OPSF while the Ministry of Finance (MOF) promulgated the necessary rules and regulations. The decision to place the OPSF under the auspices of the MOF reflected the IMF confidence in Virata as it also requested that he take responsibility for the operations of the state-owned nonfinancial agencies.[71]

With these additional policies in place, the GOP reached a tentative agreement with the IMF on an economic adjustment plan and a $613 million standby credit. Despite receiving an endorsement from the IMF negotiation team, the GOP still had to overcome the objections of de Larosière and the IMF executive board. De Larosière agreed "in principle" to the terms of the economic program but had reservations about the Philippines' ability to persuade its foreign private and public lenders to carry the balance of the GOP's $3.3 billion fiscal shortfall.[72] To ease de Larosière's doubts, the GOP needed to secure firm commitments from the private financial lenders before he endorsed the program. One portion of this condition was settled in November 1984 when the ten-bank advisory committee for the Philippines' commercial lenders agreed to supply the Philippines with new financing. The second concern was met in December 1984 when the country's public lenders, known as the Paris Club, agreed to reschedule the Philippine's debts. The final obstacle that remained came from the IMF executive board. When presented with the Philippine Letter of Intent, the board initially withheld its approval, only until a passionate plead by Virata that the GOP would live up to its commitments outlined in the Memorandum of Economic Policy (MEP).[73] His speech found sympathetic ears as the "IMF board trusted him and it knew that he always tried to put in place the right [market-oriented] policies."[74]

On December 14, 1984, the IMF executive board approved the GOP Letter of Intent. According to the arrangement, the IMF would disburse Special Drawing Rights (SDR) 615 billion ($613 million) in seven tranches over an 18 month period. The Philippines would only have access to a little over half of this facility, or SDR 335 million ($352 million), due in part to an agreement between the IMF and the GOP that the Philippines must pay off its previous IMF loans.[75] In exchange for the Standby Agreement, the GOP agreed to implement economic policies aimed to reduce the deficit, curb inflation, and address the weakness in the structure and management of the economic system.[76] Given the emergence of major burdens on the budget, the delay in support of guaranteed obligations from the major financial institutions, as well as the delayed implementation of budgetary measures in 1984, the budgetary targets of the economic program did not start until 1985.

In all, the program design was tight. Too tight, according to Virata.

> I disagreed with the IMF's formula for controlling domestic credit. Their main control point was to rein in the increase in net domestic assets (NDA). I told then IMF Managing Director J. de LaRosierre [*sic*] that their program was too tight. It went against the grain of conventional wisdom. The peso was appreciating versus foreign currencies and there was really no need for them to impose very tight controls. But he told me to just grin and bear it, saying that the IMF's credibility will suffer if they agree to modify their program immediately. So we just bit the bullet.[77]

After the IMF accepted the program, the GOP pursued other sources of foreign financing. First, the GOP renegotiated with the Paris Club. An agreement required that the club agree on common principles, but it still required each individual country to sign off on the agreement. Afterward, the GOP had to renegotiate with its commercial creditors.[78]

DOMESTIC POLITICAL ECONOMY OF THE 1984 PETROLEUM PRICE OBSERVATION

Despite its negotiations with the IMF, the GOP remained careful not to upset the already fragile domestic political landscape. Repercussions from the May 1984 *Batasang Pambansa* election, investigation proceedings from the Aquino assassination, and numerous protest movements intensified public scrutiny of the Marcos administration. Acting quickly and decisively, the GOP managed to implement petroleum price increases while simultaneously being sensitive to public needs. The GOP addressed these concerns by distributing side-payments and implementing "safety nets" designed to protect consumers against profiteering firms. Although the GOP used these strategies, it also managed to draw attention away from the GOP-IMF negotiation process and paint the necessity for price increases as a matter of economic-national security.

After the *Batasang Pambansa* elections, the GOP faced stronger legislative resistance over economic policy-making. The election results were hardly surprising, as Marcos' *Kilusang Bagong Lipunan* (KBL) party won 114 of the 183 contested seats, and the opposition coalition, the United Nationalist Democratic Opposition (UNDO), the *Lakas ng Bayan* and Pilipino

Democratic Party (PDP) coalition, took the remaining 69 seats. Despite its "success" in maintaining political support, the Marcos government received increased criticism over its management of the economy. Specifically, parliamentary opposition and business leaders questioned the utility of Virata's position as prime minister and minister of finance arguing that his position in government served as only a token gesture to lend the GOP credibility before the eyes of the country's foreign creditors.[79] Furthermore, the legislature considered filing a vote of no confidence against Virata.

Parliamentary and business criticism of Virata and the Marcos regime notwithstanding, the GOP began to implement IMF prior action measures to increase petroleum product prices. The first price increase occurred without warning.[80] Shortly after the May election, the oil companies forwarded their price increase petitions to the BOE, a process that normally goes before the public. With little or no public notification, the BOE reviewed the petitions and the next day it authorized a petroleum products price increase by an average of P0.42 per liter. The BOE justified bypassing the public hearings by claiming that the increase was necessary to indemnify the oil companies for the 3 percent import levy increase initiated in April. In other words, the price increase merely restored the DOCT ratio before the implementation of the additional levy. In a television appearance, Marcos defended the move as necessary to minimize consumer consumption of petroleum products and divert the saved foreign exchange for the import of production goods used to stimulate export production. In addition, he defined the issue to be a matter of public concern where economic growth would continue to and increase the number of jobs. Never once during the telecast did he mention the ongoing IMF negotiations.

Several weeks after the May price increase, the GOP adopted two reform policies intended to appease the IMF: the "deregulation" of the country's foreign exchange regime and an additional 2 percent increase in the ad valorem duty on oil imports. These reforms had dramatic effects on the landed cost of crude and predictable results on the price of petroleum products. Feeling the added pinch, the oil companies again petitioned the BOE for an increase in pump prices. Unlike the previous price increase observation, the BOE opened its doors to public scrutiny; two days later the board allowed the oil companies to increase the price of petroleum products by an average of P1.27 per liter. For the five kinds of petroleum products, DOCT increased by an average of P1.19 per liter while the GOP received an average of P0.06 per liter for this series of price increases.[81]

The "October capitulation" by the GOP again resulted in changes in the landed cost of crude, a move that had dramatic repercussions for many Filipinos. Days after the GOP dismantled the tacit multiple exchange rate, the oil companies petitioned the BOE for an average increase of P0.29 centavos per liter. Like the "token" hearings of the second price increase, the BOE held a one-day, scarcely attended public hearing on the request. Two days after the oil companies made their request, the BOE allowed the price of petroleum products to increase by an average of P0.87 centavos per liter, three times the amount originally requested. The GOP rationale for the increase became obvious: it took advantage of the opportunity to restore the DOCT level as well as fill its coffers through a dramatic restructuring of the price buildup. It eliminated the energy development impost on gasoline and fuel oil, an assessment that carried a value of P0.63 centavos per liter and P.0035 centavos per liter, respectively. The government also added an ad valorem tax based on the wholesale posted price of petroleum products, and it adjusted the specific tax designed to go to government coffers. Finally, the price buildup also included a P0.10 centavo per liter across the board impost designated for the OPSF. In total, the GOP slightly reduced the average DOCT from 83 percent to 78 percent of the wholesale posted price while it increased the specific tax from 7 percent to 11 percent of the wholesale purchase price and tacked on an additional 13 percent in ad valorem taxes designated to go to the GOP general fund.[82] The GOP reshuffling move was intended to increase government revenue by P3.14 billion from Pl 1.87 billion to Pl 5.02 billion.[83] Besides reshuffling the petroleum product price buildup, the GOP also followed the IMF "recommendation" of removing the distortion among the petroleum products. Diesel and kerosene fuels, used primarily by the lower-income sector, recorded an increase of 62 percent during this period. Premium and regular gasoline, used primarily by the middle and upper income sectors, incurred an average increase of 42 percent increase. Fuel oil, used by local manufactures, also increased by 53 percent from the November 1983 price levels. In the end, the price increases rationalized the petroleum products, eroding the socialized pricing policy Marcos relied on to minimize tension among the affected sectors.

Ultimately, Marcos had little choice but to follow the economic convention of cost recovery. Typically, in circumstances of costs adjustment with market forces, Marcos' first instinct was to do nothing, hoping the trend would reverse itself and wipe away the need for action.[84] Typically this first choice tactic allowed conditions to worsen, increasing the eventual costs of adjustment. Indeed, when divine intervention failed to arrive, the second choice tactic would come into play, with seemingly political effects. He allowed the circumstances to crescendo and then provided commanding leadership with viable solutions, the knight in shining armor who rides to the rescue scenario.[85] From the principals' perspective, heroes that champion their cause and resolve problems earn their political support, at least in the short term.

Riding to the rescue was well exemplified in the petroleum price increases. During these three increases, the Marcos government convened the Tripartite.

The parties involved were Marcos, NEDA, the oil companies, the Employers Confederation of the Philippines (ECOP), and the Marcos controlled Trade Union Congress of the Philippines.[86] Typically labor demanded increases in wages or cost of living adjustments, but the ECOP opposed these demands on the grounds of an already weak economy where such adjustments would drag production and eventually lead to employee layoffs. After the BOE announcement on petroleum price increases, the National Wages and Productivity Commission (NWPC), an agency of the Ministry of Labor and Employment (MOLE), agreed to set a higher minimum wage and increase the cost of living adjustment but not at the levels sought by labor. The GOP also managed to appease its employees before and during the price increase rounds. In early May the GOP announced a 10 percent government employee wage increase, an obvious attempt to garner the support of public servants. In June, the GOP announced another pay increase, the largest to date, a P8 minimum wage and allowance for nonagricultural workers. Indeed, the GOP's efforts to pass on the cost to less influential sectors of the economy contributed to the inflationary pressure of 1984, prompting the IMF to later urge the GOP to initiate wage controls. Consumer groups received some compensation for their troubles. The GOP agreed to institute safety nets that ensured that prices of daily staples would not increase accordingly, at least in the short term. Managed by the Ministry of Trade and Industry (MTI), price ceilings were maintained through networks of surveillance and threats of punishment for merchants who immediately increased the price of staples.

For those groups outside of the tripartite circle, they mobilized to demonstrate and strike. After the GOP initiated the first round of price increases, opposition from transportation associations, labor unions, and consumer groups began to mobilize. Although these groups held strikes to display their dissatisfaction with BOE decisions, they sought different economic remedies and often failed to coordinate their efforts. Still reeling from the late 1983 oil price increases, transportation associations, made up of bus and jeepney drivers, demanded commensurate fee and fare hikes with the price increases; however, after the first round of increases, solidarity among association members begun to wane as some groups refused to join the call for strikes. Marginalized during the Marcos regime, labor unions were more concerned about the inflationary impact of the price hike than the actual increase itself. To mitigate these effects, the labor unions, led by the far left leaning KMU, demanded increases in wages and cost of living allowances. Consumer oriented interest groups had the widest constituency base as they sought to protect the purchasing power of the lower-income sector and demanded that the GOP roll back petroleum product prices.

The GOP expected and received opposition to its revenue moves, and it made conscious efforts to minimize distributive conflict. The GOP agreed to hold hearings on transportation demands for increases in fares and fees; eventually, it authorized bus and jeepney associations to pass on the fuel increases to mass transit commuters. The fare and fee increases, however, were not commensurate with the petroleum price increases. This discrepancy could be due to the nature of the institutional arrangement where the level of petroleum prices were determined by the BOE while fare and fee rates are adjudicated by another agency, the Land Transportation Franchising and Regulatory Board (LTFRB), an agency of the Ministry of Transportation and Communications (MOTC).

For the most part, the GOP stayed within the confines of the political process when it came to mollifying opposition to the petroleum product price increases; yet, it also relied on distraction. During the last round of price increases, the long anticipated Agrava Report was released, a document that reported on the investigation of the Benigno Aquino assassination. The report's results covered the spectrum of the media, focusing attention away from the third and final petroleum price increase.

EPILOGUE TO THE 1984 PETROLEUM PRICE OBSERVATION

Going into the negotiations with the IMF, the GOP understood that it needed to make numerous concessions to get the economy back on track.[87] For the most part, the GOP managed to stabilize the economy at the cost of reducing economic activity to a virtual standstill. For 1985, the current account registered a positive gain of $1.03 billion, reducing the 1984 deficit from $1.11 billion to a more manageable $77 million. In addition, the overall balance of payments recorded a gain of $2.3 billion, achieved through an increase in Central Bank international reserves that climbed to $1.06 billion in 1985 from $88 million in 1984.[88] In addition, the GOP managed to contain the average Consumer Price Index (CPI) from a projected 2–5 percent to an actual 23.1 percent.[89] The GOP also increased its revenues aided in part, by the switch to ad valorem taxes in 1984 and the intensification of collection measures.

The GOP petroleum product price increases contributed to putting the brakes on the economy. As a rule of thumb, for every 1 percent increase in the real price of energy, the demand for energy drops by 0.22 percent.[90] Petroleum product consumption dropped 17.5 percent from 11.90 billion liters in 1983 to 9.81 billion liters in 1984. The use of asphalt dropped 27 percent from 1983 suggesting that road construction and repair did not pick up in 1984. Petroleum consumption of other sectors like paper processing, ceramics, sugar, and the textile mills also declined sharply by 48.7 percent, 48.1 percent, 32.5 percent, and 27.7 percent, respectively.[91] Even the biggest user of fuel oil, NAPACOR, (formerly the NPC) cut back on its use of fuel oil and slowly shifted to nonpetroleum energy sources such as geothermal, hydro, coal, and other nonconventional sources of energy.[92]

The IMF clearly wanted the Philippines to resolve its revenue problem through the use, in part, of a transparent petroleum-pricing scheme; however, economic fortunes did not favor the GOP as the OPSF drained GOP finances. In 1985, the cost of international crude dropped to $15 per barrel. In contrast to the private oil companies that purchased their supplies on the spot market, Petrophil, a subsidiary of the GOP-backed PNOC, purchased petroleum on a government-to-government contract basis to hedge against a disruption in the supply of crude. As a result, PNOC bought its crude at the agreed contract price of $29 per barrel. Given the high cost of its supplies, the OPSF reimbursed PNOC for P32.1 million out of a reimbursement total of P35.8 million for 1985.[93]

ASSESSMENT OF THE CAUSAL MECHANISMS

The causal mechanisms that produced the outcomes had various effects. The deference tactic was unclear regarding its effectiveness. At Level IIA, the evidence shows that Minister of Trade and Industry Ongpin was instrumental in moving things forward. A confidant of Marcos who became an indispensable sounding board, Ongpin was instrumental in implementing critical moves such as the naming of Fernandez as the new Central Bank governor. Indeed, the deference role he had on Marcos was clear. While Ongpin also played significant roles in securing the bridgeable loans from regional neighbors and the Binondo Central Bank scheme, evidence of his involvement in increasing the price of petroleum products would be at best circumstantial or at worst nonexistent. As a result, it would be difficult to show the application, let alone the support of the deference tactic.

This tactic served two purposes. First, with regard to the regional neighbors and Binondo Central Bank, Ongpin and his agency was a signal from Marcos to the regional and Chinese leaders of the Binondo Central Bank that their support was critical for securing foreign financing. This signal also worked in the other direction where Virata and Leung translated their thinking though a trusted instrument. Why might Virata use the deference strategy? The political noise surrounding the Philippine Central Bank, the IMF, and the selection of a new central banker required that credentials, not politics, be the guidepost. The other candidates also had impeccable credentials, but they also had close connections to Marcos, links that the IMF could not accept. Why was this tactic not used? First, the conditions have to be right; in this case, the IMF threated to derail the negotiations unless someone acceptable to the organization took the reins as the central banker. Given the shaky and corrupt political landscape that shaped the Marcos regime, the IMF could not accept someone with connections to the current regime, effectively taking Mapa and Gapud out of the running, putting Fernandez as a candidate of one, someone with familiarity with the banking industry and pro-markets. Every delay of an IMF program would delay negotiations with commercial banks, which was tantamount to increasing the financial gap between international revenues and expenditures and grinding the economy to a halt. This issue dwarfed the need to increase petroleum prices. Given the magnitude of the circumstances, it is unlikely Virata could use this tactic with frequency or its effectiveness becomes compromised.

At the Level IIB arena, Marcos clearly served as the deference vessel for securing social veto player support. Virata lacked the political acumen, institutional capacity, or even the charisma to win over Level IIB players. Much like the import episode, Virata had to rely on Marcos, who had the institutional powers to convene the Tripartite, had access to the media to marshal support via nationalism, and alter the rules via decree, even if martial law ended in 1981, to secure formal and informal ratification. Furthermore, Macros, not Virata, was the politician. Due to the broad groups affected by petroleum price increases, they would hear the Marcos call over a Virata call. At this level, the deference tactic explains much.

The external-internal tactic appears to be a more robust causal mechanism to draw in support for petroleum price increases. At Level II, the sequence of events showed the increased costs to garner domestic support for reform. Without the IMF program in place, the Philippines would be unable to restart the moribund economy, let alone bridge the balance of payments financial deficit. As the level of exports were relatively fixed and Paris Club members, especially the United States, were putting on political heat, the balance of payments problems could only be rectified with an IMF program. Intuitively, we can conclude that under these circumstances, Virata should have been able to convince Marcos that the economic costs of the absence of an IMF program outweighed the domestic political costs of reform. For Marcos, the costs associated with the United States was unclear as mixed signals emanated from Washington, allowing Marcos to initiate reforms at his own pace. The petroleum price increases in isolation are insufficient to address the price buffer fund deficit and government coffer shortages. With other mechanisms, they could have a pronounce effect. Similarly, as Washington pronounces its policy toward Marcos with more clarity, Marcos inches more and more toward reform. This mechanism explains both the decision by Marcos as well as the pace and timing.

At Level IIB, the external-internal causal mechanism appears nonexistent. Marcos clearly moves the veto players and the institutions around which they operate, but he did so without the association of the international financial climate, removing the costs associated with the IMF and financial creditors from their calculations. Instead, calls for sacrifice for the sake of nationalism become prominent. He neglected to include international negotiations as costs; instead, focusing exclusively on nationalism and economic recovery.

Both of these mechanisms have a role to play in both Level IIA and IIB. At the first arena, the external-internal causal link provides a stronger explanation of Virata securing Marcos' support. The deference link has little empirical support. At the Level IIB arena, the evidence supports a stronger cause and effect for the deference tactic over the external-internal link, only because negotiations with the IMF were omitted from the narrative. The deference explanation has a stronger causal impact.

Can another causal impact explain the GOP behavior in the 1984 petroleum price observation? The technocratic alliance is an attractive alternative explanation. According to the constructivist/epistemic community argument, the ideas of the free market, rationalization, and the Washington Consensus are transferred from the IMF to the GOP chief negotiators to the COG to the social veto players. The success or failure of this transference, and subsequently, the efficacy of the mechanism relies on the variation of the institutions. At Level IIA, Virata already informed Marcos that the IMF needed these reforms. The conversations between the IMF staff and Virata would explain the different price hikes, but it fails to explain the delay for the election and the three separate phased in periods. These maneuvers are laid out at the domestic level where both Virata and the IMF have little choice but to acquiesce. Furthermore, the IMF staff held numerous meetings with Marcos, indicating that the technocratic alliance managed to overcome any institutional obstacles to the COG; his political calculations, not institutional obstacles, served as the obstruction to reform. At Level IIB, the technocratic alliance lacks attractiveness as an explanation. First, evidence to suggest that the IMF met with businesses, labor, and consumer groups to convince them of the logic of reforms does not exists. This point alone should carry enough weight to weaken or invalidate this explanation. Assuming for a moment that the IMF managed to lobby these veto players, it would appear unlikely that the IMF could provide compelling arguments to alter veto actor preferences to absorb the costs without some guarantee of some form of side-payments, guarantees that would lack credibility. While these issues are not answered satisfactorily by the alliance, it does provide a clue to the creation of the OPSF. Ironically, the existence of the OPSF and the political cover it provided allowed for its abuse. Certainly the IMF recognized this problem. Ironically, it is also the IMF that sanctioned the creation of its successor mechanism, the OPSF, which will also suffered the same fate as its predecessor. The key to the IMF's acceptance to this mechanism seems to be tied to Virata himself as the administrator for its rules and policies. Evidently, the IMF and Virata had nearly similar preferences regarding the rational functioning of petroleum prices. They might have assumed that since Virata was also of like mind, he would have the capacity and wherewithal to manage the fund rationally.

Under this observation, we can derive several conclusions. First, the causal mechanism with the strongest explanatory power is the external-internal process at Level IIA. At Level IIB, the deference argument, tailored with an agenda setting mix, provides the best explanation. While the technocratic alliance does not have the explanatory power to explain the outcomes at Levels IIA and IIB, it can explain the creation of the OPSF.

NOTES

1. The petroleum market in the Philippines consists of numerous products: premium and regular gasoline, kerosene, diesel, fuel oil, aviation turbo, liquid petroleum gas, asphalts, thinners, and solvents. For manageability, I focus on only their first five products unless otherwise noted. Price listings reported are for the Metro Manila area only. Prices are lower in other cities and rural areas.
2. Makasiar, 1.
3. Garchitorena, 1.
4. In the short-run, taxation of petroleum products proved successful to alter consumer behavior. But in the long run this policy distorted the price gap between diesel and gasoline products which resulted in an increased demand for diesel. The shift in demand changed refinery production practices where the Philippines had to eventually import diesel and export surplus gasoline. For a good review of GOP responses to the first oil shock, see Villegas.
5. Paderanga and Paderanga, 100.
6. Interview with Senior Ministry of Energy Planning official, anonymity requested, July 13, 1995.
7. Republic of the Philippines 1983a.
8. The boundary between private and public was quite blurred. When the GOP reinstituted the forward cover in 1983, Cesar Buenaventura, president of Pilipinas Shell, also served on the Philippine Central Bank Monetary Board.
9. Center for Research and Communications Research Associate Rafael Emmanuel Macatangay, January 14, 1996, personal communication.
10. Interview with Senior Ministry of Energy Planning official, anonymity requested, July 13, 1995. It should be pointed out that this is merely one perspective on the price buffer mechanism.
11. Ibid.
12. Sciat, 529.
13. Ibid., 528.
14. Ibid.
15. Koehane 1984, 51–54.

16. Ibid., 529.
17. West 1997, 108–124.
18. Ibid., 109.
19. Ibid., 109; Lindsey, 170–171.
20. Lindsey 1993, 170.
21. West 1997, 110.
22. For excellent analyses of the Philippine economic crisis see Emmanuel S. de Dios 1984; Haggard, 1990; and Montes 1987.
23. Remolona, Mangahas and Pante, 4.
24. Haggard 1990, 240.
25. International Monetary Fund 1984, 15.
26. Lopez, 129.
27. Sicat, 491.
28. Ibid., 493.
29. Ongpin 2013.
30. Sicat, 500.
31. Ongpin 2013.
32. Karnow, 1989b, 407–408.
33. Kamm.
34. Ibid., 1989a.
35. Ibid., 1989b, 407–408.
36. In 1979, Laya was named as a World Bank executive director for Asia and Latin America.
37. Virata served as a career technocrat in the Marcos government. For details on his career, see Tandem and Sciat.
38. Joaquin, 293.
39. Interview with Senior Department of Finance official, anonymity requested, April 27, 1995.
40. Joaquin, 293.
41. Ongpin 2016.
42. Interview with Senior Department of Finance official, anonymity requested, April 27, 1995.
43. Montes 1991, 18–23 and Interview with Senior Department of Finance official, anonymity requested, April 27, 1995.
44. International Monetary Fund 1985, 3.
45. Walker, 14.
46. Interview with Senior Department of Finance official, anonymity requested, April 27, 1995.
47. Haggard, 1990, 21.
48. Tiglao, 1988, 13.
49. Sicat, 533.
50. Ibid.
51. Ibid., 503.
52. Ibid, 504.
53. For an excellent review of the IMF-GOP relationship see Thompson and Slaton 1985.

54. IMF internal document, EBS/84/ 117, l.
55. The CPSD refers to the combined budget shortfalls of the GOP and fourteen monitored government agencies.
56. *Business Day* 1984k.
57. Republic of the Philippines 1983.
58. *Bulletin Times* 1984b; and NEDA.
59. The CPEF was dismantled in June 1983 and the Special Fund was inactive during this period.
60. Bulletin Times 1984b and Sacerdoti, 84.
61. Totals do not match due to rounding.
62. *Business Day* 1984j.
63. *Business Day* 1984o.
64. Walker, 33.
65. According to a senior official, "Virata were always trying to do the right thing with regard to the economy, but he was constantly frustrated in his attempts." Interview with Senior Department of Finance official, anonymity requested, April 27, 1995.
66. In reality the Philippine exchange rate regime can be best described as a system managed by the Central Bank. Alm, Collins, Krugman and Remolona, 8.
67. *Bulletin Times* 1984d.
68. Republic of the Philippines, 1984b.
69. Interview with Senior Bangko Sentral ng Pilipinas official, anonymity requested, June 15, 1995.
70. Ibid.
71. *Business Day* 1984h.
72. Sacerdoti, 86.
73. Interview with Senior Bangko Sentral ng Pilipinas official, anonymity requested, June 15, 1995.
74. Interview with Senior Department of Finance official, anonymity requested, April 27, 1995.
75. *Business Day* 1984a.
76. *IMF Survey* 1985.
77. Yuson, 57.
78. Sicat, 519.
79. *Business Day* 1984i.
80. This section draws on *Bulletin Times 1984a and 1984b*, unless otherwise noted.
81. The components of the price buildup do not match due to rounding.
82. Republic of the Philippines, Board of Energy.
83. *Bulletin Times* 1984c.
84. Sicat, 533.
85. Ibid.
86. Sicat, 529; Kuruvilla, 131; and Hince and Lynch, 155.
87. Interviews with Senior Bangko Sentral ng Pilipinas official, anonymity requested, June 15, 1995 and Senior National Economic and Development Authority official, anonymity requested, June 27, 1995.

88. Central Bank of the Philippines 1984c.
89. Montes 1987, 23A.
90. Dimanling, 4.
91. *Times Journal.*
92. Dimanlig, 4.
93. Bureau of Energy, from the IBON Databank.

Chapter 3

1986 GOP-IMF Petroleum Price Observation

The political climate that surrounded the follow-up 1986 economic program was filled with refreshing optimism as well as continuous uncertainty. The "EDSA revolution" broke the reign of Marcos after fourteen years of authoritarian rule.[1] Afterward, Corazon Aquino, widow of assassinated leader Benigno Aquino, took power and eventually restored the democratic practices that her predecessor had all but virtually eliminated. As a reward to those who aided her political campaign, Aquino distributed cabinet positions as campaign spoils, shaping a broad-based cabinet that ranged from the far right to the middle left. This coalition proved to be the target of substantial criticism as GOP economic policy often wavered and failed to adhere to a clear-cut policy direction.

The Aquino cabinet lacked consensus on economic issues, which did not bode well for the Philippine economy as there was much work to be done. The economy was structurally weakened by years of misallocation of resources and financial drains. The savings rate had fallen from 25 percent to 15 percent of GNP. Domestic demand had collapsed and new investment had ceased. Company balance sheets had deteriorated badly and many workers were either unemployed or underemployed. Per capita income had fallen by 20 percent since its peak in 1981, yet population growth continued at the high rate of 2.5 percent. The financial system was still recovering from the IMF stabilization program and restructuring had a long way to go; the need to write off bad debt of government banks constrained options in dealing with the government deficit. Yet, despite these drawbacks, there was reason for optimism as well. Inflation had dropped and there was relative price and exchange rate stability. The current account and overall balance of payments were in better shape. Capital outflow had ceased and the Marcos drain was gone. Foreign exchange reserves were set to climb.

The key to resolving the nation's economic problems revolved around the GOP's ability to channel its resources toward economic development projects; however, government development operations lacked financial support due to insufficient domestic savings and the reluctance of foreign financing to continue lending. Again, the IMF played a pivotal role toward opening the floodgates of foreign financing, and signaled to domestic and foreign investors that the GOP had instituted the proper policies and was on its way toward economic recovery. During these negotiations, the IMF made recommendations on how the GOP could increase its revenue; suggestions that, for the most part, the GOP ignored. The GOP insisted that it would develop its own program and proceed at its own pace. On this issue, the IMF had little leverage. International oil prices were down and international Level II actors were unwilling to support the IMF. The distributive bargaining context that surrounded the 1984 negotiations over oil prices was absent in the 1986 petroleum negotiations. Toward the end of 1986, negotiations appeared on the verge of a conclusion, but they stalled as the IMF required the GOP to also reach an agreement with the World Bank, a strategy referred to as cross-conditionality.

INTERNATIONAL POLITICAL ECONOMY OF THE 1986 PETROLEUM PRICE OBSERVATION

Like the previous IMF-GOP negotiations, the 1986 negotiation process was inundated with numerous international actors striving to influence Philippine domestic policy. During its negotiations with the GOP, the IMF insisted that the government continue its commitments to economic reform, but the IMF aired doubts about the GOP ability to adhere to its reform commitments given the broad-based composition of the Aquino cabinet. Midway through the talks, the United States entered the fray as it became concerned that the GOP might repudiate its foreign loans and restrict American access to the military bases. As the IMF and GOP negotiations drew closer to a conclusion, finalization of the agreement stalled due to inability of the GOP and the World Bank to agree on the terms of an Economic Recovery Loan. During the tail end of the World Bank-GOP negotiations, the IMF authorized the economic program.

Before it entered negotiations, the IMF was pleased with the steps the GOP had taken toward a market-oriented economy, steps that were manifested in Aquino's selection of economic managers. Without question, the IMF and World Bank found themselves comfortable around Aquino's chief economic planners, Finance Minister Jaime Ongpin, and Central Bank Governor Jose "Jobo" Fernandez. In Ongpin, the international financial institutions had an ally who shared their belief in the need to minimize government interference and allow market forces to drive the economy.[2] In Fernandez, who retained his post with the strong support of Ongpin, the multilateral lenders endorsed his reappointment as a "very significant plus for the new government" and praised their "close relationship with [Fernandez] in the past and we are looking forward to its continuation."[3] In short, Fernandez's experience would be too critical to jettison. He was trusted by international commercial banks and knew the details of the Philippine international financial issues.[4] To be sure, the new administration's economic managers consisted of a similar mantle that steered the course of the economy during the country's 1983–1985 crisis years.

After the selection of Ongpin and Fernandez, the IMF felt uneasy about the economic orientation of the remaining key economic policy makers. This sentiment emerged from a World Bank memorandum that had been circulated between the IFIs.

> We found . . . a somewhat confused situation, in which most senior officials were either completely new or "holdovers" who did not expect to be retained for long. The new senior officials have many different preoccupations, among which making new appointments and exposing the misdeeds of the past regime have higher immediate priority than forming a consensus on a development plan for the future. Moreover, cabinet members have not yet adopted the discipline of an agreed policy line—on the contrary, they frequently make mutually incompatible statements based on their personal views and political constituencies. Perhaps none of this is surprising in the circumstances, but for us it means that the Minister of Finance and the Governor of the Central Bank do not have an easy time when they present "technocratic" arguments and proposals to the cabinet and this is bound to complicate our work.[5]

The IFIs concluded that this problem could potentially lead to greater dilemmas in the future, as exhibited in a July 1986 World Bank memorandum:

> While the new Government has achieved a temporary political consensus, there is the risk the political instability will increase, with the result that the Government will not be able to carry out the agreed policy program. In this situation, the Bank might be put under pressure, both by the Government and other donors, to release the second and third tranches (of the loan) despite a failure to complete the policy program.[6]

The concern over possible GOP defection raised several multilateral strategies, among other things, "to ensure a clear division of labor (in monitoring the country's compliance with the programs agreed with the Bank and the Fund)."[7] In essence, the inability of the IMF to extend the technocratic link beyond the nation's economic managers prodded the IFIs to encompass a strategy of cross-conditionality, a tactic designed to boost the bargaining power of the IMF vis-à-vis the recipient country.

Ironically, Ongpin was partially responsible for isolating the technocratic alignment from the IMF. Taking a lesson from the 1984 IMF-GOP negotiations, he instituted a critical change in the GOP negotiation format. Filipino negotiation teams usually included only Central Bank and MOF representatives, and they would often make "commitments that they would not be able to deliver since they did not have prior consultation with other agencies." To resolve this dilemma, Ongpin widened the scope of the negotiations to gamer the support of the critical departments and to ensure that the "document [MEP] was defensible to the public."[8] Ongpin accomplished this end by including representatives of the Department of Budget and Management (DBM), Department of Trade and Industry (DTI, formerly the MTI), and NEDA to participate in the IMF-GOP negotiations. Later, this process was expanded to include consultation with the forthcoming legislative branch.[9] Some members of the mission also met with Aquino.[10] As a result of Ongpin's procedural change, power within the cabinet had become diffused among numerous actors, making attempts to determine domestic oil prices on a rational basis more difficult. This situation was compounded with the approach adopted by Aquino, where she would lay back and require the cabinet to achieve a consensus rather than act as the final arbiter, as did her predecessor. Indeed, it was a rare occasion when she would solely decide an issue.

The IMF had difficulty negotiating with the GOP as it was unable to persuade the GOP to increase the price of petroleum products due to domestic considerations. As it entered negotiations, the IMF made and submitted a list of policy recommendations that included petroleum price increases to generate additional revenue. From the perspective of the GOP negotiation team, this suggestion was a nonstarter as it had both economic and political implications. First, an increase in the price of petroleum products would serve as a tax on domestic transactions and would prove counterproductive to stimulating economic growth. Second, this move would be politically unfeasible due to the March 1986 price reduction of petroleum products when Aquino ordered lower prices to reflect the lowered cost of international crude.[11] For the most part, the cabinet was in consensus on the issue to increase prices, and as such, the IMF failed to locate sympathetic domestic interlocutors. Finally, the Aquino administration rejected the IMF list of proposals and decided to follow its own course of action as described in the recovery plan promulgated by numerous University of the Philippines economics professors—a proposal referred to as the "Yellow Book."[12]

Although the GOP refused to increase the price of petroleum products, it remained open to seek other sources of revenue. Indeed, the focal point of the 1986 negotiations centered on the financing gap level that existed between the

GOP's revenues and expenditures. The IMF insisted that the GOP would not seek "new money" from commercial banks for the next two years.[13] In view of this agreement, the GOP needed to drastically slash expenditures and increase revenues.[14] The GOP optimistically estimated that improved collection techniques could raise an additional P3.2 billion during the second term of 1986. The IMF skeptically viewed this assessment. It hypothesized that the GOP's P3.5 billion revenue shortfall for the first half of 1986 would continue into the second half. Regardless of what the GOP projections indicated, the IMF staff wanted to impress upon the GOP realism in its forecast.[15] In the midst of the negotiations, the GOP announced that the GOP-IMF talks had to be set back two to three months to hammer out some solutions for the government's revenue shortfall problems.[16]

During its call for a negotiation hiatus, the GOP took the opportunity to shore up its revenue shortfall. In May, the cost of international crude dropped to $13 per barrel. Given domestic clamoring for lower prices, the GOP reduced the price of petroleum products by the same level it had reduced two months earlier. This time, the GOP seized the opportunity to make extra revenues by increasing the specific tax of gasoline by an average of Pl.00 per liter, diesel by P0.25 per liter, and kerosene by P036 per liter.[17] It also derived an ad valorem formula to slap on the price of petroleum products. Although the intensification of indirect taxation through the use of petroleum products remained inconsistent with the Yellow Book reform proposal for a greater reliance on direct methods of taxation, the GOP made commitments to the IMF to implement and maintain these levels for the purposes of generating revenue.[18] Furthermore, the GOP planned to rationalize and reorganize public financial and nonfinancial corporations with the ultimate purpose, in some cases, of privatizing them. This was a major step in restructuring public finances by reducing a major cause of the deficit. Another major move was the reform of the tax system including modification of corporate and income taxation, abolition of export duties, and withdrawal of some tax exemptions and incentives.[19]

Against the backdrop of revenue concerns, the United States began to play a larger role in the IMF-GOP negotiations. Shortly after the first cabinet meeting, Economic Planning Minister and NEDA Chief Solita Monsod publicly floated the notion that the GOP consider shoring up revenue from the other side of the equation: the reduction of expenditures used to pay the country's foreign loans, especially the denouncement of shady loans taken out during the Marcos period, a strategy referred to as "selective repudiation." In June, concern about this issue grew as US Secretary of State George Shultz paid a visit to Aquino, broached the issue, and received assurances that the GOP planned to adhere to its financial commitments. During the visit, Shultz suggested that the United States would influence the IMF to "ease up" on the Philippines.[20] Meanwhile, US President Ronald Reagan sped up appropriations for additional US financial aid to the fledgling democracy. Several times, the United States promise of an additional $200 million in financial assistance stalled in the US Senate, only to be passed finally when the GOP linked assistance with the issue of the forthcoming lease agreement on the US military installations in the Philippines.[21]

When the IMF and GOP returned to the bargaining table in August, agreement and dissent colored the final round. Both teams agreed on the technical levels regarding fiscal and monetary targets. In a bolstering statement, Monsod bragged that the targets of the economic program matched the country's economic recovery plan. "There is no question about it. We kept referring to it (the recovery plan) all the time. Our Letter of Intent is based on our recovery plan."[22] The consensus ended on the issue of the country's World Bank SAL commitments, without which the program could not continue. Resolution of the issue was eventually taken out of the hands of both negotiation teams and placed in the hands of the Chief of Governments (COG). During her first official trip to the United States, Aquino met with both de Larosière and the World Bank President Barber Conable and assured them that the Philippines would develop a consensus on its economic program and that it could live up to its stabilization and structural adjustment commitments.[23] De Larosière gave his consent to the economic program; however, he insisted that he had no choice but to await the progress of the current World Bank-GOP Economic Recovery Program talks before he could formally endorse the program to the IMF Executive Board. In the middle of October 1986, de Larosière was satisfied with the progress in the World Bank-GOP talks and he endorsed the economic program.

On October 24, 1986, the IMF Executive Board approved the use of IMF resources equivalent to SDR 422.1 million (roughly $508 million). Of this total, 198 million was available over the next eighteen months under a standby agreement and the equivalent of SOR 224.1 million was immediately available under the compensatory financing facility.[24] Access to this fund allowed Fernandez to term this program as "more generous."[25] Releases of the tranche reflected the IMF's caution regarding the GOP's commitment toward the economic program as the funding was heavily back-loaded; the agreement called for drawdowns of $6 million in October 1986, $23.05 million in February 1987, $45.90 million in April 1987, and $44.81 million in June 1987.[26]

DOMESTIC POLITICAL ECONOMY OF THE 1986 PETROLEUM PRICE OBSERVATION

The wave of euphoria that swept through the Philippines during Aquino's inception to power began to diminish slowly as interest groups and powerful elites pressed the government for the economic rewards that evaded them during most of the Marcos era. Aquino attempted to maintain the surge by molding the domestic political landscape in a fashion more advantageous for the new government. She cleaned up the remnants left behind by her predecessor, called for the resignation of many government officials, dissolved the *Batasang Pambansa*, abolished the 1973 Philippine Constitution, and initiated a constitutional convention to draft a new national charter. Sweeping away the old and instituting the new system, Aquino became more accessible to the reach of disenfranchised interest groups. In many cases, she agreed with the position of these interest groups, but she managed to placate some of their demands until the economy grew in strength. In March 1986, she saw an opportunity to both address some of these demands and give the economy a boost as she pushed for lowering petroleum product prices.

As her economic manager, Aquino selected and named Jaime Onpgin. A former president of Benguet Corporation, a gold mining company, and brother of Marcos Trade and Industry Minister Roberto Ongpin, he was a loud critic of the Marcos government, especially during the crisis years. The problem, he concluded, was the "favoritism" that industries received as it cluttered and clogged the dynamics of the economy.[27] Primarily, it was the government's interference in the economy that was the problem. To deal with this, he argued to privatize numerous industries such as PNOC. According the Lueng, Ongpin worked with haste on privatization then he could work on the debt problem.[28] State retrenchment from the economy must precede international financial issues. It made little sense to keep these revenue draining corporations and have foreign financing throwing away good money after bad. Privatization and reform traveled at a pace much slower that Ongpin anticipated. According to Leung, this was due to differences in style.

> In the broader part of government [*sic*]. There were differences of view in the Cabinet. Jimmy's style of management was strictly the corporate kind: That of the chief officer in charge of the company. But decisions that are normal in a corporation are never that simple in government, where a decision by say, the finance minister may interact or overlap with the decisions of other ministries. So right there you get a resistance the chief officer of a company doesn't get.[29]

Since People Power was a conglomeration of mixed interests to pursue one goal, the coalition had numerous representatives in the cabinet. According to Aquino spokesman Rene Saguisag, "It was 'the liberals' versus the 'militarist'; the 'nationalist' versus the 'Council of Trent'; the 'blacks' versus the 'reds' or 'pinkos.'"[30] Under this landscape, Ongpin insisted that if he were to take the finance portfolio, then Aquino must retain Central Banker Fernandez. He asked for him not only because he needed an ally in the cabinet, but also because he helped steer the Philippines past the 1983 economic crisis and had vast expertise and knowledge of the Philippine economy and debt problem.[31]

As the World Bank correctly pointed out, the Aquino administration had failed to develop a consensus on much of its economic policies as market and populist-type policies emanated from Malacañang Palace, a reflection of the diversity of the members who sat on the Aquino cabinet.[32] During meetings, dissent and squabbling were the norms, not the exception. But according to Saguisag, this dynamic was all they had. "We must remember that we had no legislature at that time. The Cabinet was the functional equivalent of a congress. So, if we had debates that lasted hours, we welcomed them, because we were committed to that kind of democratic discourse."[33] Indeed, this type of process might be preferable considering the image of the Marcos government. Saguisag, who would eventually be elected to the Senate, praised the interaction and the debate.

> Ideas competing in the marketplace: That was what the Cabinet bought back. And in hopes of distilling the best from those ideas, we were fierce, we were fractious, we were fighting all the time. It was stimulating; the President encouraged it. Ours were the sounds and noises of democracy.[34]

To limit these conflicts, Ongpin and Fernandez used their agenda setting powers to shape the debate. In cabinet meetings, in which Fernandez also sat, the agenda typically did not include debt or foreign financing. If it did, the discussion was dominated by Ongpin and Fernandez, who would often talk at such levels of complexity that the information "went over the heads of other ministers."[35] If other cabinet members tried to "enlighten" themselves of such complexities, they were allegedly either discouraged or surpassed or viewed as obstructing Ongpin and Fernandez.[36] They monopolized and jargonized information, making it more technical so that Aquino and other cabinet members had to defer.

Just as Ongpin, Fernandez, and others tried to steer the cabinet to the right, several individuals also tried to steer the cabinet to the left. Monsod constantly argued for foreign commercial banks, not the Filipino people, to bear the costs of adjustment. Similarly, Joker Arroyo, the executive secretary and confidant of both Beningo and Corazon Aquino, was a well-regarded human rights lawyer critical of the Marcos regime. According to reports, he attempted to shield Aquino from making difficult decisions, an effect that was aided by Arroyo's relatively weak administrative skills. The result of both of these characters was that he served as a gatekeeper for Aquino. He was the proverbial "go-to-guy." More on Arroyo is provided in chapter 4. Arroyo and Ongpin battled frequently, in both private and public settings.[37]

One issue of contention was petroleum prices. During the early years of the new administration, Aquino and Monsod made numerous attempts to lower the price of petroleum products that reflects the lower cost of international crude and would translate to price reductions of local goods and services. On the other hand, Ongpin and Fernandez argued that the government should take advantage of the reduced cost of international crude and maintain the current price level, thereby accruing a higher percentage of ad valorem taxes. To support their case, the economic managers relied on a hypothetical counterfactual in which they posited the grave political dilemma the GOP faced if after dropping the price of petroleum products then the price of international crude would rise. Simply put, "It is not easy to raise prices once you have lowered them."[38] Surprise support for Ongpin and Fernandez came from MTI Secretary Jose Concepcion, a protectionist whom some initially suspected of favoring a reduction in the price of fuel oil to reduce manufacturing sector costs. Instead, he argued in favor of maintaining the current price level to generate revenue to finance much needed infrastructure projects.[39]

Dissent also occurred on a debt strategy. Selective repudiation was only one of three debt strategies on the table for the GOP. This strategy, not surprisingly, served as a compromised position from the other parties to the EDSA Revolution. The Freedom from Debt Coalition, a cause-oriented grassroots organization, proposed a debt limit service of 10 percent of the export revenues and repudiation of all loans tied to corruption, such as the Bataan Nuclear Power Plant constructed by Westinghouse and financed by the US Eximbank.[40] The strategy's problem was in the "selective" portion. The government would create resource, current economic and political problems. According to Leung, the government would be looking into the past, a dangerous direction to go. "The government record on each transaction will have to be dug up. But was that where, in 1986, we should have placed the major part of our efforts? We were in crisis; we had a new government—and were we to spend our time wrangling?" Furthermore, until the issues were cleared, imports would lack financing.[41] Finally, political problems started to abound as the GOP would have to determine which assets associated with Marcos were legitimate, and then proceed through the legal process. Another strategy was debt repudiation—to put a stop to all foreign debt payments. According to Leung, the problem with this position was that "if we stop debt payments, the letters of credit from banks that enable our businessmen to buy equipment and get export contracts, would stop. We would be back to our situation in 1983, when our exporters could get no financing because we wouldn't pay our debts."[42] The final option would have been to pay off all the debts under the concept of "negotiated restructuring," a pledge voiced by Aquino during the presidential campaign against Marcos.[43] According to Leung, this was the best option for the country under the circumstances.

> So, scenario No. 3 is: using that global sympathy to win not only a rescheduling of our foreign debt [for at least three years of '86, '87, and '88] but also a reduction in the cost of that re-scheduling. And in the meantime, to welcome all cash offered in admiration of EDSA. It was not a scenario option for us: it was the *only* [emphasis in original] scenario! How to maintain the good will of the governments that had already expressed a lot of good will for the Aquino government, and how to maximize that good will and transform it into resource transfers to meet our needs.[44]

During the early stages of the Aquino administration, the GOP managed to balance both populist and economic pragmatist concerns. The international crude market softened as the cost of international crude dropped to roughly $13 per barrel. In response to this development, consumer, labor, and industry groups called for an immediate review of petroleum prices and argued that the big dip in world prices should be immediately translated into lower domestic prices. The industrial sector took this argument one step further by pointing out that a significant reduction in local prices could provide relief from labor's continued demands for higher wages and enable manufactures to reduce their operating cost.[45] To increase its ability to control prices, the GOP abolished the BOE and turned direct control of petroleum prices to the Cabinet as it established a special oil pricing review committee. After the first official cabinet meeting held in March 1986, the GOP dropped the price of petroleum products by an average of P0.41 centavos per liter, with the price of fuel oil incurring the largest reduction of 18 percent (see Table 3.1). The GOP afforded these cuts by reducing oil company DOCT by an average of P0.42 centavos per liter and by reducing the ad valorem tax by an average of P0.06 centavos per liter.[46] Two months later, the GOP again responded to calls for a reduction in the price of petroleum products. The GOP again cut the price of petroleum products by an average of P0.42 centavos per liter, and it placed emphasis on the reduction of fuel for the lower-income and manufacturing sectors (see Table 3.1).

On the surface, the domestic price reduction appeared to drain the GOP of valuable revenue. For the first round of price reductions, the GOP stood to lose an estimated P2.8 billion for the remainder of 1986. The degree of revenue loss would have been greater as Aquino wanted deeper cuts only to be restrained by her economic managers who argued that the government could not withstand such a drain. In contrast to the first round of prices, the GOP managed to retrieve lost revenues in the second round of price cuts through an alteration of the price buildup that was retroactively promulgated in Executive Order No. 21.[47] The price buildup shuffle increased specific

Table 3.1 Historical Petroleum Product Prices (Per Date of Adjustment 1-9-1985 to 8-1-1986) (Metro Manila) in Peso per liter

Dates	*Premium*	*Regular*	*Diesel*	*Kerosene*	*Fuel Oil (Wholesale)*
1/9/1985	9.01	5.58	7.04	6.95	5.02
3/7/1985	8.4	8.03	6.43	6.34	4.67
5/11/1985	8.4	8.03	6.42	6.34	4.67
1/25/1986	7.4	7.03	5.72	5.76	4.12
3/20/1986	7.15	6.78	5.26	5.31	3.47
5/22/1986	6.9	6.53	4.76	4.84	2.81
8/1/1986	6.9	6.53	4.76	4.84	2.5

Source: Republic of the Philippines, Energy Regulatory Board.

taxes by P0.64 centavos per liter while it reduced ad valorem taxes. Despite the "Yellow Book" proposal to move the government away from indirect taxation measures, the GOP continued this practice as an efficient avenue to securing revenue.

Despite appearances, all parties benefited from the second price cuts with a little GOP sleight of hand. The GOP did not inflict any significant financial damage to the oil companies when it reduced their DOCT as they merely requested OPSF reimbursements, valued at P500 million for May 1986, up from the previous month's OPSF reimbursement request of Pl8 thousand. The intensified outpour for reimbursement request dropped the OPSF monthly terminal balance to Pl.54 million in May from a first quarter average surplus of Pl.6 billion (see Table 3.2). For the remainder of the year, OPSF reimbursements averaged P262 million a month, with a peak of Pl billion in July. To help stem the OPSF losses, the GOP reinstated the OPSF impost into the

Table 3.2 Oil Price Stabilization Fund (OPSF) Monthly Terminal Balances (10-1984 to 12-1996) (in Million Pesos)

Months	*1984*	*1985*	*1986*
January		161	1,647
February		161	1,677
March		161	1,708
April		512	1,343
May		634	154
June		682	1,059
July		750	1,187
August		834	1,115
September		971	1,347
October	23	1,186	2,035
November	73	1,365	2,154
December	147	1,529	2,382

Source: Office of Energy Affairs and Department of Energy.

price buildup that Marcos had previously suspended. In essence, the public perception of the savings, for example, in gasoline, proved illusionary as consumers overpaid by contributing more money to sustain the OPSF.

To shore up its fiscal deficit, the GOP planned to reduce expenditures drastically, in particular, the forward cover subsidy given to oil companies; however, it eventually succumbed to persuasive pressure from a powerful member of the industry. Early in the Aquino administration, the GOP economic managers convinced the president that the forward cover resulted in a significant financial drain on government coffers, persuading her to drop the subsidy given to all oil companies. Despite the direct GOP savings discontinuing the expenditure would entail, it guaranteed to post added cost for oil companies as they would have to pay as much as three times the amount of their previous bill for each month the subsidy would be reduced. Likewise, the reduction or elimination of the forward cover would increase the demand for foreign exchange and possibly weaken the value of the peso in relation to the American dollar.[48] During a private discussion with Central Bank Monetary Board member and President of Pilipinas Shell Cesar Buenaventura, Aquino recanted her decision.[49] To continue the forward cover, the Aquino government put an innocuous statement in an executive order that maintained the existence of the forward cover.[50] The forward cover was eventually phased down to 180 days from the previous 360 days in 1989.

EPILOGUE TO THE 1986 PETROLEUM PRICE OBSERVATION

The October 1986 program provided the catalyst for the GOP negotiations with its other creditors. In January 1987, the GOP reached an agreement with the Paris Club to extend to ten years $870 million of loans. The successfully concluded negotiations also triggered the disbursement of $350 million in new loans by commercial banks. This money was held up since the IMF-GOP negotiations stalled in early 1986. In March 1987, the GOP and its twelve-member foreign creditor advisory committee came to terms on the rescheduling of the country's $13.2 million debt and a reduction in the interest rate.[51] This agreement received significant criticism from many circles as it did not include "new money," and under the terms, the GOP would pay out from 1988-1992 $15 billion more than it would receive.[52] Furthermore, a faction within the committee, led by Barclays Bank of Britain, insisted that the GOP guarantee the $57 million obligation of Planters Products, a privately owned firm, but one dependent on the subsidies under a government fertilizer program during the Marcos regime. This issue was not decided until December 1987.[53] The second criticism regarded the apparent secondary status of the Philippines. Argentina, which at the time had made arguments for debt repudiation, received more favorable terms 1/16 less over LIBOR than the Philippines received.[54] These criticisms were loud, but the defenses were attenuated. According to Fernandez, the critics missed the point. The point was to "buy time."

> Well, we moved the maturity term of the agreement from ten to seventeen years. And we lengthened the grace period from five years to seven and a half years. And we reduced the interest rates on long term debts to almost half, the largest single reduction in any debt restructuring of that time. All these things tended to ease pressure on the Philippines, making it easier for the new government to make a start with fewer swords dangling over its head. We knew we got a good deal. Unfortunately, what became the cynosure of everyone's attention was that Mexico won a lower rate of interests.[55]

The GOP fell short of most of the agreed targets, but overall it presented a respectable showing of keeping within the limits of the MEP. In 1987, the economy grew at 5.7 percent of GNP, missing the target of 6 to 7 percent; however, for 1988 the country managed to reach its target as economic activity increased by 6.6 percent of GNP. According to the standby agreement, the CPSD was targeted to take a dramatic drop from 1986 to the next two years, at 2.9 and 2.6 percent or GNP for 1987 and 1988, respectively.[56] In 1987, the GOP managed to reduce the CPSD as a percentage of GNP to 2.7 percent but the following year CPSD inched up to 2.8 percent, missing the program target.[57] Inflation had been kept to 3.8 percent in 1987, well within an acceptable IMF target; however, the increased economic activity carried with it additional pressure as inflation jumped to 8.8 percent in 1988. The external current account, in surplus by a greater extent than expected during 1986 because of the delay in sparking economic recovery, remained in small surplus in 1987 of 0.7 percent of GNP, and shifted into deficit, about 1.7 percent of GNP in 1988.[58]

The moderate success of the 1986 standby program was tempered by the political and economic events that transpired the following year. In early 1987, the public enthusiastically endorsed a new constitution based on the framework of the 1935 Philippine constitution aborted by Marcos in 1972, requiring the immediate election of a two chamber legislature. The follow-up elections demonstrated massive support for Aquino as her new party, Laban ng Demokratikong Pilipino (LDP), secured over 150 of the 200 contested seats in the House of Representatives and 22 of the 24 seats in the Senate. The "success" of these developments colored the nature of the economic problems that the GOP encountered.

With the plebiscite and elections in the foreground, the Aquino administration had to be careful about how it navigated through the upcoming rough economic waters. In late 1986, OPEC members reached an agreement to curtail oil production in an effort to push up the cost of international crude from $13-$14 per barrel to $18 per barrel. Since the GOP calculated 1986 domestic petroleum product prices based on $13.30 per barrel, a price increase of this magnitude would have forced the GOP to raise the cost of domestic petroleum products; but given the proximity of the elections, the GOP postponed the price increase for as long as possible. Even after the elections concluded, the GOP put off the petroleum price increase by allowing the oil companies to request additional drawdowns from the OPSF at a rate that threatened to deplete the fund by mid-1987 (see Table 3.2). Afterward, the GOP removed the price setting powers from the hands of the cabinet and set up the Energy Regulatory Board (ERB), a quasi-judicial commission that required oil companies to petition for price increases before a public forum. To cushion the impact of the inevitable price increases on the economy, the GOP restructured the tax components of certain petroleum products by scrapping the specific tax and adopting a three-tier ad valorem tax structure on petroleum products. In keeping with the "socialized pricing" policy practiced by Marcos, the GOP taxed "luxury fuels" with a 48 percent ad valorem impost; diesel, kerosene, and LPG were taxed by 24 percent and no taxes were levied on fuel oil designed to be used for energy and manufacturing purposes. Although the OPSF recorded a deficit for the remainder of the year, the terminal balance gradually reached a surplus with a P3 billion transfer from the Philippine Amusement and Gaming Corporation (PAGCOR). Fiscal problems eventually plagued the GOP, producing the problem of correlating petroleum price increases with military coups. This connection will be elaborated upon in chapter 4.

ASSESSMENT OF THE CAUSAL MECHANISMS

The deference tactic at Level IIA is a modest explanation for the 1986 petroleum price observation. Ongpin and Fernandez were the "go-to" players when it came to international negotiations, but they had limits to their reach. Two institutional processes were at work, the wide network of GOP negotiators set up against the IMF and the consensus process within the cabinet. While the IMF negotiations shifted the bargaining leverage toward the GOP against the IMF, the consensus requirement allowed one principal to stall reforms. Monsod and other populists leaning members limited the viable range of policy choices. Furthermore, Arroyo served as the gatekeeper of information to Aquino. The vast differences in preferences of Ongpin and Fernandez with Arroyo made the transfer of rationalization policies highly unlikely. In addition to this mix of preferences, cabinet rules of consensus amplified the variation of preferences among the cabinet members. Under this circumstance, the deference tactic to secure Aquino's support proved limited.

The limits likewise applied to the external-internal tactic. The price of international crude dipped significantly, limiting the cost recovery urgency of the GOP. The drop also gave the government the room to maneuver, and reduce the domestic price of petroleum products. This latitude limited the tactic as it reduced the cost of a nonagreement on this issue. In addition, the US political landscape in the Congress and the Reagan administration appraised the Aquino government with both favor and caution. The signals coming out of Washington were quite optimistic, even glowing when considering the pledges of financial support. The Reagan administration, however, was cautious especially when it came to the mixed GOP signals on selective repudiation. As a result of these two factors, Ongpin had limited capacity to play off the international Level IIB arena against the domestic Level IIA. Furthermore, the IMF did not have a high preference for petroleum price increases given the OPSF surplus, but it did serve as a mechanism to generate revenue in a fiscal institutional arrangement that cried out for more revenues in an environment of expanding expenditures. Reshuffling the price buildup without raising the domestic price of petroleum was a significant tactic to address both IMF insistence and domestic Level IIA resistance. This explanation, unfortunately, is less than compelling.

In comparison to the deference and external-internal process tactics, does the technocratic alliance provide us with a clearer picture? The alliance hypothesis would also assume that IMF preferences for IMF policies transfer to the COG via well-positioned domestic interlocutors. Drawing out its logic, we can apply it to the 1986 Aquino cabinet and predict that it would fail to deliver the predicted outcome. Much of the IMF preferences fall in line with Ongpin and Fernandez's preferences. The GOP chief negotiators were stymied by those whose preferences clashed with theirs. Their inability to have a meeting of the minds with others due to the institutional arrangement demonstrated their inability to overcome the institutional arrangement constructed for under the Aquino cabinet. Furthermore, Ongpin was unable to manipulate the rules and norms that surrounded the cabinet, attempting to shift preferences. In line with the reshuffling of the petroleum price scheme, the technocratic alliance tactic is also silent on this development. The absence of an alliance explains the inability of Level IIA ratification. Furthermore, if the tactic did explain the outcome, why does it provide only a partial explanation? If the technocratic alliance succeeded with price reshuffling, why did it fail with increasing prices? Given this empirical incongruity, we can conclude that the external-internal tactic does a marginally better job at explaining the cabinet decision.

From an institutional-equilibrium perspective, we can show that Ongpin and the cabinet managed to pursue their preferences to alter the rules. The norms and process required the BOE to hold hearings, assess the need for a price hike, and arrive at a decision. Bringing in the petroleum price decision-making process to the cabinet insulated it from domestic pressures. They would have complete control. Furthermore, they used the OPSF surplus to buy them more revenue, more public legitimacy, and more political stability.

NOTES

1. EDSA is the acronym for the Epifania de Los Santos Avenue highway running between the two military camps where the revolt took place.
2. Gigot 1986b.
3. *Bulletin Times* 1986a.
4. Joaquin, 293.
5. World Bank International Finance Corporation, Office Memorandum, "Visit to the Philippines for Operational Discussions," May 15, 1986.
6. This excerpt was dated July 9, 1986. A copy of it was later leaked to the *Manila Chronicle* 1986c.
7. Ibid.
8. Interview Senior Budget and Management official, anonymity requested, July 5, 1995.
9. Ibid.
10. International Monetary Fund 1986a, 2.
11. *Philippine Daily Inquirer* 1986b. The GOP had earlier dropped the price of petroleum products as a result of the first cabinet meeting of the new Aquino administration.
12. Economic Recovery and long-Run Growth: Agenda for Reforms. This proposal argued that GOP policies be directed toward the goals of economic recovery and sustainable economic growth. According to the report, such goals could be achieved by stimulating aggregate demand, but the magnitude of the foreign debt created problems in both the short and long runs. To overcome these problems, the GOP should engage in pump priming activities that concentrate on export expansion and expenditure on rural infrastructure. In addition to these suggestions, the proposal advocated a series of relatively conservative, market-oriented development policies, designed to reduce the scale of government intervention, encourage development in rural areas, and promote equity through land reform and a reorientation of government expenditures and move to the reliance on a method of direct taxation from indirect taxation.
13. Solita Collas-Monsod 1989, 480.
14. Ibid.
15. *Manila Bulletin* 1986k.
16. Ibid., 1986e.

17. Republic of the Philippines, Board of Energy.

18. Collas-Monsod 1989, 480.

19. Interview Senior Budget and Management official, anonymity requested, July 5, 1995.

20. *Philippine Daily Inquirer* 1986c; and *Manila Bulletin* 1986c. Shultz did not public link the two issues. The *Philippine Daily Inquirer* reported that the United States would use its influence to move the IMF. The *Manila Bulletin* reported that after his meeting with Aquino, Shultz was relieved to learn that Aquino bad no intention of applying selective repudiation toward the foreign creditors.

21. *Manila Chronicle* 1986a and 1986b.

22. Ibid., 1986m.

23. Import liberalization proved to be an important component of the IMF economic program, not for its own sake but due to the cross-conditionality strategy applied by the IFIs. In tum, the World Bank required the GOP to reform three large government financial institutions as prior conditions to its economic recovery program.

24. *IMF Survey* 1986.

25. Joaquin, 294.

26. International Monctary Fund Press Release 1986.

27. Joaquin, 289.

28. Ibid., 291.

29. Ibid., 290.

30. Ibid., 278.

31. Ibid., 293.

32. Dohner and Haggard, 69. Not surprisingly, Aquino admitted to the squabbles in her cabinet. "It wasn't a secret that we all came from different factions, and the common denominator was that we were all against Marcos. It would be asking for a miracle again if, at this early stage, we would all be of one mind." Aquino 1986.

33. Joaquin, 278.

34. Ibid.

35. de Dios 1990, 129.

36. Ibid., 133.

37. Mydans, 1987.

38. Interview with Senior Department of Finance official, anonymity requested, April 27, 1995.

39. Ibid.

40. de Dios 1990, 129.

41. Jaoquin, 292.

42. Ibid., 291.

43. Timberman 1991b, 336.

44. Jaoquin, 292.

45. *Philippine Daily Inquirer* 1986a.

46. Republic of the Philippines 1986c.

47. Ibid., 1986d.

48. Phone interview Senior Energy Resources Supply Administration Division official, anonymity requested, June 6, 1995.

49. Interview Senior Budget and Management official, anonymity requested, July 5, 1995.
50. Ibid.
51. Dolan, 181.
52. de Dios 1990, 128.
53. Dolan, 181 and Dohner, 9.
54. Dohner, 9.
55. Joaquin, 297–8.
56. Republic of the Philippines 1987.
57. Central Bank of the Philippines 1988.
58. Republic of the Philippines 1989b, and Collas-Monsod, 480.

Chapter 4

1989 GOP-IMF Petroleum Price Observations

The following observations were clouded by the August 1987 military coup against the Aquino government. It had raised petroleum prices, producing protests, strikes, and demonstrations, which was subsequently followed by a military coup attempt. The coup failed, but it left a residual effect on the Aquino and Ramos administrations regarding increasing prices of petroleum products with military coups. The former increased the possibility that the latter would occur. This connection made it difficult for the governments to increase prices.

Unlike the previous observations, the 1989 petroleum observation consists of three distinct stages. In the 1989a observation, favorable international and domestic conditions, a low international price of crude oil, and an OPSF surplus allowed the IMF-GOP negotiations to pass over petroleum price increase as an agenda issue. While revenue issues still remained prominent for the GOP, the military coups of the past months ruled out any price increases. The parties eventually sealed an Extended Fund Facility (EFF) agreement in March 1989, an arrangement designed to promote economic growth, not stabilization. With increased government spending, the economy began to heat up, pushing the GOP past the initial EFF targets. The GOP attempted to increase prices in September 1989, but it backed off. Subsequently, the GOP submitted an addendum in November that required reducing the deficit though an increase in petroleum prices, among other actions. This provided the 1989b petroleum observation. By mid-1990, the government damped the heated economy, but other developments, which without GOP policy adjustments, would have forced the GOP to miss the addendum targets. As a result, the GOP in September 1990 increased petroleum prices, which provides the 1989c Petroleum observation. The three observations are distinct, but at the same time, linked.

The three observations pivoted around the Philippines' foreign commercial creditor problems. The GOP's foreign debt obligations outpaced revenue generating measures, a development common to many highly indebted nations in the mid-1980s. The international financial community had to devise a new strategy to provide new monies with structural adjustment measures. Toward this end, the US-initiative Baker Plan was hatched, but it failed to gain any traction.[1] Its replacement strategy, the Brady Plan, attempted to ease the debt burden of these nations provided that both the IMF and the World Bank would coordinate their efforts and each institution would defer to the other in macroeconomic policies and structural adjustment, respectively.[2] But the IMF and World Bank programs (1989a) would have to be in place before the Brady Plan would kick in.[3] The addendum (1989b) was critical as the GOP and its commercial creditors concluded in January 1990 a buy back agreement. Finally, the IMF required adjustments be made to fiscal deficit practices to continue the program (1989c), which commercial creditors required to continue financing.

Foreign financing and debt servicing had to be compatible with domestic consensus. It was not. Since 1987, the Philippine economy achieved impressive growth rates. By mid-1988, the abounding economic activity began to demonstrate signs of weakness that led experts to conclude that much still needed to be done to resolve the country's ability to sustain economic growth.[4] To continue the country's economic progress, the GOP required $3.1 billion to finance its "Yellow Book" programs; however, the GOP refused to increase taxation measures that might place a drag on the economy. Similarly, the GOP could not secure other domestic sources of revenue as the financial system remained fragile. The remaining option for the GOP was to tap official foreign donors and commercial banks for approximately $2 billion, of which $1.8 billion was expected to shore up international reserves that fell below the equivalent of two months' imports.[5] To secure access to external financing, the GOP turned again toward the IMF, where only an endorsed IMF agreement could signal to foreign lenders that they were not throwing away good money after bad policies. The choice to turn toward the IMF was not made without carefully considering the alternatives.

Before its outset, GOP-IMF negotiations were surrounded by controversy. First, the GOP restructured its commercial bank debt in 1987, an agreement that excluded access to "new money" and included interest rates that were deemed by domestic critics as unfair.[6] Second, the GOP suffered a significant revenue shortfall in 1988 due to changes made in the taxation of oil, leading to lower petroleum receipts and the slowdown of foreign finance. Revenues for 1988 were predicated to be P10.7 billion behind the targeted receipts and could expand the projected 1989 budget deficit by 13 percent to P25 billion.[7] Third, Congress held deliberations on cutting back payments on the country's foreign debt to 1–0 percent of annual current account receipts. The IMF expressed concern regarding these discussions and the effect they would have on the outcome of the negotiations.[8] The IMF vigorously opposed this congressional initiative and recommended that the GOP undertake measures to raise more revenues to finance development projects.[9] Fourth, external issue linkages played a pivotal role in the negotiations as the ongoing Military Bases Agreement talks and foreign financing negotiations helped to close existing rifts and speed up the IMF-GOP negotiation process. These issues were prominent under the first phase of the oil observation, and finally subsided under the second phase of the oil observation. In fact, the IMF was virtually pushed out of the picture by this time.

1989A GOP-IMF PETROLEUM PRICE OBSERVATION

International Political Economy of the 1989a Petroleum Price Observation

During the GOP-IMF 1989a talks, the negotiation process was tempered by bargaining power that swayed back and forth between the parties. The GOP seemed to have the upper hand as it had successfully completed the 1986 Standby agreement and was in line for access to increased IMF funding. Bargaining power appeared to shift as dissent broke among the GOP negotiation team. Finally, the momentum appeared to swing back to the GOP when participation into the Brady Plan and an international consortium of the Philippine Assistance Program (PAP) required an IMF approval before the pledging session could continue.

When the parties began bargaining in November 1988, the GOP based its enthusiasm on the positive trends in the economy. Successful completion of the 1986 Standby agreement gave the GOP the option to tap higher levels of IMF funding, the EFF, which would have provided the government with access to $900 million for economic development. Furthermore, the GOP could also request access to the Compensatory Contingency Fund (CCF), giving the country access to $400 million for protection from exogenous shocks. Having reached this status, the GOP felt it could quickly work out a Standby agreement with the IMF on its own terms. For the GOP, the issues centered on the economic growth target for the next three years and the source of financing necessary to achieve these levels. The GOP believed that measures that raised revenue through increased taxes, reduced current expenditures, and controlled or reduced interest payments remained unattractive options given the need to maintain a stable political environment following the 1987 coup attempt. This development shortly followed public protest and demonstrations. To finance expenditures for necessary infrastructure projects critical to sustained economic growth at the proposed 6.5 percent of GNP, the GOP looked toward its foreign donors and creditors.[10]

The IMF took the more cautious position. The IMF staff, led by Ulrich Baumgartner, wanted to keep the level of external financing down to lower and more manageable levels.[11] In its view, the bottom line came down to achieving this goal while other targets became residuals.[12] The IMF staff suggested a lower growth target of 6 percent, higher pubic savings, lower private savings, lower foreign savings, and lower total investments, both public and private.[13] Included in these lower targets, the IMF asked the GOP to slash its budget deficit to P19.7 billion for 1989; Pl2.5 billion for 1990; P8.9 billion for 1991; and Pl0.9 billion for 1992. The IMF based its position on two suppositions. First, they doubted the GOP's ability to generate sufficient revenue to match its planned expenditures, which would lead to creating an even wider deficit. Second, they projected that the GOP's plan of sustained growth at these levels would eventually result in higher levels of inflation.

For the most part, the GOP chief negotiators sided with the IMF and shifted the bargaining strength toward the IMF. Holdover Central Bank Governor Fernandez, a co-chair of the Philippine negotiation team, had for all practical purposes "already sided with the IMF" before negotiations began.[14] Similarly, the other co-chair, Secretary of Finance Vicente Jayme also agreed with the IMF position. Going into the negotiations, the chairmen adopted and continued Jaime Ongpin's approach of a "nonconfrontational" debt strategy that would "allow us to continue working with our creditors and trading partners, a position that rejects the radical proposals emanating from official circles, the Congress and even the academe and cause-oriented groups."[15] The strategy exhibited a certain amount of caution, and had reflected Jamye's intentions of setting modest and sustainable targets, taking a "success breeds success" framework that followed a clear logic: "If you can sustain these levels, then you can go to the next level and the next level after that."[16] But that is where the Ongpin line stopped. Toward the end of 1988, Jayme and Fernandez recognized that the financial gap would be larger than they initially anticipated ($1.7 billion)[17] and new money would be needed.[18] Given this reality, the GOP's approach to its debt issue was to "reduce net resource outflows."[19] As a result, the completion of the IMF negotiations was viewed as urgent.[20]

Within the Philippine delegation, Economic Planning Minister and NEDA General Secretary Solita Monsod expressed opposition to the IMF's proposal of a 6 percent GNP growth target. She anchored her position on the government's fragile legitimacy, which could only be strengthened with higher levels of economic growth.[21] As a result, she pressed the GOP's chief negotiators to stand firm with an average 6.5 percent GNP growth rate target for 1989 through 1992, which would carry with it a projected deficit of P3l.9 billion for 1989; P32.1 billion for both 1990 and 1991; and P27.5 billion for 1992.[22] Furthermore, Monsod insisted that the GOP should shoot for the higher target rates; abandoning these targets meant leaving more people in unemployment as well as the perpetuation of poverty conditions. Monsod eventually acquiesced to the IMF position on the deficit figures,[23] but she maintained her stance on the target growth rate.[24] Despite her isolation from other GOP delegates, Monsod managed to gain the support of one critical individual: Aquino.

The second round of the talks resumed in early 1989 as a compromise was reached under the direction of Aquino. In a meeting with IMF Asian Department Director Prabhabar Nervekar, Baumgartner, and the Filipino chief negotiators, Aquino convinced the IMF delegation that the financing gap would be sufficiently covered by the official loans received from the launching of the multilateral PAP.[25] Afterward, Aquino announced that the parties reached an agreement that included several key positions for the GOP: a 6.5 percent growth target set by the government for 1989, no new tax measures in 1989, flexibility in the timing of current expenditures to keep deficits within targeted levels, and large investments in key areas such as power.[26]

For both the IMF and the GOP, the PAP served as a bargaining chip to speed up negotiations. Ironically, the PAP was also a carrot used to speed up negotiations on the US military base leases. Both sides agreed to the principle of the bases, but disagreed on the terms; as a result, the GOP dragged its feet in negotiations.[27] The PAP was the brainchild of the Reagan administration in May 1988 to induce the GOP to support economic reforms and speed up negotiations on the bases.[28] Originally known as the Multilateral Assistance Initiative (MAI), the PAP was regarded as the mini-Marshall plan, designed to solicit a total of $10 billion in pledges from numerous countries in the form of grants and aid distributed over five years.[29] According to the US interpretation of the financing scheme, the United States would determine how to distribute the loan while Japan would pledge most of the money for the plan; however, Japan was adamantly opposed to linking aid to the bases issue and had pressed for the scheme to be managed by the World Bank, a viewpoint that eventually gained acceptance.[30] Ironically, both the IMF and the GOP used the PAP to further their negotiation positions. Aquino relied on the PAP to convince the IMF that foreign revenues would be forthcoming. Similarly, the GOP chief negotiators argued to their domestic Level IIA constituents that the EFF should be rushed through before the PAP session begins, as the pledging session could not continue without an IMF "seal of approval." In the end, PAP pledged $3.5 billion for 1989/1990.[31]

PAP produced several purposes. First, it served as a reward for the "good behaviour the country showed[s] in the negotiations."[32] Its financial stream arguably offsets the fiscal and political sting associated with IMF programs and commercial bank debt obligations. This offset, however, was shown to be illusionary as net outflows would outpace net inflows.[33] Second, it streamlined the normal Official Development Assistance process as it bypassed government bureaucracies, in this case, NEDA. Through the Investment Coordinating Committee, as Head of NEDA Monsod had absolute authority to negotiate with donors and vet projects paid from aid money. Selecting projects and contractors and distributing the funds was often the source of the bureaucratic logjams. NEDA severed several functions. First, it filtered out project requests from both public and private organizations. Frequently, NEDA found itself clashing with the finance ministry and the Central Bank.[34] Second, NEDA had to select contractors through an open bidding process as a means to ensure "efficiency." This process was continuously thwarted by US representatives who intervened on behalf of their constituents.[35] Finally, NEDA would require numerous studies on these development projects, slowing the process. In contrast, PAP is administered through a private organization, at the time, a construction and development company that received private requests. Other arguments for the presence of PAP lied with the feeling to bypass the Monsod led NEDA or that the government cannot absorb and implement the additional ODA.[36]

The American military bases were also a constant theme that remained intertwined with the PAP issue. From 1986 onward, the bases were a subject of contention between the United States and the Philippines, with Washington eager to renew the leases. At the outset of the IMF-GOP talks, the IMF strongly encouraged the GOP to conclude the negotiations with the United States.[37] In October 1988, the United States and the GOP agreed to extend the Military Bases Agreement and increase the compensation from $180 million per year to $481 million for the remaining two years of the lease.[38] For the GOP, the increased compensation meant additional revenues that it could count toward the country's financing gap. It also included a $160 million grant from the United States.[39] In contrast, the United States and the IMF argued that the funds were to be designated for outlying areas of the bases and therefore were excluded from being counted as general revenue. In the end, the United States agreed to increase its military assistance to $962 million: $400 million was targeted toward military assistance; $320 million from the Economic Support Fund; $192 million toward development assistance and food aid; $50 million in housing investment guarantees; and other loan guarantees from the Export-Import Bank of the United States.[40] While the military assistance was an important component of the agreement, the Economic Support Fund was also critical. It provided funds for broadly promoting political and economic stability in areas of US interests. More specifically, it financed recipient governments' capacities to transition from centralized political and economic systems toward liberal systems, civil engagement building, balance of payments issues, and budget issues.[41]

Its applications are broad in scope and at the discretion of the recipient government, with the only limitation that it not be used for military or paramilitary purposes. Military financing proved to be a constant and significant issue for the Aquino government.

Foreign financing levels, from both public and private sources, proved to be a key consideration to the conclusion of the negotiations. In the midst of negotiations, the GOP stopped servicing its debt obligations with official creditors due to the expired debt restructuring agreement that ended in September 1988. Furthermore, the GOP had failed to make progress on the 1987 debt restructuring agreement with the commercial banks, due to expire in 1989. In contrast to the previous IMF-GOP agreements, this time the IMF agreed to the terms of the Letter of Intent before the commercial creditors and the Philippines managed to negotiate new terms on a debt restructuring agreement and with the Paris Club. IMF Managing Director Michel Camdessus had stressed that the IMF management was satisfied with the talks and that the Philippines restructuring agreement between the Paris Club and the GOP could not proceed without an IMF agreement in place. Later, Camdessus had to plead with a dissenting bloc of European nations to abstain in a vote on the IMF-Philippine program request rather than block it completely. In view of the deadlock between the GOP and the commercial creditors, the European nations sought to have the IMF approve the economic program "in principle" and wait for the banks and the GOP to reach agreement on debt restructuring.[42]

In March 1989, the IMF executive board approved the GOP's economic program. In exchange for SDR 946.9, the GOP would target real GNP at 6.5 percent for the next three years. The GOP would keep inflation under control by keeping it to 8.0 percent in 1989 and capping it to 6–7 percent in 1990 and even lower targets of 5–6 percent through 1991 and 1992. The CPSD percentage would be carried through 1989 at 3.4 percent of GNP, but by 1990 the government would slowly reduce its deficit to 2.9 percent by 1990, 2.5 percent by 1991, and 2.3 percent by 1992.[43]

The GOP deficit target was scheduled to be minimized, but petroleum product prices are far from front and center, but it did make several commitments in this area. The GOP intended to continue with the policy of timely adjustment of wholesale pump prices of petroleum products in line with import price developments, and it pledged that the OPSF would not be permitted to borrow or accumulate arrears and would not receive subsidies. Furthermore, the GOP agreed to examine the forward cover provision of the OPSF. This eventually led to the 1989 reduction of the forward cover period from 1 year to 180 days.[44] In short, the GOP promised to keep an eye on petroleum prices. Other products, however, were not immune to price rationalization. Rice and water were scheduled for price increases.[45]

The inflows of PAP also coincided with the outflows of foreign debt, addressed primarily under the Brady Plan. An agreement with commercial creditors was eventually reached in January 1990, a ten-month gap between the IMF agreement and the creditors. The delay hinged around the amount the country was able to pay and in part due to the uncertainty of the IMF and World Bank role under the plan. Even after the EFF was finalized by the IMF, it allowed the GOP to use IMF ($125 million) and World Bank ($150 million)[46] financing to buy back $1.31 billion of government debt at a 50 percent discount that was being sold on the secondary market. This strategy reduced this country's debt by approximately $650,000 million. It also permitted debt due from 1990 to 1994 to be rescheduled. The Paris Club also had difficulty with the GOP, but this was soon pushed forward.

Domestic Political Economy of the 1989a Petroleum Price Observation

During the IMF-GOP 1989 EFF negotiations, the GOP constantly faced fiscal constraints. An atmosphere of political uncertainty and tension already existed and predated the August events.[47] The trigger was the early August increase to the international cost of crude. To adjust, the ERB authorized a price increase in mid-August (see Table 4.1). Student organizations, consumer groups, moderate and far left labor unions and transportation associations threatened to and initiated strikes to pressure Aquino to issue commensurate side-payments and to rescind the ERB judgment. The threats turned to action when the opposition produced a countrywide broad-based *Welgang Bayan.* The effect was devastating as the economy almost grinded to a halt.[48] Riding

Table 4.1 Historical Petroleum Product Prices (Per Date of Adjustment 8-15-1987 to 10-5-1990) (Metro Manila) in Peso per liter

Dates	*Premium*	*Regular*	*Diesel*	*Kerosene*	*Fuel Oil (Wholesale)*
8/15/1987	8.2	7.85	5.64	5.69	2.82
8/26/1987	7.5	7.15	5.25	5.3	2.82
5/3/1988	7	6.65	5	5	2.82
8/18/1988	6.8	6.43	4.7	4.7	2.82
8/16/1989	5.8	5.43	3.7	3.7	3.42
9/1/1989	5.8	5.43	3.7	3.7	4.91
9/21/1989	8.87	8.5	6.24	6.24	2.82
11/1/1989	5.8	5.43	3.7	3.7	3.42
11/30/1989	7.06	6.69	4.96	4.96	4.91
9/21/1990	8.87	8.5	6.24	6.24	4.91
10/5/1990	8.87	8.5	6.24	6.24	4.98

Source: Republic of the Philippines, Energy Regulatory Board.

on this result, the student organizations and far left labor unions wanted to ramp up the demonstrations and called for an August 26 nationwide and more intense *Welgang Bayan*. Consequently, the ERB agreed to a partial average rollback of P0.50 centavos per liter made possible by the reduction of customs duties on oil imports from 20 to 15 percent. The reduction was set to go into effect on August 26. The protests and demonstrations proceeded regardless, intensifying, and crippling eleven cities including Metro Manila, Cebu, Davo, and Bacolod.[49] These produced clashes with government forces. The coup plotters initially scheduled its coup for August 29 but moved it up a day to catch the peak of the *Welgang Bayan*. Then, the Reform the Armed Forces Movement (RAM) troops entered Metro Manila and attacked Malacañang Palace in an attempted coup d'etat.[50] Forces loyal to President Aquino eventually neutralized the coup attempt but could not remove the stigma left by this experience, a residual that correlated petroleum price increases with attempted coups.[51] The coup also intensified criticism of and dissent among the cabinet members, forcing Aquino to reshuffle her cabinet and release Ongpin and Arroyo from their duties.[52] Although some observers noted that the cabinet reshuffle eliminated the source of controversy and disagreement within the Cabinet, steered the cabinet to the right, friction continued into the upcoming IMF-GOP program talks.[53]

The 1987 August coup attempt constrained cost-recovery efforts, which ultimately had an effect on attempts to secure international financing. The GOP agents started to find their principals reining in their slack. Opposition originated from the 1987 debt restructuring agreement with the nation's commercial lenders as Congress, a delegate member of the negotiation team, and academicians constantly reminded the chief negotiators to take into greater consideration the development requirements of the Philippines in crafting an economic program. Congressional leaders called for the resignation of both Fernandez and Jayme. Monsod repeatedly questioned the positions taken by Fernandez and Jayme and persuaded Aquino to support her position.

Congress also constantly called to lower the percentage of payments on the country's foreign loans. Handling all these issues, Aquino managed to soothe opposition and allow the process to play out, culminating in the 1989 IMF-GOP EFF agreement. Although criticism of the nation's economic managers remained constant after the 1987 debt restructuring agreement, the calls for their ouster became louder as the IMF-GOP negotiations approached. Academicians began the assault with a critical analysis of the Philippine debt situation and echoed for the removal of the chief negotiators whose views coincided with foreign bankers.[54] Always eager to be in the spotlight, Congress picked up on this issue and began its efforts to reshape the upcoming negotiations by calling for a program that placed a ceiling on the debt service ratio of 10 to 20 percent of the country's export receipts where the "outflow must be tailored to the Philippines' capacity to pay."[55] In February 1989, Congress also tried to create a Joint Executive-Legislative Debt Council that would determine debt policy, a measure that Aquino vetoed. Congressional opposition had the numbers to override the veto, but Aquino marshaled support to prevent it. Apparently, Congress lacked the taste for having to deal with a contentious and controversial issue.[56] The congressional calls grew louder in volume and number, but Aquino maintained her original position and retained the chief negotiators. In declaring her vote of confidence in them, Aquino emphasized "the continuing need to address our serious external debt problem in a manner that will provide effective debt relief, without derailing our continuing growth or detracting from the international confidence we have achieved."[57] Without question, the IMF staff became concerned over congressional threats to usurp the country's foreign debt payments and the IMF-GOP negotiations.[58]

Cleavages between the chief negotiators and their constituencies were compounded by dissent within the GOP negotiation team. The GOP followed the outside-inside negotiation pattern. The principal debate within the team centered on the pace of the country's economic growth. On the one hand, Jayme and Fernandez agreed with the IMF staff on the assessment that the Philippines should proceed with moderate growth GNP targets. They lacked a "go-to-person" to secure Aquino. On the other hand, Monsod argued that the new economic program should set a growth target of 6.5 percent of GNP, a growth level, she argued, that could be met with minimal inflation and with a reduction in the government deficit.[59] The source of this internal controversy derived from different interpretations of the same data arrived at by both NEDA and the Central Bank.[60] Some of the differences in interpretation concerned the government's budget deficit, projected revenues, growth rate of domestic savings, and other policy areas. After the first round of the IMF-GOP negotiations, the negotiating team regrouped and ironed out most of its differences. Later Aquino sided with Monsod and refused to drop the GNP growth target below 6.5 percent of GNP. With that issue out of the way, negotiations continued between the IMF and GOP. The ensuing rounds were held in secret and Monsod was not informed of the place or time of the scheduled talks. After the negotiations had concluded, Monsod could "do nothing as the terms of the agreement had already been set."[61]

Aquino also played a significant role in reducing tension between Congress and the domestic negotiation team. Aquino originally planned to sign and submit the Letter of Intent to IMF, but left and right winged blocs from both legislative chambers called for its deferment to allow time for an inquiry into questionable IMF terms and conditions. In response to this call, Aquino allowed the negotiation panel to field questions from Congress. The reward of the PAP assuaged the opposition to the MEP.[62] Afterward, the Senate endorsed "with reservations" the MEP by a vote of thirteen to seven. The thirteen senators voted to accept the MEP provided that a review program would be conducted at every six-month interval, and constant consultation would be made with senators.[63]

Meanwhile, world oil prices fell to $11 per barrel by November 1988, allowing the ERB to lower prices. Soft international petroleum prices allowed the GOP to avoid making difficult choices. By early next year, international prices jumped back up to $16 per barrel, but the ERB did not make a corresponding adjustment. As a result, oil companies continued to make reimbursement requests against the OPSF. At the end of 1988, the fund had a balance of P1.8 billion. By November 1989, the OPSF incurred a deficit of P8.3 billion (see Table 4.2).[64] The pattern of behavior was to be expected as the GOP was "gun-shy" from the correlation with price increases and military coups. The incapacity to reintroduce price increases proved to be the undoing of the GOP.

Epilogue to the 1989a Petroleum Price Observation

The terms of the MEP were ultimately unsustainable. The GOP "abstained" from increasing petroleum prices, but soon the economy began to heat up as domestic and international equilibriums as moved to deficits and inflationary pressures started to kick in. The GOP no longer had the luxury of sitting out a petroleum price increase. The following next episode required the GOP to rationalize prices.

ASSESSMENT OF CAUSAL MECHANISMS

The external-internal tactic succeed, but not for the IMF, but for the GOP. This pattern of negotiations was no different from the previous episodes in that the IMF served as the catalyst for foreign financing. The IMF program came first, then the substantial levels of financing from commercial banks started. In this observation four considerations differed. First, PAP funding was based on the "good will" of the donors, who wanted to see the Philippines develop economically. Second, PAP governments were not only interested in the economic development of the country, but also the funding met the needs of other interests such as the military bases or other domestic industries. Third, the Aquino government had earned the reward for good fiscal behavior, noting that an EFF not a stabilization fund was sought. Finally, the 1987 August coup attracted much sympathy from the international community, garnering much need to shore up the GOP legitimacy. At Level IIB, the strategy also worked. Congress put pressure on both Aquino and her lead

Table 4.2 Oil Price Stabilization Fund (OPSF) Monthly Terminal Balances 1-1987 to 12-1990 (in Million Pesos)

Months	*1987*	*1988*	*1989*	*1990*
January	2,354	−50	2,480	−7,604
February	2,257	140	2,422	−7,684
March	1,715	337	841	−8,114
April	1,212	583	299	−8,860
May	438	689	−818	-9219
June	−256	1,096	1,096	−6,103
July	−703	1,358	−3,103	−7,123
August	−719	1,592	−4,204	−8,423
September	−595	1,990	−5,043	−11,077
October	−39	2,345	−5,806	−13,950
November	−290	2,368	−7,133	−16,575
December	−123	2,523	−7,068	−15,482

Source: Office of Energy Affairs and Department of Energy.

negotiators, but any semblance of resistance or extension of the negotiation process would have signaled an absence of needed funding for specifically domestic investment, and second for economic growth.

The deference tactic provides a weak explanation. Drawn out to its logic, the deference tactic should have proven no results. The hypothesis does support the evidence. Jayme and Fernandez were unable to secure a "go-to" person on which to extend their preference to Aquino. The connection is clearer on at the international arena where Camdessus was used as the "go-to" guy to secure the support of the executive board.

Could there be another plausible alternative explanation to the outcome? The technocratic alliance would prove to also achieve the same conclusion. Whereas Monsod had her voice heard on growth target rates, her voice was silence by Jayme and Fernandez when they outmaneuvered her out of the meetings. This change of schedule/place permitted the chief negotiators to manipulate the rules to derive their outcome. The alliance and the deference hypothesis both arrive at the same conclusion for nearly similar reasons.

1989B GOP-IMF PETROLEUM PRICE OBSERVATION

The GOP dodged a petroleum price increase in the 1989a observation; it was not as lucky in the 1989b observation. After the IMF accepted the terms of the MEP, the GOP lobbied its foreign donors and creditors, both public and private, for new money, debt relief, and debt restructuring. It successfully negotiated agreements with its public donors and creditors, but negotiations with its private creditors were drawn out. The GOP pro-growth policies succeed, but the economy began to heat up, pushing the economic indicators out of line with the MEP targets. Since foreign financing hinged on a sustained MEP and its tranche disbursements, the GOP could either amend the targets or the policies. The dire conditions required the GOP to do both.

In November 1989, the GOP submitted to the IMF an addendum to its MEP.[65] Under this revision, the GOP reassessed its commitment to a growth level of 6.5 percent of real GNP and reevaluated its prospects by targeting economic growth to an average of 6 percent over the medium term. Furthermore, the GOP also pledged to reduce the CPSD to 2.9 percent of GNP, which had leaped to 4.5 percent of GNP in comparison to the original target of 3.6 percent. To come to terms with the increased domestic demand and to return the economy on track, the GOP pledged to take the following actions: revenue mobilization, expenditure curtailment, and adjustment of prices of petroleum products and the exchange rate. The petroleum price increase led to the most devastating military coup of the Aquino administration.

International Political Economy of 1989b Petroleum Price Observation

The robust growth levels of mid-1986 1989 were coupled by the expected outcomes. The slack in the economy tightened as inflation hit 13.5 percent in the fourth quarter. This increase was fueled by stronger aggregate demand that was pushed by congressional legislation on increased wages, expanded fiscal policies, and the subsidies for petroleum demand.[66]

In addition to PAP financing, the GOP began to readdress its debt servicing. In June 1989, it sought and received debt relief when it reached an agreement with the Paris Club, referred to as the Agreed Minute of May 26, 1989.[67] This arrangement allowed the GOP to reschedule all of its past due principal and interest in addition to all of the principle and interest due for the next two years. Joe Ledesma, board member of the Philippine Freedom from Debt Coalition, notes that "the country is still required to pay all of the debt and the crisis is simply postponed until 1992 when the principle is due."[68]

The GOP simultaneously negotiated with its commercial lenders.[69] In November 1989, it secured a commitment for $600 million in new money, significantly short of the GOP's target of $1 billion. The money was to be dispersed into three tranches, 60 percent would be available immediately, 20 percent in June 1990, and the reaming 20 percent in December 1990. The two later disbursements coincide with the EFF disbursement tranches. The agreement also had other provisions included a waiver on debt and debt servicing operations, and future debt restructuring that excluded the new money. The network of foreign donors and creditors was of utmost importance for the GOP. If it could not maintain the targets, then the entire foreign donor and creditor funds would surely be in jeopardy. The increase in petroleum prices was worth the risk if the GOP wanted to continue financing.

Domestic Political Economy of the 1989b Petroleum Price Observation

While domestic level negotiations between the GOP and its constituents were absent under the first phase of negotiations, it was strikingly present in the second phase of the observation. In 1989, the economy had begun to beat up. Aggregate demand targets agreed under the EFF exceed the designated level. Increased import levels and inflation also began to show signs of pushing the economy toward an undesired course. To correct this problem, the GOP submitted an addendum to the IMF, reasserting its commitment to the terms of the EFF targets though sound fiscal and monetary policies. As the CPSD had begun to grow, the GOP promised to initiate a number of course correction policies, including an oil price increase. The price increase had devastating repercussions as it triggered another military coup against the Aquino government, a coup that left deep scars.

The disaster centered on an unfavorable change to the landed costs of crude and the GOP attempt to insulate its constituency from its effects. In January 1989, Arab Light cost $14.61 per barrel and increased to $16.28 per barrel in a span of two months. To cover their losses, domestic oil companies accelerated their OPSF reimbursement requests, a move that widened the OPSF deficit (see Table 4.2) and accounted for 0.9 percent of the CPSD in 1989.[70] In a similar fashion, the National Power Corporation (NAPACOR, formerly NPC) balance sheet also continued to deteriorate as consumer prices remained the same, but the government corporation suffered from cost increases. Undeniably, the price buffer scheme and the government corporation began to show the strains of diminishing rates of returns. To shore up both of these facilities, the GOP funneled P5 billion to the OPSF, which in turn diverted P3 billion to NAPACOR, a maneuver necessary to restore a favorable balance sheet to attract investors needed for fast-track energy projects designed to resolve the nation's brownout problem.[71]

For the GOP, changing oil prices had become a more structured process. After the 1987 oil price debacle, the GOP created a dual arena for the discussion of oil prices. Regarding the social actors, the process continued to rely on the ERB as a distributive conflict mitigating device. As petroleum price increases would have a reverberating effect on other issues, the GOP added a second layer decision-making process. It created a joint-agency cabinet committee that addressed the geopolitical implications of price hikes. The committee was composed of department heads from finance, labor and employment, PNOC, NEDA, the ERB, and the Office of Energy Affairs.[72]

The Department of National Defense (DOND) was conspicuously absent from this list of agencies. It would be later included among the joint agencies and would be given veto rights over all cabinet proposals. The rules and punishment for violating the rules were simple. The finance minister had institutional powers to set the agenda, but he always needed to thread two needles. According to new Secretary of Finance Jesus P. Estanislao, "since I was in the hot seat I had to be the one to convince the President that petroleum prices could NOT (emphasis in the original) be subsidized and that we would have to let market forces determine price levels."[73] In addition, he also had to convince the security cluster, which approved any and all proposals presented before the cabinet.[74] Parties that failed to secure the subcommittee's approval received the ire of Aquino.[75] To secure this consent, finance ministers would attempt to increase the military budget[76] or use the prospect of foreign loans as leverage to bypass resistance.[77] The military's role in politics during the democratization process never wavered; furthermore, it also pressed the Aquino administration for budget increases.[78] On the mission and the budget, both the military and the government saw eye to eye. Then General Fidel Ramos succeeded in lodging the military's role in politics, at first on a de facto basis,[79] but eventually institutionally through its inclusion with the Cluster for Political and Security Affairs, later called Cluster E.[80] This standing committee had a gatekeeping function on petroleum price adjustments, among other things, and frequently frustrated numerous finance secretaries' attempts to increase prices. To secure this consent, finance ministers would attempt to increase the military budget[81] or use the prospect of foreign loans as leverage to bypass resistance.[82]

In addition to these institutional arrangements, the GOP also formalized wage determination. In July 1989, it created the National Wages and Productivity Commission (NWPC) and regional tripartite wages and productivities boards. The commission was empowered to call for a corporatist arrangement among government, labor, and business that would from time to time assess, review, and determine regional wages.[83] The commission, in the end, depoliticized wage increases as it insulated the legislature and the executive decision-making processes. Much like how the ERB serves as a compensation mechanism for oil companies, the NWPC also provides the same function as labor group's petition for increases. Almost immediately after its creation, the NWPC authorized a wage increase by P35 from P54 to P89 in July.

Shifting funds to the OPSF and NAPACOR proved to be a short-term solution as the soaring OPSF deficit eventually forced the GOP to test the domestic waters and increase the price of petroleum products. In September 1989, the ERB allowed the oil companies to adjust their prices by an average of P2.54 per liter, with the price of gasoline accounting for the bulk of this increase while the price of fuel oil increased by Pl.49 per liter (see Table

4.1). The ERB price authorization process was kept short and the GOP did not have the opportunity to use the IMF as a "lightning rod." At this stage, the wage compensation process through the NWPC proved to be no longer salient. It had already authorized a wage increase in July; businesses and industries would have protested loudly and vehemently another round of increases. This mechanism remained silent. Key opponents to the price hike were the KMU, Trade Union Congress of the Philippines (TUCP) and transportation associations, which did not request side-payments to compensate for the increases.

They had denounced the decision, and initiated protests and strikes against the GOP. Feeling the heat, Aquino later requested that the ERB reassess its earlier proclamation. Later, the ERB had "reconsidered" its earlier decision and returned the price of petroleum products to their pre-September levels. The temporary surge in domestic petroleum prices did little to halt the sliding OPSF deficit. Meanwhile, international crude oil prices continued to rise. In January 1989, international crude oil costs $12.40 per barrel. By November 1989, the price increased to $16.00 per barrel. As they were prohibited from legally increasing the domestic price of petroleum products, Philippine oil companies continued to accumulate claims on the OPSF. By end-November, the oil companies had claims against the OPSF in the amount of P6.5 billion. To rectify their balance sheet situation, the oil companies petitioned the ERB. With a quick turnaround time, the ERB announced increases in domestic wholesale prices of petroleum products by an average of about 30 percent in December 1989, creating a flow of funds for the OPSF. The issue of the IMF or the Addendum to the MEP did not surface. Except for fuel oil, which increased P0.65 per liter, the price of petroleum products across the board had risen by Pl.26 per liter. To cushion the impact of the price increase, the GOP maintained the current ad valorem excise tax rates on petroleum products as it reduced the OPSF impost by an average of P0.73 per liter.[84]

After the price increase, in early December the Aquino administration faced an unprecedented level of domestic turbulence. Transport associations, the KMU, the TUCP, and the Citizens Alliance for Consumer Protection (CACP), staged numerous strikes that crippled several key cities. After several days, the strikes were contained. The GOP offered the veto actors concessions to offset the price increase. Before the compensation process could be started, the Philippines was hit with another devastating incident, a military coup attempt. The coup plotters scheduled the ouster attempt to take advantage of the confusion racked by the oil price strikes.[85] RAM forces managed to sustain a siege on Malacañang Palace for one week. Aquino made public appeals to unify the nation against the coup. Pushed into the background of a destabilizing political-military situation, the oil price veto actors dissolved their protests and reluctantly accepted the price increase. With the aid of the United States Air Force, military units loyal to Aquino quelled the rebel faction's attempt to topple the government.

The coup attempt inflicted numerous civilian and military casualties, but it also reinforced the psychological residual regarding petroleum price increases: coup attempts were correlated to price increases, making it more difficult to authorize price increases. As Estanislao pointed out: "Whenever we brought up the issue of increasing oil prices, someone would say 'military coup,' and that would be the end of it."[86] Indeed, the issue of security became of paramount importance. "The real problem is security," a Malacañang source says, explaining what he calls the government's "step-by-step" approach. "Anytime you change anything, wages, oil prices, anything, you get into a spot of instability and you realize there is someone out there who will exploit it. We are always afraid it will hit the government when it is vulnerable—and that will be the end of it."[87] Along with economic and political considerations, the GOP now had to contend with another variable in determining the prices of petroleum products: security.

Epilogue to the 1989b Petroleum Price Observation

The turmoil suffered by the Philippines in December significant political and economic scars. On the political side, the Aquino administration recognized that it held a position far from secure vis-à-vis its domestic constituency. Petroleum price veto actors, countryside peasants, communists, and the Muslim faction from the south combined to place demands on the government. In addition, the government opened up another problem when it received pressure from factions within the military. As a corollary to this dilemma, the GOP realized that its legitimacy heavily relied on the country's economic performance, which had begun to show signs of slowing down. To restructure the economy, the GOP attempted to institute dramatic reforms. This effort, however, had met considerable opposition and, in conjunction with pure misfortune, the GOP found itself again knocking at the IMF door. Indeed, the GOP was in a very precarious position.

ASSESSMENT OF CAUSAL MECHANISMS

The processes at domestic Levels IIA and IIB were complex. The vertical interaction between the levels was entangled, just as the horizontal interaction of events in July, September, and December, which both created a synergy that would reverberate through the GOP for years. At Level IIA, the deference tactic went through the finance minister, who had the capacity to tap international financial sources to receive new monies as GOP targets were linked to the EFF disbursements, which were tied to commercial lenders new monies. Aquino formally accepted the price increase, but after the prices were instituted, she reversed course. We can then assess that the deference tactic did not matter. This conclusion, however, is clouded by the December price increase, where the security cluster recommended the increase at a cabinet meeting and Aquino stuck to it. The difference between the two events is that the addendum was present in December, not in September. The GOP had to implement the price increases successfully, with no reversal. We might conclude that the deference process at Level IIA in isolation yielded unexpected results, but in combination with the process at Level IIB, the evidence suggests that the application at both levels made reform irreversible.

Deference at Level IIB played a critical role. In the September increase, the GOP appeared to rely upon the wage increases as a means to marshal support for policies yet to be implemented. In contrast, the December increase had Aquino generate political support for the survival of the government. She managed to provide a rally around the flag sentiment. We must view the causal link of this mechanism with some caution, however, as it was the military coup that she tried to quell, not the price increase. The increase faded into the woodwork with the coup attempt. The presence and absence of the deference tactic explanation at Levels IIA and IIB does provide strong support for the zigzagged process.

Can the external-internal sequencing provide a clearer and simpler connection for the outcome? This causal mechanism also traces a clearer link at Level IIA, where GOP chief negotiator Jayme can use his capacity as the gatekeeper of finance to provide an increase in costs, via the reduction/absence of international finance to keep the economy moving, and a decrease in benefits, via belt tightening fiscal policies. This clear explanation, however, has limits when applied to Level IIB. While the connection between the IMF program and petroleum prices are linked in domestic news sources, the causal mechanism is either lost or forgotten. In the September attempt, the GOP chose not to link the two, producing a backlash that it could not overcome. The December attempt also provided a different set of circumstances as the military coup took center stage, pushing the petroleum prices out to the wings, and no attempt was made to link the IMF program with the price increase. Instead, the agenda was changed from one of a fiscal crisis to one of national security.

Does the technocratic alliance provide the best explanation at both levels? This tactic posits that the IMF negotiation staff and the GOP chief negotiations share and subscribe to the same information and through the institutional arrangements are able to expand the information network to others in the cabinet including COG Aquino. The institutional cabinet rules become critical as the security cluster serves as the gatekeeper of all proposals. Much like Joker Arroyo in the early Aquino years, the group vets information coming not only to the attention of the full cabinet, but also to Aquino. Unlike Arroyo, the cluster filters out only recommendations that do not meet its security threshold, which tend to be left leaning policy recommendations. The distinction is nuanced, where the information is one of degree (i.e., level of security) versus type (i.e., left or right). Under this circumstance, the technocratic alliance tactic is specious as it is limited to explain the Level IIA acceptance. Furthermore, its application to Level IIB was not applied in either the September or December situations. The September price increase occurred before the addendum, and the December price increase occurred after. In neither case was price rationalization an issue of prior-conditions. Taking the counterfactual, we can claim that IMF technocrats inundated Level IIB institutions and in the first case failed and in the second succeeded. Our assessment is inconclusive.

The proximity of the attempts to increase petroleum prices gives us an opportunity to examine the learning process. The August 1987 coup attempt had reverberating effects through the Level IIA decision-making process. It created an expectation of coup attempts with petroleum price increases; as a result, a consensus on price increases proved inadequate process to interpret adequately the security ramifications, requiring a final vetting the lens of preventing coups. The institutional arrangement was adjusted to pass a security test, depoliticizing the context of price increases. In a similar fashion, the GOP removed from the cabinet the price increase decision-making process to the ERB, from Level IIA to Level IIB, giving social veto actors input, at least the appearance of input, into the process. The GOP expected that Level IIB actors would view the price increase process as "fair," taking into consideration the interests of all involved parties. These institutional changes attempted to remove politics from policy and in the case of Level IIA, exchange politics for security. In providing a security test, the cabinet focused on two issues: the impact price increases have on social veto actors and the disruption they may cause and the opportunities these disruptions may provide RAM. These two points may account for why Aquino quickly asked the ERB to reconsider the increase. Diffusing the social veto actors expected reaction muted any opportunity for RAM to launch another attack. Furthermore, perhaps the absence of a coup in September allowed the GOP to learn in December. In comparison to the September price increases, the December increases were roughly half, perhaps suggesting that the GOP could get through a smaller increase. Social veto actors (Transport associations, the KMU, the TUCP, and the CACP) responded at unexpected and unprecedented levels. Under the cloud of these demonstrations, RAM took advantage of the situation and launched its most devastating military coup against the government. The GOP learned, but it did not foresee the outcome.

1989C IMF-GOP PETROLEUM PRICE OBSERVATION

In the first three quarters of 1990, the GOP attempted to take advantage of its international opportunities and assuage its domestic constraints. According to the 1989 MEP targets, the economy had gone off track. Aggregate demand expanded, but it stemmed from government and consumer spending, which was aided by raises in imports without a corresponding rise in exports.[88] The economy began to overheat. Wages increased by approximately 40 percent, pushing inflation to 10.5 percent clearly surpassing the MEP target rate of 8.8 percent. Since foreign donors and creditors tied their disbursements to the MEP tranches, the GOP tightened monetary and fiscal policies but the latter was done in an uneven manner. This imbalance stemmed in large part from the fragile perception of Aquino's legitimacy, triggered by the military coups and the increasing levels of poverty. Coupled with these concerns, the GOP also dealt with natural disasters, forcing the GOP to expand expenditures without commensurate revenue generating measures, primarily keeping petroleum prices stable. In short, the GOP traveled the neoliberal line, the populist line, and back to the neoliberal line, with some exceptions. The August 1990 Iraq invasion of Kuwait sealed the GOP fate. It had to raise petroleum prices.

International Political Economy of the 1989c Petroleum Price Observation

In contrast to economic improvements shown in 1986–1989, the economy in 1990 began to deteriorate dramatically. In the first half of the year, the economy expanded by only 3.4 percent of GDP, 1.4 percent below the GOP minimum expectation.[89] The 1986–1989 pump priming policies had much to do with it as the economy began to reach full capacity, contributing largely to inflationary pressures, which by 1989 had reached 10 percent and by the first semester of 1990 reached 12.5 percent.[90] To keep to its 1989 MEP targets, the Central Bank tightened monetary policy, increasing the interest rates of treasury bills, the rediscount rate, and the reserve requirement.[91] To compound this problem, a drought devastated agricultural production; power shortages hampered industrial production; an earthquake struck the Luzon region; and a typhoon struck Cebu.[92] These misfortunes forced the GOP to increase expenditures by 5 percent over the MEP target.[93] These domestic problems were paralleled by macroeconomic imbalances. The 1989 coup attempt hindered foreign investment and tourism. For the first semester of 1990, foreign investments grew by only 13.5 percent, down from 80 percent of the same period a year prior.[94] Similarly, tourist arrivals in the first semester dropped by nearly 70,000 in the same period a year prior.[95] Furthermore, the Philippine Peso depreciated by 7.5 percent, which exacerbated the trade deficit and the foreign debt servicing outlays. Debt servicing, both domestic and foreign, accounted for approximately 40 percent of the national budget.[96] These declining indicators were aggravated by the August 1990 Iraq invasion of Kuwait.

The Persian Gulf crisis made bad economic conditions worse. The aggregate trade deficit widened from $2.6 billion in the previous year to $4 billion.[97] By the time of the invasion, the Philippines imported approximately 83 percent of its oil from the Middle East.[98] The oil imports increased by $1,787 million from the previous year where oil accounted for approximately 25.1 percent of the increase.[99] One part of this increase resulted from cost of crude jumping from an average of $15 a barrel to $30 by October. Another part came from an increase in import volumes as the government encouraged oil companies to increase their inventories to protect against oil shortages.[100] On the revenue side, exports decreased from the previous years due to a slack in international demand, primarily from Middle Eastern countries, and a softening of traditional exports due to the drought. Furthermore, Middle East Overseas Contract Workers (OCW), about 79 percent of which were in Saudi Arabia, decreased their remittances by $24 million.[101]

Relations between the IMF and the GOP remained on solid footing as Aquino's new economic managers took center stage. After December coup attempt, Aquino reshuffled her cabinet putting in Jesus Estanislao as finance, Jose Cuisia Jr. as the Central Bank governor, and Cayetano Paderanga as the head of NEDA.[102] Estanislao understood that the key to obtaining economic growth and stability was keeping the CPSD within a prudent range of 2–2.5 percent of GNP.[103] The deficit served as the linchpin for bringing down and maintaining low levels of inflation, which both Cuisia[104] and Estanislao treated as "public enemy number one."[105] A CPSD of greater than the desired range would only temporarily push real growth upward. Afterward, inflation and other undesired macroeconomic imbalances would take effect, and the adjustment of this situation could only be corrected through painful economic stabilization programs.[106] Estanislao's fears materialized as the OPSF portion of the deficit began to increase from 3 percent in 1988 to 4 percent in 1989 and ultimately to 5 percent in 1990.[107]

Shortly after Iraq invaded Kuwait, Estanislao had to marshal support for foreign financing. He received support from the GOP's main donors and creditors, Japan, the United States, the IMF, and the World Bank, which agreed to provide additional financial assistance. Estanislao's credibility was supported by Aquino's policy to repay all the foreign debts. Indeed, Aquino supported Estanislao on many issues. "I was lucky. I had a President who had steel for what is reasonable. She gave me support all the way."[108] Furthermore, commercial bank creditors increased the releases of more financing.[109]

Estanislao used an external-internal strategy, securing support of the foreign creditors and donors to get domestic players to participate in the GOP's plans. Furthermore, the Congress was deliberating plans to suspend or waive foreign debt, but in light of the country's foreign support, such plans would undermine its "relations." Aquino informed Estanislao that she was in contact with numerous senators on this matter.[110]

While Cuisia tried to rein the money supply of the Central Bank in March 1990, Estanislao tried to also increase revenues to stay within the 1989 MEP targets. Under the program, the inflation target was 7 percent; the economy had 14 percent inflation. The CPSD target was 2.5 percent; the actual deficit was 5.2 percent.[111] In mid-September, the GOP tried to rein in the OPSF portion of the CPSD by increasing petroleum products pricing. The ERB authorized the oil companies to increase the price of petroleum products by an average of Pl.40 per liter. Under normal circumstances, a price increase of this magnitude would have eradicated the OPSF deficit. However, on September 19, two days before the price increase took effect, Congress had passed and Aquino signed into law Republic Act No. 6965, changing the tax formula on premium and regular gasoline, aviation turbo fuel, diesel and asphalt from ad valorem to specific taxes rates and exempted liquid petroleum gas and fuel oil from taxation.[112] Because of the tax change, the GOP stood to lose an estimated Pl.7 billion in revenue for the remainder of the year.[113] To illustrate the reduction in GOP revenues, ad valorem taxes derived from premium gasoline averaged P3.75 per liter from September 1989 to September 1990. With the change in taxation rules, government collection was reduced to P2.52 per liter.[114]

The GOP's ineffective adjustment of local petroleum prices brought a stern response from IMF lead negotiator Ulrich Baumgartner. To pressure the GOP into making the proper adjustments, Baumgartner issued a memorandum to Estanislao and Cuisia stating that the GOP must continue to increase the price of petroleum products as well as devalue the peso to receive consideration for a 1989 EFF drawdown.[115] In the first quarter of 1990, the Central Bank maintained an overvalued peso at P21.35 to the dollar as a result of its efforts to mop up excess liquidity. In October the GOP devalued the peso by 24.8 percent from a year earlier as the peso fell from P25.75 to P28 to the US dollar. According to Cuisia, Baumgartner's memorandum had nothing to do with the devaluation, rather "it was something we knew we had to do anyway."[116]

The October devaluation had two significant repercussions for the GOP budget deficit. First, it increased the landed cost of crude, and thereby expanded the OPSF reimbursement request and exacerbated the price buffer's deficit to nearly P14 billion. Second, it increased the country's debt service cost by an additional P4 billion.[117] Due to the IMF and GOP economic managers' insistence that debt payments continue without hindrance, the government had to reduce other expenditures and increase revenues to trim down its deficit. At this juncture, both parties realized that the GOP could no longer stay within the limits of the 1989 EFF and time had come to draft a new Standby Agreement.

Domestic Political Economy of the 1989c Petroleum Price Observation

In early 1990, the Aquino government proceeded to develop a program to ensure economic growth. In a secret cabinet meeting, the Aquino administration committed itself to rely more on market forces to develop the economy. Several proposals were made and accepted that called for the privatization of several government owned corporations, increasing the relaxation of trade barriers, and the deregulation of the oil industry and its prices.[118] The government's desire to deregulate oil originated from several sectors. First, particular government officials continued to call for the deregulation of the industry to release the GOP from its fiscal constraints. The OPSF owed oil companies about 27 percent of the 1990 GOP budget deficit. Second, the government received tremendous criticism for its late 1989 decision to adjust oil prices. Third, the government had a clear preference to divorce itself from the politically sensitive and highly controversial process of petroleum pricing.[119] Despite the best intentions of the GOP to initiate reform in the oil industry, it found that economic conditions were not conducive to deregulation as it would only serve to incite public fears on price increases. To mitigate these fears, DBM Secretary Guillermo Carague proposed that oil prices be deregulated gradually over a 6 month period, and to allow oil prices to rise at only 1.2 percent a month. In addition to deregulation, the meeting also launched Aquino's *Kabisig* movement, shorthand for the institutionalization of executive power with the regions.[120] Eventually, the GOP postponed these plans due to the country's natural disasters and the Iraqi invasion.

During this period, the GOP worked tirelessly with the Congress to address fiscal matters. In July, before the Iraqi invasion, the two parties convened and Estanislao submitted a package of revenue generating measures, including increases in sin taxes, road taxes, and affluence taxes. These measures had to be passed by the lower chamber, which was amenable to increases in sin taxes, but stopped short of passing a tax on wealth. The leadership had concerns about keeping the *Laban ng Demokratikong Pilipino* party intact and protecting its members from a "no new taxes" pledge it had made earlier. Furthermore, the 1992 elections made taxing the rich politically unpalatable.[121] Just as revenues were hard to collect, so would expenditures be difficult to cut. Aquino ordered her departments to slash 20 percent of their budgets in an effort to reduce the CPSD. Members of her cabinet objected to this measure, which required Aquino to readjust the budget strategy.[122] Members of Congress subsequently called for a "Budget Summit" to reconcile fiscal differences among the branches and bureaucracies. The results were inconsequential, forcing both chambers to agree on the use of an executive order to produce a temporary import levy, shielding legislators from public scrutiny.[123]

The Iraqi attack pushed the GOP into crisis mode; it had to address not only petroleum supply and price issues, but also the lives and remittances of OCWs. Almost immediately, the cost of oil jumped from $15 per barrel to $21 per barrel. In turn, this development produced two effects. First, the Philippine demand for foreign exchange increased, especially from domestic oil companies.[124] Second, it also increased the uncertainty regarding Philippine access to oil, which prompted the GOP to urge domestic oil companies to increase their stocks from 60 days to 90 days.[125] The increase in demand contributed to a rapid depreciation of the peso from the end of July until the end of September, a development that proved costly for the GOP as it remained committed to support the forward cover. To prop up the peso, the Central Bank participated in the trading floor as a net seller of foreign currency. Despite these developments, the GOP continued to maintain the current price levels of petroleum products, accelerating the OPSF deficit to P8.4 billion by September (see Table 4.2). The combination of the devaluation and the increase in the price of crude pushed the Philippine cost to $25.60 a barrel, necessitating a price increase.[126]

Bracing for the worse, Aquino went through a list of contingencies. She instructed the cabinet to begin consultations with Congress, examine conservation measures, and reduce the demand for energy.[127] Furthermore, Cluster E, the eventual formal title of the cabinet security committee, gamed out political and security scenarios. To provide international cover for its policies, Aquino also instructed Estanislao to ensure continued access to new money from foreign creditors. The groundwork for international and intergovernment cooperation was just the beginning. Commitments had to be made on the public side as well as social groups began to grumble about the day-to-day implications of the invasion. As a response, the GOP called for a two-day multisectoral meeting consisting of Congress, business, labor, the media, and NGOs. Criticism about the crumbling conditions was rampant, but in the end, these groups rallied around the calls for nationalism. According to Estanislao, "They knew we were in a state of emergency. They all wanted to do their part to help."[128] A week after the meeting, the ERB held hearings.

The hearings were short. The ERB received the oil company petitions, bypassed the formal hearings to include opponents to increases, and ruled in favor of the oil companies.[129] On September 21, 1990, it raised prices of kerosene and fuel oil from P4.96 to P6.24, a 31 percent increase. Several senators later filed petitions regarding the unfairness of the process. Later, the ERB announced that it would reassess its decision in several weeks, when it could either increase, decrease, or suspend its initial decision. By the next session in early October the ERB maintained the price hike, making only a modest hike to fuel oil (see Table 4.1). To assuage the impact, Aquino instructed DTI to monitor prices and ensure supplies.[130]

The GOP was prepared for the worse. Some of the anticipated responses materialized but in muted form. Rebel soldiers occupied a military installation in Northern Mindanao; their hold fizzled after a few days. Twenty days after the ERB announcement, the *Bagong Alyansang Makabayan* (BAYAN) and KMU staged a *Welga ng Bayan*, but its impact was minimized due to the tripartite process, fractured opposition, and anticipated planning. The multisectoral meeting helped to assuage domestic opposition. Furthermore, attempts to increase wages to offset the petroleum increases were just around the corner. The acceptance of emergency contingencies was temporary as DTI Secretary Jose Concepcion announced that the merchants and manufacturers suppressing of prices could no longer be sustained.[131] Prices of commodity goods would have to increase. Some cabinet members suggested that the GOP absorb these increases, a move Estanislao rejected.[132] Opposition to the price increase also had little traction. BAYAN and KMU supported the strike, but other groups such as transportation associations received authorizations from the LTFRB to pass on the price increases to consumers. The TUCP did not participate in the strikes, nor was it inactive. Indeed, it, along with the KMU and the Labor Advisory Consultative Council, threatened to initiate their own strikes if the GOP did not dismantle the ERB and replace it with a multisectoral agency that included labor and consumer interests, rescind the price increases, and an across the board increase in minimum wages.[133] The reduced strikes, however, did produce an uneven effect around the country as some regions were disrupted and others were muted. Finally, the GOP suspended school classes, mobilized the military to use its vehicles as commuter transportation, and arranged to ration petroleum through the creation of the cabinet level Energy Operations Board with Executive Order 422.[134] Despite the successful price increase, the GOP failed to gain sufficient revenues due to its alteration of the petroleum tax scheme ad valorem to a specific rate.[135] The GOP managed to sustain the price increase, but it needed another round to restore its fiscal balance. It chose not to go down this road again.

Instead, the GOP implemented a 5 percent tariff import designed to raise P16 to P17 billion in government revenues, resulting in an additional P0.53 centavos loss per liter for oil companies. The financial conditions had severely deteriorated for domestic oil companies, prompting Dutch Shell to seriously consider pulling its operations out of the Philippines.[136] For the GOP, the financial cost of no action proved just as devastating, since delays would cost the GOP P90 million per day or P2.7 billion per month in subsidies.[137] Another round of price increases was necessary.

Epilogue to the 1989c Petroleum Price Observation

As with their many dealings, the IMF and GOP agreed on the principles of stabilization but dissented on the details. Both parties recognized that the burgeoning CPSD was in large part responsible for the economy's problems. The IMF strongly suggested drastic expenditure reductions and revenue enhancing measures, in particular, a petroleum price increase, recommendations with which Estanislao and Cuisia already agreed. Prices were increased. A heterogeneous set of domestic veto actors, however, vehemently had opposed the move and pressured Aquino to recant the decision. Later, the ERB rescinded part of the price increase. Having lost the price increase battle, the GOP resorted to another measure, an import levy. This time the IMF initially balked at the idea, but later reconsidered its position when the chief negotiator provided critical information to shift IMF costs perceptions.

ASSESSMENT OF CAUSAL MECHANISMS

At domestic Level IIA, the external-internal tactic has the clearest effect. Estanislao relied on the international Level IIB actors (public and private financial creditors) to move domestic opposition. The critical component of this observation, similar to the 1989b observation, lies in the established EFF. The program had already been in place and the international Level IIB actors have bought in to provide immediate funding. The cost of missing the agreed upon IMF targets would be extremely devastating for the GOP as it would not only mean cutting back government wages or jobs and development projects need to promote economic activity, but also limit the capacity of businesses to engage in international transactions, bringing the economy to a crawl. The benefit of hitting the targets was that new monies would be forthcoming. These targets were going to be overshot due to the current economic trends, the Iraq invasion of Kuwait accelerated and increased the imperative, serving as a catalyst for action.

The tactic also seemed to push through a successful domestic price increase Level IIB ratification. The price increase process embraced many veto players; many right and center-right leaning organizations participated and reluctantly accepted the costs associated with the price increase and pursued offsetting remedies, if even partially. They would either pass along or bear some of the costs. In contrast, some left leaning veto players refused to participate. The unevenness of Level IIB resistance to the price increase demonstrated that laying the ground work for the cost proved successful.

Under the prevailing institutional arrangement, Estanislao was required to rely on the deference tactic. The circumstances regarding international financing, the Iraqi invasion, and the increase in international oil prices forced Estanislao to make the price increase recommendation. A majority of the cabinet principals knew it was inevitable, absorbing the costs could not be an alternative as the CPSD would continue to surpass the IMF targets. Cost recovery was the only viable option. At Level IIB, the GOP also deferred the process to the multisectoral group. The outcome was as predicted, but perhaps for a different logic. After the cabinet decision, the GOP used the deference tactic at Level IIB and threw a wide net around Congress, business, labor, the media, and NGOs opening up the participation process. The consultations occurred a week before the ERB hearings, showing the GOP willingness to provide transparency. The board hearings themselves and the decision was quick, giving the ERB the responsibility to revise its decision should the expected demonstrations and strikes reach a threshold level with which RAM can take advantage. The deference tactic fails to "get the guy to get the guy"; rather, it "gets the guy to ensure that the other guys don't play." These mechanisms can explain the developments at Level IIA and Level IIB, but can the technocratic alliance framework explain the developments better?

This tactic can explain the outcome at Level IIA. The meeting of the minds by the IMF staff and the GOP chief negotiators predated the negotiations. This outcome is unremarkable. Given the consensus institutional arrangement of the GOP inner circle, the ideas of rationalization easily transfer from Baumgartner to Estanislao, or the other way around, and to Aquino, receptive to US and IMF ideas about liberalization. Similarly, no evidence exists to suggest that IMF interlocutors interacted with Filipino citizens. To support this hypothesis, the IMF would have to "infiltrate" Congress, transportation and labor unions to assist their issue area counterpart. This tactic can suggest that its effectiveness is more pronounced at Level IIA rather than Level IIB.

The GOP learned from its past policies. The 1989b observation (both September and December) reflects a series of changes and adjustments that culminated in the 1989c Observation. Fearful of the August 1987 coup experience, the GOP treaded carefully when it increased prices in September 1989. Surveys, NGOs, and government commissions, which culminated in the Davide Commission Report, provided the GOP with information about the state of the military, the economic, social, and political circumstances that precipitated the coup attempts.[138] These circumstances included the fragile nature of the constitution and the parties, players, and possible political aspirations concerning the 1992 presidential elections. Furthermore, the turmoil associated with the oil price increases provided an opportunity for coup plotters.[139] These mechanisms were instrumental in providing information for the executive branch. It subsequently made alterations to its decision-making process to "depoliticize." Within the formal authority of the executive branch, the GOP institutionally funneled the decision-making process through a security gate, delegated the price level process to an ERB, and formalized the wage increase process to the NWPC. Petroleum prices in September were raised with caution and only as a last resort. The IMF program targets were off-line and required adjustment. The NWPC authorized the wage increases; the GOP agreed to increase petroleum prices; the ERB then determined the appropriate levels. Once social veto players reacted, the GOP understandably got cold feet and then backed off. The ERB subsequently restored the original prices. The fiscal problem remained and the IMF targets could not be met.

The GOP altered the process of the second round of price increases. Any attempt to employ the NWPC would be futile given the response by businesses in September. Instead, the oil companies served their petitions, the ERB quickly answered, and authorized the increases. Transportation, left leaning labor, and consumer organizations expected the outcome and quickly responded with debilitating strikes. The subsequent military coup attempt forced Aquino to appeal to the country's patriotism. The fast-paced process allowed social veto players to marshal significant support to stage devastating and turbulent strikes. The process succeeded to increase and maintain petroleum prices, but certainly this process would not be duplicated in the future.

From the December 1989 devastation, the GOP adjusted its process. This time, the prices had to increase and stick. It called for a multisectoral meeting with Congress, private veto players, and NGOs, and pushed for a fast ERB process that gave it the capacity to reassess its decision. In addition, it tried to hold down the pass through of prices to consumers for as long as it could, during which a *Welga ng Bayan* was staged. Finally, the LTFRB and the DTI allowed the pass through of prices. The multisectoral meeting appeared to temper social veto player concerns and shrink the veto player coalition and also the possibility of a military coup.

NOTES

1. Boughton, 1002.
2. Ibid.
3. Ibid., 508.
4. See Krugman, Alm, Collins, and Remolona on the problems with transforming the Philippine economy.
5. Galang 1988.
6. Dohner, 8–9.

7. Galang 1988b.

8. Interview with Senior Bangko Sentral ng Pilipinas official, anonymity requested, June 15, 1985.

9. *Business Bulletin* 1988a.

10. Collás-Monsod, 476.

11. Ibid.

12. Ibid., 477.

13. Ibid., 478.

14. Interview with Senior National Economic and Development Authority official, anonymity requested, May 23, 1995.

15. *Business Bulletin* 1988c.

16. Interview with Senior Department of Finance official, anonymity requested. April 27, 1995.

17. Broughton, 509.

18. de Dios 1990, 130.

19. Ibid., 131.

20. Ibid., 130.

21. Ibid., 130.

22. *Business Bulletin* 1988f.

23. Ibid., 1988e.

24. Ibid., 1988d.

25. Tiglao 1989a.

26. *Manila Bulletin* 1989a.

27. Thompson 1992, 47.

28. Ibid.

29. The contributors were expected to be Japan, West Germany, France, Italy, Canada, the World Bank, and the Asian Development Bank. $10 billion was the figure originally proposed, but no one could really point to how the figure was derived. This figure was merely used as a reference and stuck throughout the planning phase. Interview with Senior Department of Finance official, anonymity requested, April 27, 1995.

30. According to Collás-Monsod, if the military base agreement was signed, then the PAP could proceed. Collás-Monsod, 484.

31. International Monetary Fund 1989a, 27.

32. de Dios 1990, 136.

33. Ibid., fn. 13, 144.

34. Tiglao 1991b.

35. Thompson 1992, 50.

36. de Dios 1990, 137.

37. *Manila Bulletin* 1988b.

38. The Military Bases Agreement was to end in 1991 and was subject to continuous ongoing negotiations.

39. International Monetary Fund 1989a, 28.

40. U.S.—Philippine Military Bases Agreement Review, 1988.

41. US Foreign Assistance Reference Guide, 6–7.

42. Tiglao 1989b.
43. Republic of the Philippines, 1989b.
44. Ibid.
45. Ibid.
46. Broughton, 509–510.
47. Davide Commission Report, 176.
48. Ibid.
49. Ibid.
50. RAM represents approximately 10 percent of the officer corps. Its announced goal is to build an effective and fair minded armed forces in the service of the country and people, "weeding out corruption while improving training and moral." Wurfel, 290–291.
51. Interviews Secretary of the Finance Department Jesus Estanislao, April 20, 1995 and Acting Secretary of Finance Department Ramon del Rosario, May 11, 1995.
52. A key member of the cabinet also released was Executive Secretary Joker Arroyo, a left leaning member who commanded considerable influence with Aquino and had fought bitterly with Ongpin within the cabinet. Arroyo chided Ongpin for "agreeing to impositions to the World Bank and IMF." Ongpin reportedly criticized "serious delays: in implementing financing and tax programs and lack of coordination by Arroyo." Dohner and Haggard, 69.
53. Timberman 1991, 188–193.
54. Alonzo, Canlas, Fabella, Medalla, and Paderanga Jr.
55. *Manila Bulletin* 1988a.
56. de Dios 1990, 135.
57. Galang 1988a.
58. Interview with Senior Bangko Sentral ng Pilipinas official, anonymity requested. June 15, 1995.
59. *Business Bulletin* 1988d.
60. Interview with Senior Department of Finance official, anonymity requested, April 27, 1995.
61. Interview with Senior NEDA official, anonymity requested, July 28, 1995.
62. de Dios 1990, 136.
63. Romulo.
64. International Monetary Fund 1989a, 8.
65. International Monetary Fund 1989a; and Republic of the Philippines 1990b.
66. International Monetary Fund 1989b, 5.
67. Ibid., 27.
68. Smith, 1.
69. Addition to the IMF Memorandum of Economic Policy, 27.
70. International Monetary Fund Circular 1991, 90.
71. Interview with Senior National Economic and Development Authority official, anonymity requested. June 27, 1995.
72. Republic of the Philippines. Administrative Order 91, Creating a Cabinet Committee on Matters Affecting Oil, September 12, 1988.
73. Secretary of Finance Jesus Estanislao, email message to author, August 30, 2016.

74. Chambers, 108.
75. Interview with Senior National Economic and Development Authority official, anonymity requested, June 27, 1995; and interview with Senior Budget and Management official, anonymity requested, May 23, 1995.
76. Timberman 1991, 259.
77. Chambers 2014, 105.
78. McCoy 1999, 261.
79. Timberman 1991, 259.
80. Chambers 2014, 108.
81. Timberman 1991, 259.
82. Velasco 1991, 191; Chambers 2014, 105.
83. Republic of the Philippines 1989a.
84. Office of Energy Affairs.
85. In larger part, some extreme right-wing military leaders were displeased with the overtures the Aquino government had made with the communist party in its efforts to end hostilities.
86. Interview with Secretary of Finance Jesus Estanislao. April 20, 1995.
87. Tiglao 1990a.
88. Benjamin E. Diokno, "Philippine Macroeconomic Policies Affecting Households," 1990, 2–3, and Kruger, Alm, Collins, and Remolona.
89. Timberman, 1991, 160 and Yap, 7.
90. National Statistics Coordination Board and Central Bank of the Philippines as cited by Yap, 3.
91. Ibid., 13.
92. Ibid., 7.
93. Ibid., 14.
94. Ibid., 12.
95. "Visitor Sample Survey," Department of Tourism, as cited in Lambrete and Yap, 50.
96. Timberman 1991a, 160.
97. Lamberte and Yap, 38.
98. Ibid., 9.
99. Ibid., 38.
100. Ibid.
101. Ibid., 36.
102. Estanislao first replaced Monsod as Economic Planning Secretary and head of NEDA. When Jayme left the cabinet, Estanislao took over that position, and he was succeeded by Paderanga.
103. Estanislao, 1.
104. Interview Senior Philippine Central Bank official, anonymity requested. May 29, 1995.
105. de Dios 1992, 98.
106. Estanislao, l.
107. Lamberte and Yap, 19.
108. Secretary of Finance Jesus Estanislao, email message to author, August 30, 2016.

109. The Aquino Management of the Presidency, the Gulf Crisis.
110. Ibid.
111. Lambrete and Yap.
112. Republic of the Philippines. 1990d.
113. Tiglao 1990b.
114. Department of Energy. n.d.
115. Ibid., and Collás-Monsod 1991.
116. Interview Senior Philippine Central Bank official, anonymity requested, May 29, 1995.
117. Tiglao 1990c.
118. Interview with Senior National and Economic and Development Authority official, anonymity requested. June 27, 1995.
119. Gamboa, 2–3.
120. The Aquino Management of the Presidency, the Gulf Crisis; and Clarke, 39.
121. *Manila Standard* 1990d.
122. Ibid., 1990e.
123. Ibid., 1990d.
124. Ibid., 1990a.
125. Lamberte and Yap, 38.
126. Interview with Secretary of Finance Jesus Estanislao. April 20, 1995; and The Aquino Management of the Presidency, the Gulf Crisis.
127. The Aquino Management of the Presidency, the Gulf Crisis.
128. Interview with Secretary of Finance Jesus Estanislao. April 20, 1995.
129. *Manila Standard* 1990b.
130. Ibid.,1990a.
131. The Aquino Management of the Presidency, the Gulf Crisis.
132. Ibid.
133. *Manila Standard* 1990g.
134. Republic of the Philippines 1990a.
135. Office of Energy Affairs 1990.
136. Interview with Senior Pilipinas Shell Corporation Finance official, anonymity requested, June 27, 1995.
137. *Manila Bulletin* 1990h.
138. The Davide Commission Report. The Davide Commission was started in January 1990, and its findings released in stages throughout the year. It was formally released for public consumption in October 1990.
139. Ibid., 83 and 93–94.

Chapter 5

1991 GOP-IMF Petroleum Price Observation

The economic, military, and political developments of 1989–1990 proved too much for the GOP. It straddled the line between its international creditors that required neoliberal policies and domestic pressures that pushed for populist-leaning policies. The attempt to walk the line proved catastrophic. The GOP attempted to adhere to the targets set by the 1989 MEP, but it was apparent that the GOP could not reach them. As a result, the IMF suspended the program's disbursements. In January 1991, the IMF abrogated the EFF and it negotiated with the GOP to develop a more manageable stabilization program. Two factors proved critical to the negotiation process: the ability of the GOP to demonstrate its commitment to the stabilization program by formulating and implementing revenue-generating measures prior to the approval of the formal program and the military and economic linkages that existed between country members of the IMF executive board and the Philippines.

INTERNATIONAL POLITICAL ECONOMY OF THE 1991 PETROLEUM PRICE OBSERVATION

From the outset of the negotiations, Estanislao kept one eye on the IMF and the financing that a concluded agreement would accrue. For the most part, the IMF agreed with Estanislao that unless the right steps were taken, the CPSD would spiral out of control. The hemorrhaging OPSF was the culprit. Although Estanislao agreed in principle with the IMF, he sought some political breathing room by suggesting "looser targets."[1] Specifically, he wanted to set the CPSD target to be P43 billion for 1990 and P31 billion for 1991.[2] Regardless of the targets derived, Estanislao grew more concerned over the benefits that an IMF agreement would bring. It would restart public and private credits that were held up due to the impasse between the two parties. In particular, Estanislao was interested in ensuring the flow of PAP financing, where donor countries would hold another pledge session in early 1991.[3]

As Estanislao sought to meet the revenue requirements of the IMF staff, he also played the "executive board card." As early as August 1990, Estanislao secured commitments from central donors and creditors to continue additional assistance and from commercial banks to accelerate the release of new money, pledges made easier by Aquino's position on full repayment of debts.[4] Estanislao also lobbied critical member countries on the IMF executive board. Still the tacit leader of the PAP, Japan consented with the Philippines' other major aid donors that the GOP should first meet the IMFs directives to meet the next round of pledges. According to one Japanese official, "We do not wish to undermine the efforts of the IMF to persuade the Philippine Government to correct fiscal and monetary policies."[5] Despite its reluctance to interfere in the IMF-GOP negotiations, Japan pledged to help the Philippines with $200 million in "special commodity loans," under the heading of earthquake-damage repair loans and assured the GOP that it would urge the international organizations to recognize problems caused by the Gulf crisis when deliberating loans to the Philippines.[6] Along with securing Japanese support, Estanislao also heavily lobbied Singapore, Germany, and Saudi Arabia to pressure the IMF. German support for the Philippines came as no surprise since it had vested trading and financial interest in the country. Singapore and Saudi Arabia's special interest lay with the country's proximity to critical shipping lanes and the reliance on a steady supply of OCW as laborers.[7]

Even with these countries' special interests in the Philippines, their cooperation did not come without a price. In exchange for their support, Estanislao promised to press for the realization of three issues: increases in the price of domestic petroleum products, increased taxes, and the continued presence of American bases stationed in the Philippines. In effect, Estanislao informed the representatives of the member countries: "You must show support for me so that I can push for you."[8] With more ammunition in hand, Estanislao again attempted to sway the GOP cabinet. This time, his arguments for an oil price increase were more persuasive, as he managed to apply leverage in the form of committed foreign financing.

Like the 1989 IMF-GOP EFF talks, these negotiations were influenced by linking them to the upcoming PAP pledging session. Estanislao used the PAP donor meeting as his trump card to push the IMF into endorsing the economic program much faster than it wanted.[9] Indeed, IMF officials felt pressured to secure an agreement before the meeting or it would have most likely been postponed.[10] Under the PAP, the Philippines received another $3.3 billion in loans.

In early December, the Fund and GOP agreed in principle to an economic program, a plan that would become official when the GOP increased petroleum prices.[11] To ameliorate the deficit, the GOP needed to reduce its expenditures and raise additional revenues as well as make further adjustments in the exchange rate. Since Congress was reluctant to pass the necessary legislation, the GOP initiated a 5 percent levy on oil and other imported crude, a move that increased the landed cost of crude. Changes to the landed cost of crude and the magnitude of the OPSF deficit forced oil companies to petition for price increases. The only question remained was how much the price should be increased. In a December cabinet meeting, two price figures were placed on the table before Aquino where the IMF had suggested the higher of the two values. Aquino chose the lower.[12] This decision on how much to raise the price of petroleum products did not take effect immediately as the GOP still had to go through the ERB and hold public hearings.[13] Later, the ERB authorized the price increase, but Aquino succumbed to private and public opposition and requested that the ERB reconsider its decision (see Table 5.1).

Estanislao's "involuntary defection" from his promise to the IMF to increase petroleum prices forced him to resort to harsher measures. Immediately, the GOP had to address the fiscal shortfall before the negotiations could go on further. Estanislao had traveled to Washington, DC and explained the circumstances regarding the marginal increase in petroleum prices to IMF Managing Director Michel Camdessus, securing his support and convincing him that he had "done all he could when it came to oil levies."[14] Having temporally assuaged the IMF, Estanislao still faced the problem of insufficient revenues. To resolve this problem, he proposed to the IMF and Aquino that the GOP alter the 5 percent levy on imported crude and petroleum products to an 9 percent across the board levy.[15] Initially, the IMF staff resisted this option since it violated its free trade position; however, the IMF accepted the measure only after agreeing that without the PAP loan package, the Philippines would be pushed into a balance of payments crisis.[16]

Table 5.1 Historical Petroleum Product Prices (Per Date of Adjustment October 4, 1990 to August 14, 1991) (Metro Manila) in Peso per Liter

Dates	*Premium*	*Regular*	*Diesel*	*Kerosene*	*Fuel Oil (Wholesale)*
10/5/1990	8.87	8.5	6.24	6.24	4.98
12/5/1990	15.95	15	9.35	9.35	5.16
12/6/1990	20.7	19.56	7.75	7.75	5.16
12/10/1990	15.95	15	7.75	7.75	386
8/14/1991	12.7	12	7.3	7.3	3.45

Source: Republic of the Philippines, Energy Regulatory Board.

The decision to use the import levy to generate revenue did not come without exhausting other options. Indeed, as early as October, Estanislao attempted to persuade Congress to push through a series of new taxation measures, but house leaders balked at his recommendations. During the negotiations, the IMF took a turn at attempting to persuade Congress of the necessity of new revenue legislation, convincing both chambers on the need for revenue, but failing to move them into action as the house insisted that intensification of current taxation practices would suffice to reduce the GOP deficit. The outside-inside interaction strategy was more than just an effort at coalition building; it was also an attempt to demonstrate how difficult it was to move Congress. "Now they (the IMF) know what we are up against," said Estanislao.[17]

In February 1991, the IMF and the GOP eventually agreed on an economic plan to get the Philippine economy back on track. The centerpiece of the GOP's economic plan was a strong effort to bring down the overall fiscal deficit through measures to raise revenue, a reduction in expenditure appropriations accompanied by tighter controls on spending, and the gradual elimination of the deficit of the OPSF (see Table 5.2). At the same time, monetary policy was put on a tight and steadier course. Furthermore, the country was to depreciate the peso to improve the Philippines' external competitiveness and facilitate adjustment to the adverse movements in the terms of trade over the previous two years.

The plan expected the economy to move at a slow pace in 1991 and begin recovery in 1992. Inflation was to be brought down to 9 1/2 percent by end of 1991 and to 7 percent by end of 1992. During 1990, the fiscal position deteriorated sharply, with the GOP deficit increasing to 3.6 percent of GNP and

Table 5.2 Oil Price Stabilization Fund (OPSF) Monthly Terminal Balances January 1990 to December 1991 (in Million Pesos)

Months	*1990*	*1991*
January	−7,604	−15,104
February	−7,684	−14,998
March	−8,114	−12,877
April	−8,860	−10,579
May	−9,219	−7,240
June	−6,103	−4,327
July	−7,123	−1,980
August	−8,423	−49
September	−11,077	1,791
October	−13,950	2,387
November	−16,575	2,927
December	−15,482	3,271

Source: Office of Energy Affairs and Department of Energy.

the CPSD rising to 5.2 percent of GNP. In recognition of the paramount importance of a reduction in the overall fiscal imbalance for stabilizing the economic situation, the GOP and the leaders of Congress would try to reduce the deficit of the GOP to 2.0 percent of GNP in 1991 and to 0.2 percent of GNP in 1992. The CPSD was targeted to decline to 3.7 percent of GNP in 1991 and to 2.5 percent of GNP in 1992. Cuts in government expenditures were to be made across the board, with the exception of debt service, which would remain at the same level.[18]

To address the OPSF deficit, the GOP sought to introduce two measures. First, the government promised to follow a pricing policy that would ensure the attainment of zero balance by August 1991. In particular, the GOP would maintain the present price level despite the projected world price decline until a cumulative surplus in the OPSF had been generated. Second, a budgetary transfer of P5 billion would be provided in 1991 to settle outstanding claims on the OPSF.[19] The agreement also marked the first time the GOP promised to look into the deregulation of oil prices. Once calm had been restored to world markets, the GOP would begin to explore this issue and lay down the technical and legal groundwork with a view toward deregulation as soon as possible.[20]

DOMESTIC POLITICAL ECONOMY OF THE 1991 PETROLEUM PRICE OBSERVATION

The price increase occurred in the context of severe financial difficulties for the oil companies.[21] The cost of international crude had jumped from P23.99 per barrel in September 1990 to P30.33 per barrel in November 1990. The ERB reference rate for determining domestic prices was exacerbated by fluctuations to the peso-to-dollar exchange rate as it deteriorated from P24.45: $1.00 in September 1990 to P25.55: $1.00 in November 1990. Oil companies petitioned the ERB for an average price increase of Pl.42 per liter.

While the GOP attempted to generate revenue, labor groups sought to offset any adjustment measures costs. Organized labor, including the moderate TUCP, used the tripartite system to petition for an across the board P38 wage increase. The NWPC sat on the request, pushing labor to find remedies in the halls of Malacañang or the Congress, to no avail.[22] On November 1, it eventually authorized a wage increase from P89 to P106 an hour. Surprisingly, business leaders supported the call to increase energy prices, calling on their need to reflect the landed cost of crude. "Local prices of petroleum products must reflect the current import cost of crude oil and the current peso-dollar exchange rate to make energy conservation effective."[23]

The impending petroleum price increase triggered an onslaught of proponents and opponents to the ERB hearings.[24] During the two-week session, a new deliberation policy initiated by the ERB to hear adequately private interest on the matter, Estanislao testified in favor of the oil companies and reportedly claimed that oil price hike was necessary as it was one condition given by the IMF.[25] The link was reinforced by constant media barrages. Several legislators also were present at the hearings who pushed for the ERB to decline the petitions, stating that such a move would produce undue hardship for the "lower-income" sector. Opposition to the price hike was also expressed by social players. Midway through the hearings, these groups prepared to stage another *Welga ng Bayan* should the ERB decision favor the oil companies. Not surprisingly, the ERB ruled in favor of the oil companies and the GOP as it authorized an increase in prices by an average of P4.00 per liter, with gasoline prices incurring the largest proportional increase and the "low-income" sector fuel prices receiving a smaller proportion. These increases were announced in two stages. On December 5, 1990, for example, premium gasoline increased from P8.87 to P15.95. On the following day, it increased to P20.70 (see Table 5.1).

The ERB decision drew considerable attention from both private and public sectors. Transportation associations and labor unions called for commensurate increases in fares, fees, and wages. The backlashes from these players, together with cabinet opposition, prompted Aquino to ask the ERB to reconsider the price increases on the "lower-income" sector fuels. Cabinet opposition sprouted as several critical departments were isolated from the internal decision-making process and their interests were severely threatened. Among these instrumental departments were the Department of Transportation and Communication (DOTC, formerly the MOTC), the Department of Justice (DOJ), the Department of Resources and Environment (DORE), the Department of Labor and Employment (DOLE), and, surprisingly, the Department of Defense (DOD).[26] Oscar Orbos, Transportation and Communication secretary, spearheaded the move to revise the prices as he argued that the price increase would exacerbate problems inherent in the transportation industry and he called for a price revision of P7.75 per liter for diesel. Similarly, the Department of Public Works and Highways expressed concern that the price increase would push up the cost of public work projects related to the recent calamities.[27] These agencies were concerned about the level of the price increase, not the increase itself. In response to Aquino's request, the ERB revised its first price increase by upwardly adjusting the average price of gasoline by an average of P4.65 per liter and dropping the price of lower-income fuel by an average of P0.43 centavos per liter (see Table 5.1).

Revision of the ERB price increase failed sufficiently to assuage dissent as the GOP continued to establish "safety nets" for the public. Aquino had held meetings with various interest groups and she agreed to address their concerns if they agreed not to strike, an agreement to which several labor unions and transportation associations adhered. Eventually, the LTFRB and the NWPC agreed to make suitable concessions to wages and fares and fees in addition to the November increases. Despite these assurances, the *Welga ng Bayan* had engulfed Metro Manila and several key cities. DTI issued price controls and managed to secure the manufacturing sector's support for the price increase provided that the increases were moderate. In a television appearance, Aquino asked the nation to band together and cope with the price increase because of the economic crisis. She did not mention the ongoing IMF-GOP negotiations. The call for national unity provided an increase in the costs of social actors as they would be seen as anti-nationalistic.

Furthermore, Aquino also attempted to minimize the distributional conflict by appealing to the middle class to bear the burden of the economic crisis: "If the manufacturing sector were asked to bear the burden, then the cost would translate to the poor."[28] The partial success of the December petroleum price increase received little admiration from the IMF. To resolve the continued revenue deficit, Estanislao persuaded Aquino to increase the 5 percent temporary import levy by an additional 4 percent.[29] Under normal practice, the president can set up tariffs under 10 percent without congressional authorization. The levy was designed to be kept in place until June 1992. However, by August 1991, the levy was reduced to 5 percent. The levy will be discussed in more detail in chapter 6.

EPILOGUE TO THE 1991 PETROLEUM PRICE OBSERVATION

For the final three quarters of 1991, the Philippine economy traversed along the path charted under the negotiated IMF economic stabilization program. Real GNP dropped by 1.2 percent from the previous year. Toward end of 1991, however, the economy gradually recovered lost ground and registered positive developments, particularly on the external front. In other developments, the country's balance of payments yielded a surplus of $1.4 billion, a marked turnaround from the $185 million deficit registered in 1990. This was made possible by the narrowing of the merchandise trade gap, a notable increase in OCW remittances, positive developments in medium and long-term loans, foreign investments, and short-term capital accounts. In particular, export growth was bolstered partly by government incentives. On the other hand, the contraction in imports reflected the slowdown in domestic activity and the imposition of the import levy during the year.

The levy completed two objectives for the GOP. First, it succeeded in generating revenue for the government and reducing the GOP fiscal deficit. The levy was estimated to contribute Pl6.4 billion to GOP coffers.[30] To complement the revenue-generating measure, the GOP also drastically cut back on government expenditures. As of August 1991, the GOP budget deficit amounted to PS.33 billion, or 50 percent lower than the target it eventually negotiated with the IMF.[31] Second, the levy restored stability to the economy. The country's balance of payments yielded a surplus of $1.4 billion in 1991. This surplus helped boost the gross international reserve level to a record high of $4.5 billion. The high level of foreign exchange reserves coupled with the low demand for imports caused the exchange rate to settle down to P26.65 by December 1992.[32] In April 1992, Aquino repealed the levy, except for petroleum and petroleum products, which continued to be assessed at 9 percent ad valorem levy.[33]

In the fiscal sector, the GOP incurred a deficit of P26.3 billion in 1991, slightly lower than the programmed level during the same period as the increase in revenues more than offset the excess in expenditures. During the year, the deficit of the OPSF was successfully controlled and wiped out by mid-August. By September, the OPSF had accumulated a positive balance aided with the continued stability in the peso-dollar exchange rate and the decline in international crude prices.

The economic measures put in place by the IMF-GOP program, in combination with domestic political developments, contributed to the stabilization of the economy. With the implementation of the temporary import levy, the Philippine economy creped along at a snail's pace in 1991 and began to demonstrate some life within the next two years. During these years the GOP managed to initiate several structural reforms that paved the way toward increased economic activity and eventually more optimistic economic projections. With this optimistic outlook, the GOP looked for ways to sustain the economic boom. The most practical avenue, as envisioned by the GOP leaders, again led to the doorstep of the IMF.

The size of the GOP foreign debt changed little under the Aquino administration. In 1985, toward the end of the Marcos years, the country owed $26.92 billion. In June 1990, the debt was at $26.97 billion. The debt service also had a marginal change. At the end of 1985 it was $2.57 billion; in 1990, it was $2.35 billion.[34]

Ultimately, deregulation of petroleum prices is contentious. Both the left and right have valid arguments for the role of states and markets. Commentators, observers, and scholars make easy calls from the sidelines. Government officials, who face daunting political and economic challenges, can only reflect on their decisions. Estansialo made this call years later.

> In my view with the privilege of hindsight, I may have overdone the price increases at one point, and eventually we scaled them down a bit. But we did send the signals that we were committed to market forces and not to bureaucrats determining the price of oil. Too messy.[35]

ASSESSMENT OF THE CAUSAL MECHANISMS

The external-internal tactic proved effective. Securing the support of international Level IIB players, the principals of the IMF, the GOP chief negotiator Estanislao managed to produce two effects. First, he managed to push the IMF into an agreement without the levels of commitments or through the mechanisms it wanted. Second, IMF principal support plied open Cluster E, subsequently leading a path toward an agreeable Aquino. The same tactic failed to produce similar results at Level IIB. For many of the social veto players, the costs of an agreement or a shift from the status quo puts them at financial disadvantages, making the choice between the two positions a nonstarter.

The deference tactic proves effective. Estanislao used international Level IIB as the "go-to-guys," securing their support first before pursuing the support of actors at international Level I and domestic Level IIA. Much like the external-internal effectiveness, this tactic was able to produce two dramatic effects with one variable. Fortunately, their interest coincided in many cases with the GOP interests. At domestic Level II, this tactic proved marginally effective as the ERB reversed course. Their first "go-to guy" was the moderate veto players that used the NWPC tripartite system and the LTFBR. Capturing the ratification of these groups, the GOP was able to deal with the groups to the left, who expected the rules to favor the oil company petitions To secure their price increase consent, Aquino held their audience and appeared on national television calling for particular groups to bear the cost of adjustment. The Aquino consultations, however, had the opposite effect—reversing partial course on the price increases. Much like the September 1989b process, this tactic failed to stick. Why was this the case?

Military coup memories still reverberated strongly with Aquino. Most mainstream veto players that ratified their consent to the price increases had already received their compensation. The residual groups were politically marginalized, but still held mass power. Aquino bargained with them to address their concerns in exchange for a pledge to not strike. This arrangement required her to make some concessions. Breaking her word could increase the intensity of the strike and produce a possible military coup, the outcome they feared the most. This development was instrumental to reversing the ERB decision because the left leaning veto groups negotiated with Aquino last, a sequence that had dire consequences for the GOP. Would the outcome be different if the Aquino meetings were sequenced at the beginning? The outcome may have been different as peripheral groups would have given their consent first as they are the ones more likely to engage in *Welga ng Bayans*. This counterfactual assessment may not be definitive as strikes still occurred even after Aquino met these organizations, but they lacked the intensity of the December 1989 strikes.

Can a technocratic alliance causal mechanism best explain the 1991 petroleum outcome? At Level I, the IMF had a meeting of the minds and "secured" the GOP chief negotiators. This assessment assumes that the latter had slightly different preferences and the former was able to help them steer their course. This assessment overemphasizes the capacity of the IMF and under evaluates the preferences of Estanislao and Cuisia. Both individuals knew what had to be done. The question remained on how much they need to do and what mechanisms they can use to loosen their constraints. The alliance tactic also assumes that a Level IIA "meeting of the minds" also applies to the COG assuming the prevailing institutional arrangements are favorable. Given the role that Cluster E plays, it is dubious that the cluster would accept the virtues of the Washington Consensus; instead, it is more plausible that the economic and political costs of a nonagreement outweighs the cost of an agreement, an argument for the external-internal tactic. The unconvincing technocratic alliance application cannot explain the successful formulation of the petroleum price increase at Level IIA. The tactic also cannot explain the reversal of the price increase at Level IIB. The IMF lobbied members of Congress, but only reached those who were in the choir, finding few converts. This tactic may explain why Congress did not legislate new revenue-generating measures, but it cannot explain the role that social veto players in the observation. As an explanation at both Level IIA and IIB, the technocratic alliance tactic falls short.

What lessons did the GOP learn? Military coups were still etched in the minds of the Aquino cabinet. The Davide Commission Report was released not two months prior, giving the cabinet sufficient time to digest its implications, but not implement its recommendations. Treading on eggshells, the Aquino government needed to mute the strikes to reduce the possibility of a coup. Retrenching the ERB decision may have been the proper choice to make. Another lesson can be made regarding the actual prices. Cabinet principals and business accepted the principle of a price increase, but disagreed with the terms indicating that freeing petroleum prices from regulation may have saved the GOP heartache. Finally, sequencing the tripartite after the ERB released permits interested veto actors the opportunity to participate in the decision-making process, an inclusion that pushed veto players into the conventional side of participation rather than the unconventional.

NOTES

1. Interview with Senior NEDA official, anonymity requested, June 27, 1995.
2. Rogoberto Tiglao 1990b.
3. Interview with Secretary of Finance Jesus Estanislao, April 20, 1995.

4. The Aquino Management of the Presidency, The Gulf Crisis.
5. Anthony Rowley 1990.
6. Interview with Secretary of Finance Jesus Estanislao, April 20, 1995.
7. Ibid., and interview with Senior Department of Finance official, anonymity requested, April 27, 1995.
8. Interview with Secretary of Finance Jesus Estanislao, April 20, 1995.
9. Ibid.
10. Tiglao 1991a.
11. Tiglao 1991b.
12. Interview with Senior Central Bank of the Philippines official, anonymity requested, May 29, 1995.
13. This interpretation differs with the account recorded by Lito Zuleta, who asserts that while Aquino knew of the impending price increase, she did not realize that the severity of the increase would be as large. See Zuleta and *Manila Chronicle* 1990o.
14. Interview with Secretary of Finance Jesus Estanislao, April 20, 1995.
15. Interview with Senior Central Bank of the Philippines official, anonymity requested, May 29, 1995.
16. Tiglao 1991a.
17. Interview with Secretary of Finance Jesus Estanislao, April 20, 1995.
18. International Monetary Fund 1991. Attached to the document is the 1991 Memorandum of Economic Policy.
19. Ibid.
20. Ibid.
21. Sources for this observation are derived from the *Manila Bulletin* (1990d, 1990e, 1990f, and 1990g) and the *Manila Chronicle* (1990d, 1990e, 1990f, 1990g, 1990h, 1990i, 1990k, 1990l, and 1990m) unless otherwise noted.
22. *Manila Standard* 1990c.
23. Ibid., 1990f.
24. The following observation is derived from *Manila Bulletin* (1990h) and *Manila Chronicle* (1990n, 1990o, and 1990p).
25. *Manila Chronicle* 1990n.
26. Doronila 1990.
27. Orejana.
28. *Manila Chronicle* 1990p.
29. Collis-Monsod 1991, 9.
30. Antonio, 1.
31. de Dios 1992, 100.
32. Republic of the Philippines. Central Bank of the Philippines.
33. Republic of the Philippines 1992a.
34. Dolan, 184.
35. Secretary of Finance Jesus Estanislao, email message to author, August 30, 2016.

Chapter 6

1994 GOP-IMF Petroleum Price Observation

The 1994 IMF-GOP agreement grew out of the Fidel Ramos platform to transform the Philippines into a "newly industrialized country" by the tum of the century. To meet this challenge, referred to as the "Philippines 2000" program, the GOP would lead the way to revitalize the country's economic environment through intensive structural adjustments in the areas of education, infrastructure, and energy.[1] Qualitative improvements in these areas would resolve the environmental constraints that had continuously handicapped the private sector, and allow entrepreneurs to take the lead in the country's drive toward industrialization. The GOP sought to fund the program through increased foreign capital, as domestic public investment hit low levels in 1992 and the government was reluctant to levy taxes for fear of placing a drag on the economy. The GOP sought to increase the flow of foreign capital through securing an IMF standby agreement, intended to provide the key toward funding the "Philippines 2000" program.

Ramos sailed smoothly through his honeymoon period but met his first taste of political defeat during the 1994 IMF-GOP negotiations. Indeed, after narrowly winning the 1992 presidential election to succeed Aquino, Ramos had numerous successes implementing market-oriented economic policies. The GOP achieved many milestones but failed to meet IMF expectations of deeper market reforms and additional revenue generating measures. For the most part, Ramos succeeded in pushing these demands through congress; however, his efforts to push through passage of critical revenue generating measures stalled. Ramos resolved this problem by circumventing congress and using his powers of executive privilege to impose a levy on petroleum imports that led to the eventual increase in petroleum prices. Dramatic domestic opposition mounted in resistance to the price increase, forcing Ramos to rescind the increase and seek other means of revenue accumulation.

INTERNATIONAL POLITICAL ECONOMY OF THE 1994 PETROLEUM PRICE OBSERVATION

Like the 1989 IMF-GOP standby agreement, the new IMF-GOP negotiations were clouded by an atmosphere of doubt and dissent. The parties reached a consensus on some issues, but the first round of talks ended with a wide gulf between them. Indeed, the IMF negotiation team felt that the GOP engineered its data to dress up the country's sanguine economic prospects. The parties delayed continuation of the negotiations until the GOP instituted additional prior requirements to support the economic data. Despite these steps, the IMF insisted that these prior conditions were inadequate and pressured the Philippines to deepen its reforms. One of the central reforms demanded by the IMF was, again, that the GOP raise additional revenues. The GOP attempted to generate additional revenue through an increase in petroleum products but failed. In the end, the GOP managed to increase the nation's value-added tax to appease the IMF.

At the onset of the negotiations, the GOP felt optimistic that it had more bargaining leverage vis-à-vis the IMF. In 1992, the GOP successfully instituted, at a rate that surprised even the IMF, full liberalization of the foreign exchange market.[2] In addition, the peso registered a 5.8 percent appreciation from the previous year and strengthened significantly against other major currencies. In early 1993, the Philippines participated in the Eurobond market after an absence of ten years.[3] The country's official international reserves reached an unprecedented high of $5.2 billion at the end of 1992. The country demonstrated signs of increased foreign investment and increased foreign demand for its exports.

Having reached these milestones, the GOP felt it was in a favorable position to request an EFF package that would afford Manila access to more IMF financing. In March 1993, the GOP unveiled its economic proposal to the IMF. The GOP sought to lift the Philippine growth rate from 2.3 percent to about 6 percent by 1997, consistent with its ASEAN neighbors and simultaneously bring down inflation to about 4 percent and to achieve balance of payments viability. The GOP intended to use a strategy that included financial consolidation, financial reform, macro stabilization, and liberalization of numerous sectors of the economy, in particular, oil.[4] The IMF was not as optimistic as it had reservations and doubts on the optimistic results of the proposal.[5] The Philippine delegation, led by acting Minister of Finance Ramon del Rosario and holdover Central Bank Governor Jose Cuisia, pressed for an economic program with a growth target of 4.5 percent of GNP and a CPSD level of P46.2 billion or 3 percent of GNP for 1993.[6] The IMF feared, however, that a 4.5 percent growth rate would produce added inflationary pressure on the economy. It suggested that the GOP reduce this risk by aiming for a lower growth target of 3.5 percent of GNP and trim its projected deficit to P37.6 billion, or about 2.7 percent of GNP. According to a senior NEDA official, "the IMF always wants our inflation to be comparable to the inflation rate in other ASEAN countries. They use that target as a barometer."[7] As a reference point, the IMF relied on the country's 1992 economic indicators to illustrate that while GNP growth was minimal, inflation managed to reach 9.2 percent for the first half of the year, slightly higher than the expected 7 percent rate for the period.[8] In 1993, inflation had reached 8.2 percent. Furthermore, the IMF also pointed out that a 4.5 percent average growth may lead to a balance of payments crisis, where historically Philippine real GNP growth above 3 percent fueled balance of payments crises.[9]

To meet the nation's goals, both parties developed a program around minimizing the CPSD.[10] This variable would eventually determine the context of the GOP fiscal policy. Initially, the GOP projected a relatively optimistic outcome where it could raise P282 billion in revenues, leaving it with a budget deficit of P25.6 billion for 1993, but the IMF countered that the GOP revenue instruments could only raise P262 billion, which would result in a deficit of P56 billion.[11] GOP optimism for generating revenue was shattered when first quarter figures had shown that the Internal Revenue and Customs bureaus failed to meet their collection targets. To close the fiscal gap, the IMF insisted that the GOP shore up its revenues and cut expenditures. The IMF pointed to three areas where the GOP could make up for the fiscal shortfall: expand the government's coverage of the Value-Added Tax (E-VAT) system by decreasing the number of exemptions and broadening its base, reduce the incentives granted by the Board of Investments, and institute a fuel tax that would increase the excise tax from the current Pl per liter.[12] According to del Rosario, the GOP already knew it had to make inroads in these areas, in particular with the petroleum excise tax.[13] But he pointed out that the government needed "time to consult with our friends in Congress to see to what extent we can get support for these measures."[14] In the end, both sides agreed to the general principal of reducing subsidies; however, like other issues, the convergence ended there. According to del Rosario, "this balancing act is very critical. We can't just cut expenditures indiscriminately because it would be unfair to those recipients that need it. Thus, we have to look very closely into this."[15]

GOP reluctance to follow the IMF's advice on fiscal policy brought the first round of negotiations to an uneasy conclusion. Differences remained on several key issues, but the two parties managed to come to a compromise on one key issue. Del Rosario proposed that the GOP institute more flexibility with the economic program by adopting a growth target with a high-low cap of 3.5 to 4.5 percent growth rate of GNP, a proposal that would allow the GOP to meet other targets demanded by the IMF. In exchange for this proposal, the IMF insisted that the GOP keep its 1993 inflation target under 7.5 percent.[16] Both parties managed to reach a consensus on these targets, but this issue failed to end. In the end, del Rosario put the negotiations in proper perspective for both sides when he asserted "if you (the IMF) are wrong (about the economic targets) then you simply go to the executive board and apologize; if we are wrong, we carry the fate of 66 million people."[17] In addition to these targets, the IMF requested another consideration: the GOP must restructure the Central Bank by dissolving the Monetary Board and replacing it with a new Central Monetary Authority (CMA) and resolving the Central Bank's financial problem.[18] The implementation of this prior condition led to the establishment of the CMA and the Bangko Sentral ng Pilipinas (BSP).[19]

After the first round of negotiations, both parties experienced changes that altered the complexion of the negotiations. The IMF replaced its chief negotiator and Resident Representative, Bruce Smith and Christopher Browne, respectively, at the request of Ramos.[20] Kadhim Al-eyd and Howard Handy took over the positions. Similarly, the GOP replaced del Rosario and Cuisia with Undersecretary of Finance Ernest Leung and Senior Deputy Governor for the Central Bank Gabriel Singson.[21] At the controls of the negotiation, Leung opted to change the conditions of the talks when he postponed the negotiations to permit the GOP to make some fiscal adjustments according to the IMF position on revenues. In June 1993, the GOP placed an additional P1.00 per liter tax on oil products and 95 centavos per liter for crude petroleum imports, which became known as the "Leung Levy." The Department of Finance estimated that this measure would increase GOP revenues by an additional P10 billion.[22] The domestic price of the petroleum products, however, remained the same. To compensate for the reduction in DOCT, the oil companies increased their reimbursement request from the OPSF, which at the time registered a surplus of P5.2 (see Table 6.1).

At this juncture, the GOP began efforts to shift the domestic slack of both its own domestic constituency and the IMF. Almost immediately, Leung met with IMF Managing Director Michel Camdessus and informed him of the change in strategy. Camdessus understood and sympathized with the problems faced by the GOP and readily agreed to Leung's suggestion.[23] While Camdessus' preferences closely matched those preferences of Leung, the managing director also facilitated the ratification process as he scheduled to meet with several GOP department heads, congressional leaders, and nongovernmental organizations and impress upon them the importance of an IMF program.[24] For the most part, Camdessus managed to sway more congressional leaders than the country's anti-IMF interest group coalition.[25] In addition, the IMF even turned the strategy around by sending several Philippine senators to Washington, DC to receive a "crash course" on the principles of IMF economics. This strategy proved marginally effective as the IMF managed to reinforce the value of

Table 6.1 Oil Price Stabilization Fund (OPSF) Monthly Terminal Balances January 1992 to December 1994 (in Million Pesos)

Months	*1992*	*1993*	*1994*
January	3,531	5,970	−1,723
February	4,229	5,899	508
March	5,620	6,663	1,345
April	7,827	6,308	1,883
May	7,923	5,780	2,633
June	8,295	5,272	3,259
July	8,176	4,641	3,416
August	8,215	4,086	3,641
September	8,595	2,761	3,717
October	8,520	1,442	3,938
November	8,543	229	4,158
December	5,532	−884	4,071

Source: Office of Energy Affairs and Department of Energy.

IMF prescriptions to Senators who already agreed to the principles of a free market but it failed to persuade those Senators who resisted IMF polices.[26]

With the Philippine Senate's refusal to endorse the renewal of the Philippine military bases in late 1991, IMF executive board members looked at the Philippines in a different light. Saudi Arabia saw the Philippines as economic opportunity. In early 1994, the GOP sold 40 percent of Petron, an affiliate of PNOC, to the Saudi Arabian Oil Company (Saudi Aramco).[27] The shares sold for $503 million, of which $251 million were be directed toward the GOP while the remainder would be targeted for energy development projects.[28] In addition to the cash outlay, Saudi Aramco would also provide advanced petroleum refinery technology to Petron. Clearly, Saudi Arabia increased its interest in the economic health of the country. Similarly, Japan, France, and Germany were interested in the economic development of the country, but they insisted that the GOP secure an IMF economic program before they would continue to disburse concessional aid.[29] Again, the IMF was the catalyst for the start of foreign financing.

Talks resumed six months after the break-off of the first round and resolved around the same issue of revenue generation but under a different set of circumstances.[30] The Leung Levy responded to the IMF recommendation made in April 1993 about increasing levies; however, this time the domestic conditions had changed as the precarious OPSF balance, the forward cover, and oil price adjustments came into the picture. The IMF insisted that the GOP reduce its CPSD level from 2.9 percent to 2.2 percent of GNP, and suggested that the OPSF deficit would be a good place to start trimming. In addition to the OPSF, Al-eyd asked the GOP to enforce more permanent revenue streams rather than relying on its present strategy of selling government assets to beef up national coffers, a one-time cash increase. Al-eyd pointed out that the Leung Levy did not provide sufficient revenues to improve the cash position of the GOP and suggested that a petroleum price hike be adopted. In response to this suggestion, the GOP insisted that such a measure could only be completed through public hearings and that it was outside the purview of the negotiation team. In January 1994, the ERB allowed the oil companies to post an increase in the price of petroleum products by an average of P1.55 per liter (see Table 6.2). Ramos later renounced this announcement when intense domestic pressure mounted to oppose the increase.

The petroleum price hike debacle established a clear message to the GOP: it could not increase revenue measures without the consent of congress. To this end, the IMF met with important members of the Senate to explain the importance of generating more revenue. The IMF explained that without a solid revenue scheme, negotiations could not continue and expressed its preferences for the implementation of an expanded E-VAT system.[31] According to a high ranking official, the E-VAT would probably not have gotten off the ground but for the IMF: "they (the IMF negotiation team) visited with Congress because they understand that if the GOP is going to have reforms, then they must demonstrate to Congress the rationale behind the reforms."[32] This perspective has been challenged. The IMF's role was not definitive as the Philippine Senate already understood the need for reforms and was proceeding to expand the VAT system.[33]

In June, the Congress passed a 10 percent expanded VAT measure and President Ramos signed the bill. Several hours later, Malacañang and the IMF agreed on a new Memorandum on Economic and Financial Policy (MEFP). Under the terms of the program, the Ramos administration laid out a series of structural adjustment commitments to the IMF. These promises included deregulation of the oil industry, liberalization of the banking sector, removal of import restrictions, and completion of its plan to sell state-owned companies to the private sector. The government also aimed to boost growth

Table 6.2 Historical Petroleum Product Prices (Per Date of Adjustment October 14, 1992 to October 9, 1996 (Metro Manila) in Peso per liter

Dates	*Premium*	*Regular*	*Diesel*	*Kerosene*	*Fuel Oil (Wholesale)*
10/14/1992	10	9.5	7	7	5.65
1/28/1994	11.55	11.05	8.55	8.55	2.65
2/8/1994	10	9.5	7	7	2.65
8/10/1994	9	8.5	7	7	2.65
1/2/1996	9.08	8.63	7.03	7.03	
2/1/1996	9.5	9	9.83	7.53	
8/14/1996	10	9.94	8.1	7.87	
10/9/1996	10.25	9.44	6.6	7.67	

Source: Republic of the Philippines, Energy Regulatory Board.

from 2.3 percent in 1993 to 4.5 percent continuing upward every year to about 6 percent by 1997. The government also pledged to rein in inflation, which ran at a 10 percent annual rate early in 1994, to about 4 percent.[34]

Critical to government's economic program was the schedule to deregulate the oil industry, in particular the gradual reduction of the forward cover. According to the program, the GOP would reduce the forward cover from the present 180 days to 90 days on March 1, 1994; to 60 days on June 6, 1995; to 30 days on September 1, 1995; and a complete phasing out by December 1, 1995. The gradual phase out of the forward cover would have a significant impact on the country's exchange rate. For every 30 days the forward cover was reduced, an increase in demand for the dollar by $150 million would take effect; also, the peso exchange rate value would depreciate by roughly four centavos for every $150 million demanded, which would translate to an increase in petroleum products by P0.3 to P0.4 per liter.[35] With the phasing down of the forward cover, the Philippines would be in a circle of countries with Article IV agreement of the IMF where currencies are fully convertible. The IMF viewed the forward cover as an impediment to full convertibility of the peso and an indirect form of subsidy to a particular sector.[36]

In the MEFP, the GOP also promised to address the continuous problem with the OPSF. The GOP planned to increase the flexibility of oil prices by building up the OPSF balance up to Pl billion and then freeze it at that level by the introduction of an automatic price adjustment formula. The GOP also intended to provide some flexibility with the OPSF by setting a fund floor of zero and a ceiling of P2 billion.[37]

DOMESTIC POLITICAL ECONOMY OF THE 1994 PETROLEUM PRICE OBSERVATION

The 1994 GOP oil policy originated from a desire to generate additional revenue for the "Philippine 2000" project. Initially, Congress' reluctance to act swiftly toward passing legislative measures that increased revenues, forced the executive branch to take drastic actions. These measures culminated in the domestic increase in the price of petroleum products. However, political opposition, both public and private, forced the GOP to recant on its decision and seek other means of generating revenue. Through persistent pressure, the GOP managed to force the hand of congress and establish additional revenue-generating measures.

At the outset of the IMF-GOP negotiations, the GOP faced several difficult choices regarding the IMF's prior condition to generate additional revenues. First, the GOP scrapped the plan to reduce the time phase of the forward cover given to oil companies. This plan would have increased the foreign exchange risk to oil companies and they would have pressed their claims against the OPSF in the amount of P2.5 billion.[38] Second, and perhaps most importantly, the GOP increased the levies posted on oil imports, referred to as the "Leung Levy," by another P0.95 per liter and petroleum product imports by P1.00 per liter. Given the respective levies already imposed under the Aquino administration, the levy of oil imports totaled P1.95 per liter for petroleum product and P2.00 per liter for petroleum imports.[39] The GOP estimated that the new levies, which went in effect in September, would generate and additional PIO billion; without this revenue, the CPSD would exceed 3 percent of GNP for 1993.[40]

Instead of resolving the nation's revenue problems, the "Leung Levy," together with other major developments, had drastic fiscal, and eventually political, ramifications. First, the peso slid against the dollar by 16.9 percent from the end of the first quarter of 1993 to the end of the third quarter of 1993. This development, in addition to the increased levy, altered the landed cost of crude. To compensate for these changes, domestic oil companies increased their OPSF drawdowns. As the drawdowns continued, the OPSF registered a P884 million deficit by the end of 1993 (see Table 6.1). Besides changes in the landed cost of crude, Congress also contributed to the OPSF depletion. In 1993, NAPACOR again experienced financial problems and seriously considered increasing in the price of electricity. Congress attempted to spare consumers from the increase by diverting P3 billion from the OPSF to NAPACOR.[41] Domestic oil companies were unable to recover their losses from the financially troubled OPSF and were forced to petition for an authorization for petroleum product pricing increases.

Level IIA consent more complicated than meets the eye. The Ramos administration was perceived to be impervious to a coup given his background. This was not the case. His Finance Minister de Rosario relayed, "Every time we tried to raise the issue of oil prices, someone in the (cabinet) meeting would just make a short statement that considering continuing threats of coups, it was untimely to adjust the prices. And that was it."[42] The Philippine military was in dire need of modernization and cash was in short supply.[43] The United States contributed a significant portion to the GOP, which is not included in budget figures.[44] While the United States increased its expenditures and military assistance during the Aquino years, the GOP began to draw down its own expenditures where during the last two Aquino years, the GOP military budget as a percentage of GDP dipped to an average of 2.02 percent.[45] US spending replaced GOP cutbacks. To maintain military support, Ramos used financial instruments, promises of modernization, and appointments; yet, financing was in short supply as the military base rents accounted for 90 percent of the military budget.[46] This shortfall had to be filled either via domestic resources, which were limited, or foreign financing, which gave more leverage to the CNs. Ramos could authorize a modernization deal, done in 1996, only after he secured an IMF program in June 1994. Ironically, the road to military modernization traveled through the IMF.

A petroleum increase would entail some pass through to consumers. The NWPC authorized a wage increase from P118 in January 1991 to P135 in December 1993. Wages also increased another P10 in April 1994. The setting was set for the oil price increase. The oil companies' move to restore its DOCT levels resulted in forcing together an unusual opposition coalition. In early December, the oil companies petitioned the ERB for the necessary price increases. In the public hearings, their request was met with considerable opposition from veto actors of mass power: transportation associations, labor unions, and consumer groups; however, this time, the mighty Federation of Philippine Industries (FPI), a coalition of domestic-oriented manufacturing firms, threw its business power muscle into the ring. The FPI objected to the price increase proposal due to the potential financial pressure the manufacturing sector given the December across-the-board-wage increase.[47] Given the considerable weight FPI added to the veto coalition of mass power, the outcome of the public hearings became uncertain.

The public hearings spanned approximately one month under a media orchestrated microscope. Since its creation in 1987, the ERB always attempted to maintain its independence from the GOP, an effort that was never more apparent than during this observation. In fact, the ERB chairman continuously issued reassurances to the public that the GOP's preferences on the issue were removed from consideration, and no mention of an IMF link was publicly made by a GOP official; however, constant media exposure hammered this point throughout the ERB hearings. In contrast to these assurances, a member of Congress noted the link between Petron and the GOP, where the government-owned oil company could not have filed a petition for a price increase without clearance from Malacañang Palace.[48] In late January, the ERB authorized an increase in the price of petroleum products by an average of P1.55 per liter (see Table 6.2).

The price hike was certain to have a dramatic impact on the cost of living, production, and transportation. Attempting to assuage the price hike's pass through to consumers and other domestic products, the GOP immediately initiated a series of "safety-net" measures. First, DTI had led numerous government agencies in a surveillance campaign of merchants and threatened imprisonment to merchants who immediately increased the price of their goods. Second, the LTFRB authorized jeepney and taxi unions to increase their fares by P0.50, thereby enabling the sector to "ride out" the increase. Finally, the GOP increased the salaries and wages of GOP employees, including police, firefighters, and the armed services, by 22 percent. Although these measures were intended to mitigate the impact of the increase in prices, opposition continued to build.

After the ERB announcement, public opposition continued to build and intensify, creating the potential for massive domestic turmoil. Although additional interest groups joined the fray in opposition to the price hike, none carried more political clout than the Catholic Bishops Conference of the Philippines and the Inglis de Crista (Church of Christ). The broad-based heterogeneous coalition, known as the *Kilusang* Rollback, attempted to roll back prices through a scheduled *Welga ng Bayan*.[49] After staging debilitating strikes and demonstrations in the city of Manila, Ramos agreed to meet with the coalition. During these talks, the coalition had expressed its objectives and offered the government suggestions on other revenue alternatives.[50] Ramos took these alternatives seriously, but perhaps the most persuasive maneuver was made by the church as it assured Ramos that a coup attempt would be imminent should the situation escalate.[51] Ramos conceded to the coalitions' demand and asked the ERB to temporarily rescind the petroleum price increases, promising to permanently remove the price increase should congress pass significant revenue-generating measures to replace the P1.00 oil levy. The reversal proved to be a permanent move.

During the entire observation, many congress members publicly supported the Kilusang Rollback, but in all likelihood, they secretly supported the ERB decision. Congress again moved into the spotlight with the reversal of the ERB decision. Congress had addressed the numerous issues, but despite the sense of urgency, failed to take concrete steps toward resolving the revenue shortfall.

Congress eventually managed to pass a series of measures that liberalized the entry of foreign banks, increased taxes on stock sales, and enlarged the scope of the 10 percent VAT system. IMF meetings with key legislative figures contributed significantly to the push congress on these issues, but eventually these measures stalled in congress only to be finally passed after being assured by Ramos that he would release P9.5 billion in so-called congressional initiative funds.[52] This fund, sometimes referred to as countryside development fund, can be used by the president as a tool to sway members of congress. The process is shared by both branches. Members of congress identify the project and designate the beneficiaries. After an open public bidding process, the DBM dispense the funds.[53] However, disbursement or withholding of these funds remains at the president's discretion, as he authorizes the release of funds.[54]

The result of congressional action proved to be the catalysis toward economic growth. Afterward, Ramos signed the legislation. Several hours later, Malacañang Palace announced that it agreed to an IMF MEFP. Under the program, the Ramos administration laid out a series of other tough steps.

These reforms consisted of deregulation of the oil industry, liberalization of the banking sector, removal of import restrictions, and completion of its plan to sell state-owned companies to the private sector. The government also aimed at boosting growth from 2.3 percent in 1993 to 4.5 percent continuing upward every year to about 6 percent by 1997. The government also pledged to rein in inflation, which ran at a 10 percent annual rate early in 1994, to about 4 percent.[55]

EPILOGUE TO THE 1994 PETROLEUM PRICE OBSERVATION

In its efforts to realize the "Philippine 2000" program, the GOP bowed to both international and domestic pressures. The IMF managed to apply its strategies of technocratic alignment to push the GOP toward increasing petroleum product pricing. In an indirect manner, the IMF managed to use critical information from International Level II to shift GOP perceptions of its aspiration points. In contrast, the GOP lacked the international leverage that a financing program such as the PAP and the military bases previously provided to the country. In its efforts to enlarge Level II constraints, the GOP similarly lacked success. The prolonged ERB process and the composition of public political opposition to the price increase in petroleum products nullified the effectiveness of the GOP's strategies. The GOP fortunately managed to move congress into taking action necessary to meet the conditions of the IMF.

The GOP had difficulty meeting the commitments laid out in the 1994 MEFP and faced challenges that threatened to derail the entire program. The cases of the E-VAT, the sale of Petron to Saudi Aramco, increase in the price of electricity, the automatic adjustment mechanism, and even the issue of the location of electricity generating plants eventually found themselves before the Philippine Supreme Court. The Ramos administration fought hard to ensure passage of these policies that were deemed critical to the revitalization of the Philippine economy. At times, the GOP would secretly lobby the Supreme Court, a tactic that was exercised with great care so as not to invite separation of power violation challenges.[56] The GOP realized the delicacy of this tactic, as a Philippine Central Bank official recalled, "We couldn't do this for every case that went to the Supreme Court, thus we had to pick and choose our cases."[57] The GOP also weighed the political considerations of implementing oil deregulation, and postponed the deregulation schedule to avoid severe political repercussions in the May 1995 congressional and local elections.[58]

The GOP itself was responsible for slowing the phasedown of the forward cover. While deregulation was considered beneficial for the GOP in the long run, in the short run, it threatened to put dents in GOP coffers. Based on calculations of the BSP, the gradual phase down of the forward cover for oil imports beginning January 1995 would require an additional $290 million and $320 million correspondingly for 1995 and 1996.[59] However, planned initiatives for the phase-down of the forward cover also stalled due to the Mexican foreign exchange crisis of 1994. The GOP feared the reduction in the forward cover, along with the Mexico peso crisis of late 1994 would put severe strains on the country's foreign exchange reserves. The decision to forestall the phase down also took into consideration the close of the national stock market due to the Pope John Paul's visit to the Philippines. GOP officials had noted that increased foreign exchange activity in fellow ASEAN nations and they feared that when the stock market resumed operations the demand for the dollar would increase and that the Philippines might travel the same path as Mexico. As a result, the GOP delayed the reduction of the forward cover.[60]

ASSESSMENTS OF THE CAUSAL MECHANISMS

The absence of the urgency in the 1994 observation was manifested in the long duration of the negotiations. The 1984 IMF-GOP negotiations were under a cloud of international financial urgency and of domestic political constraints and uncertainty and also long and drawn out process. In comparison, the 1994 program was also long and drawn out and under a cloud of domestic constraints, but the atmosphere had a more optimistic air as the GOP was looking for an "exit" to the IMF tutelage. The external-internal tactic provided the GOP negotiators with sufficient leverage to move Level IIA opposition. Going to IMF principals and emphasizing their delegation to their IMF agent moved any cabinet opposition to increasing revenues, primarily petroleum prices. The tactic, however, was limited to the Level IIA as Congress and social veto players at Level IIB were almost deaf ears when it came to revenue and petroleum. As a result, this tactic demonstrates its limits.

The deference tactic also had its limits. The IMF principals deferred to the IMF before they would initiate promises of financing, giving bargaining leverage to the IMF staff. The economic conditions under which the GOP operated were not as constraining as in previous years, removing the air of urgency that typically surrounds these negotiations. GOP negotiators were cautious about increasing petroleum prices as they would always face a counterargument of "coups." The promise of more foreign financing, however, moved Cluster E, as modernization could come only through an established IMF program. At Level IIB, Ramos was the "go-to-guy." He served as the mechanism toward securing political support for the price increase. The broad based *Kilusang* Rollback coalition, unprecedented in its scope and mass power, staged a devastating *Welga ng Bayan* that forced Ramos to grant them an audience. The argument of the religious leaders about the possibility of a coup reversed the increase. Much like the 1991 observation, the deference person that was to sway veto players instead reluctantly relents. Again, the possibility of the military coup became the turning point, even for Ramos, a former military general.

The technocratic alliance also provides some insights and mix conclusions to explain the outcomes. At Level I, the IMF staff and the GOP chief negotiators had a meeting of the minds where the interaction between them extended the IMF position where more revenue building measures were required. The GOP CNs agreed in principle, but not in the terms, expressing the difficulty of securing such measures. In this context, the tactic can explain the outcome, but not the timing. This issue is addressed by the Level IIA institutional arrangement and Cluster E, but the tactic fails to go far enough to explain the outcome at this level satisfactorily. Petroleum price increases were always an option, but only when other less costly alternatives were already fleshed out, alternatives that could be done with only executive instruments. Furthermore, there is no evidence to suggest that the IMF staff had meetings with critical Cluster E members. Such evidence would substantially strengthen the attractiveness of this causal mechanism. The application of this tactic at Level IIB does explain the outcome. The IMF had multiple meetings with legislative members and social veto players. The relay of information can derive from two types: an agreement with the principles of the Washington Consensus and an agreement with the benefits of a rational petroleum pricing system. The IMF would have problems with the former as many of the veto players are dependent upon the shielding of the state from market forces. On this score, the IMF would be unlikely to attract converts. Similarly, rationalization costs and benefits could not be convincingly accepted by groups with little or no disposable income. The limitation of the tactic to capture social veto players explains the reversal of the price increase.

The GOP did not adjust the decision-making process at Level IIB. The NWPC increased prices before the ERB hearings. This month-long session was an attempt to demonstrate its independence from the GOP and to illustrate process transparency. After the ERB announcement, the *Kilusang* Rollback gained numerous supporters as other coalitions jumped on board. A *Welga ng Bayan* ensued that was followed by a Rollback audience with Ramos. The difference between the 1994 and 1991 observations was that the scale of the strikes were more debilitating in the former than the latter and the scope of the *Kilusang* Rollback was diverse. Industry and the religious organizations may be anticipated as joining the coalition, but it probably came as a surprise. Quite simply, the process to prepare for the increase was overwhelmed.

NOTES

1. This program is spelled out in the Medium-Term Philippine Development Plan. Republic of the Philippines 1995.

2. Interview with Senior Central Bank of the Philippines official, anonymity requested, May 27, 1995.

3. The GOP floated $150 million in Philippine bonds in the Euromarket, where 40 percent was absorbed by European investors, 40 percent by Asian investors and the remainder by US institutional investors.

4. International Monetary Fund 1994.

5. Del Rosario summed up the IMFs views on the GOP proposal: "They thought we made up these numbers . . . that we pulled the program out of thin air." Interview with Secretary of Finance Ramon del Rosario, May 11, 1995.

6. Originally, Cuisia wanted 3.5 percent and del Rosario wanted 5.5 percent economic growth of GNP. *Philippine Daily Inquirer* 1992.

7. Interview with Senior National Economic and Development Authority official, anonymity requested, July 5, 1995.

8. *Philippine Daily Inquirer* 1992.

9. *Business World* 1993.

10. Interview with Secretary of Finance Ramon del Rosario, May 11, 1995.

11. *Philippine Daily Inquirer* 1993c and 1993d.

12. Browne, 1–3; and *Philippine Daily Inquirer* 1993e.

13. Interview with Secretary of Finance Ramon del Rosario, May 11, 1995.

14. *Philippine Daily Inquirer* 1993e.

15. Interview with Secretary of Finance Ramon del Rosario, May 11, 1995.

16. *Philippine Daily Inquirer* 1993b.

17. Interview with Secretary of Finance Ramon del Rosario, May 11, 1995.

18. For the financial problems of the Philippine Central Bank, see Lamberte, Chapter 2.

19. The CMA would replace the Monetary Board and act as an independent policy-making body. The composition of the Monetary Board consisted of two members from the private sector and five members of government, one being the Central Bank Governor. The reform changed the composition by instituting five private sector members and two public sector members. However, independence of the board remained dubious due to the way the Ramos administration stacked the board that had three of the five private sector members who were former GOP officials. Also, the reform measure allowed the Central Bank to rid itself of its past debts, most of which still remained from the Marcos era, by wiping them clean. Interview with a Bangko Sentral ng Pilipinas official, anonymity requested, June 13, 1995.

20. This request came at the behest of Ramos after Browne publicly made public statements regarding the lack of GOP leadership. Browne also charged that the GOP economic program put on the bargaining table was fraught with erroneous assessments that could only lead to a continuation of the boom and bust cycle that the government was trying to escape.

21. The reason for the change was primarily domestic in nature. The Senate Committee on Appointments failed to confirm del Rosario's appointment to the position.

The changeover of the Central Bank to the BSP required a new governor, and Cuisia decided to step-down.

22. *Philippine Daily Inquirer* 1994b.

23. Interview with Senior Department of Finance official, anonymity requested, April 27, 1995.

24. Interview with Senior National Economic and Development Authority official, anonymity requested, July 28, 1995.

25. Interview with Freedom from Debt Coalition staff member, anonymity requested, July 7, 1995.

26. Interview with Senator of the Philippines, anonymity requested, May 16, 1995.

27. For a good synopsis of the deprivation of Petron, see Haggard 1989.

28. It is also interesting to note that Saudi Aramco became the principle supplier, up to 90 percent, of Petron's foreign oil needs. *Manila Bulletin* 1994a and 1994c.

29. Browne, 2.

30. This paragraph is derived from the *Manila Bulletin* 1994c, 1994d, 1994e, 1994f, 1994g, and 1994h.

31. Interview with Senator of the Philippines, anonymity requested, August 7, 1995. This conclusion was collaborated by an interview with a Senior International Monetary Fund official, anonymity requested, July 11, 1995.

32. Interview with Senior National Economic Development Authority official anonymity requested, July 28, 1995.

33. Interview with Senator of the Philippines, anonymity requested, May 16, 1995.

34. *Philippine Daily Inquirer* 1994c.

35. Interview with Energy Resources Supply Administration Division, Department of Energy official, anonymity requested, June 6, 1995.

36. *Business World* 1994; and interview with a Senior International Monetary Fund official, anonymity requested, July 11, 1995.

37. IMF confidential aide memoir as published to the *Manila Chronicle* 1990l.

38. *Philippine Daily Inquirer* 1993f.

39. Republic of the Philippines 1993.

40. *Philippine Daily Inquirer* 1994b.

41. Ng.

42. Author interview with Secretary of Finance Ramon del Rosario. May 11, 1995.

43. Dolan, 256–257.

44. Ibid.

45. World Bank 2017.

46. Chambers 2014, 115.

47. *Manila Bulletin* 1994b; and *Philippine Daily Inquirer* 1994a.

48. *Philippine Daily Inquirer* 1993g.

49. Kilusang Rollback cannot be directly translated into English. For all intents and purposes, it means Action Rollback.

50. These alternatives were made public in various issues of Philippine newspapers.

51. Interview with Secretary of Finance Ramon del Rosario, May 11, 1995. It should be noted that del Rosario was out of government during this observation, and these accounts were based on his observations and communications with members of government.

52. Tiglao 1994.

53. Interview with Senator of the Philippines, anonymity requested, May 16, 1995.

54. Interview with a Senior Department of Budget and Management official, anonymity requested, May 23, 1995.

55. *Philippine Daily Inquirer* 1994c.

56. Interview with Secretary of Finance Ramon del Rosario, May 11, 1995.

57. Interview with Senior Central Bank of the Philippines official, anonymity requested, May 29, 1995.

58. Interview with Senior Caltex (Philippines) Corporate Planning official, anonymity requested, May 31, 1995.

59. See the Proposed Downstream Oil Industry Deregulation Action Plan. Republic of the Philippines 1994, 24.

60. Interview with Senior National Economic and Development Agency official, anonymity requested, July 28, 1995.

Chapter 7

1984 GOP-IMF Import Observation

Like many developing countries, the Philippine economy has a myriad of weakness that contributes to the country's poor economic performance. Among these shortcomings, the structure of the trade regime continues to plague the country by creating incentive distortions among various economic sectors.[1] Since the late 1970s, these distortions, their effects and their removal, have attracted significant attention from the IMF and the World Bank. The IMF interest in the Philippine trade structure provides a fascinating and illuminating exploration into the factors that drive trade reform in the Philippines, a story that this chapter narrates.

This chapter focuses on the IMF-GOP bargaining process as each party negotiates the terms of an economic program and applies particular bargaining strategies to shift its counterpart's aspiration points. This chapter is divided into eight sections. Following this introduction, the next section paints a brief overview of the government controlled Philippine trade environment. Trade consists of both imports and exports, and certainly, exports are also critical to the growth equation, but I focus primarily on imports. Of further note, trade obstacles consist of both tariff and nontariff barriers to trade, specifically, quotas and licensing procedures. This chapter addresses both forms of barriers and distinguishes between tariff and import (nontariff barriers to trade) liberalization. The following five sections narrate the story behind the politics of trade reform in the Philippines, with each division corresponding to an IMF-GOP economic program. These divisions are broken down into five segments: the introduction, negotiations between the IMF and GOP, ratification of the agreed course of action at Level IIA and IIB, an epilogue to the negotiations, and an assessment of the causal mechanisms. Afterward, the conclusion briefly summarizes the main points of the chapter.

POLITICAL ECONOMY OF IMPORTS IN THE PHILIPPINES: 1946–1983

Trade restrictions have been a prominent feature of the Philippine trade structure during the postwar period.[2] Balance of payments difficulties, government revenue requirements, and economic development strategies prompted Philippine authorities to use a complex system of tariff and non-tariff barriers to trade. The system often vacillated in intensity as domestic and international forces exerted pressures on the GOP to change its grip on the inflow of foreign products. Eventually, import control became used as an instrument of corruption and protectionism for a raising manufacturing base.[3] Import controls, in addition to exchange rate and tariff controls contributed to strengthen the politically powerful import-substituting industries, such as textiles, non-tradable agricultural and consumer durable products that were shielded heavily from foreign competition. In addition, these industries received investment incentives and access to credit, which helped eventually to produce a Latin American style dualistic economy. The manufacturing sector employed approximately only 10 percent of the labor force while agriculture and the service sector provided 65 percent.[4] This favoritism also had other effects on the economy. First, it created market distortions that discriminated heavily in favor of investments in these industries, providing profit incentives for areas where the Philippines did not have a comparative advantage. Second, the protectionism created market distortions that discouraged private actors from investing in agriculture and exports, areas that had little allure for profit seeking individuals.

The twin oil shocks of the 1970s were instrumental in causing changes to the country's import policies. Following the 1973–1974 energy crisis, the GOP recognized that greater efforts were needed to generate more foreign exchange earnings to cope with the external upward pressure on prices of the country's major imports. In addition, foreign exchange would be required for development purposes. Thus, the GOP became concerned with export orientation as a major focus in the development of trade and industrial policies and programs. The second energy shock of 1979 required a more determined GOP effort to resolve certain structural weaknesses and thus enable the economy to be more resilient in the face of such external shocks. Policies that favored export-oriented industries were intensified. This is where the World Bank came in to play. According to Virata, it was to give the Philippines a kick-start it needed.

> So we asked the World Bank again to continue (from the 1970s programs) because—well, reducing our protection, making our economy more outward looking because we were suffering from these highly import substitution type

> of industrialization. And the way we look at it is that there was really no way to stabilize the Philippine economy unless we had more for exports because every time we improved the economy internally it sucks in a lot of imports and it does not push out the exports. So that is why Philippine development has been undulating like that: every time you improve you slow down because you run out of foreign exchange. Now this year we are in a build-up stage of our foreign exchange reserves because we have the slowdown. So with reserves higher, people get ideas again.[5]

In the early 1980s, the GOP and the World Bank joined as partners in a joint effort to overhaul the highly distorted Philippine trading system. Years of government protection left the country with effective protection rates (EPR) that ranged across sectors from 49 to 200 percent.[6] By 1980, the country had regulated 2,883 import commodities.[7] The first Structural Adjustment Loan (SAL) program, executed in 1981, aimed to reduce tariff rates of raw materials and intermediate producer goods destined mainly for export-oriented, labor intensive industries, such as textiles and apparel, electronics, leather and footwear, and food processing, and establish more uniform tariff levels during the period 1981–1985.[8] The program helped to cut the country's average nominal tariff protection from 43 percent to 28 percent by 1985.[9] The SAL II program, instituted in 1983, required that the system of industrial incentives be simplified and stripped of its capital-cheapening bias, and for energy policy to be reformed with a view toward the elimination of various subsidies, in particular, on power pricing policy.[10] In conjunction with the SALs, the GOP also adjusted its system of nontariff barriers to trade. Under the government's concurrent Import Liberalization Program (ILP), import restrictions were targeted to be reduced under a five-stage process. In January 1981, the GOP pushed the ILP into motion, with the liberalization of 263 nonessential consumer and 610 under-classified consumer items.[11]

The SALs had significant benefits that the GOP could not match. While the benefits of structural adjustment were clear, the costs were dramatic. The economic and social cost would somehow have to be absorbed. The political costs could be crippling to governments. By itself, the GOP could make the changes; it would have to provide safety nets for manufacturing industries and various types of labor. These programs would require more fiscal trimming of expenditures as increasing revenues would most certainly be a political, not to mention bureaucratic, nonstarter. This dilemma becomes reduced significantly by the World Bank and its financing, which offsets the monetary costs borne by the government. More importantly, the benefits of structural adjustment could be derived without the "threat of an early reversal."[12] Furthermore, the World Bank financing could fill macro-equilibrium financial gaps. Since it disperses loans quickly, it could use the funds to pay for the sudden rise in imports and debts that the transition phase would certainly produce. Arguably, if a balance of payments deficit surfaces, the IMF would be in place to provide stabilization financing given the deference of the World Bank to the IMF.[13]

Despite the apparent gains, the GOP achieved relatively mixed results under the auspices of SAL I and II. Broadly speaking, the GOP complied well with non-key conditions, such as reorganization of the Ministry of Trade and Industry (MTI) and import duty refunds for exporters,[14] but in terms of the items liberalized, products deregulated under the ILP were either of little trade value or not locally produced.[15] Part of the blame lay with the World Bank as it was ambiguous in its guidance as to which imports should be liberalized and failed to make provisions regarding macroeconomic framework, assuming this to have been dealt with by the current IMF Standby agreement.[16] More specifically, trade reform programs often failed to reduce or even exacerbated domestic distortions between particular sectors, as it ignored not only the agricultural sector with its lagging performance, its export taxes and monopolies, but also the inefficiency of the public financial sector, an industry laced with "cronyism."[17] Furthermore, gains made by exports were illusionary as the import content of exports was high, failing to provide sufficient value-added materials to the products.[18] In short, exports failed to provide any sufficient backward linkages toward the rest of the economy.

Resistance to the World Bank typically occurred Level IIA as a first line of defense. To gain leverage, the World Bank was used as a carrot. According to Bienvenido Noriega, then a National Economic and Development Agency official, the link with the World Bank pushed reforms in the government. "We're using the World Bank (structural adjustment) load as leverage to do things NEDA has wanted to do, but hasn't been able to. With this World Bank backing, NEDA has more strength vis-a-vis other ministries."[19] Indeed, Virata would use this leverage with ministries that typically dragged their feet on reform. By providing the promise of money, it made bureaucracies cooperative where they ordinarily would not be. Being in the first chair in negotiations with the IMF and World Bank was part of the finance portfolio; however, when it came to matters regarding trade in the Uruguay Round or ASEAN trade discussions, Viata and subsequent finance ministers held seats by virtue of their membership in certain economic committees, but they were participants, no longer principals.[20] Subsequently, trade and industry ministers and their preferences become key players in IMF negotiations, where GOP CNs—the holders of the finance and Central Bank portfolios—would defer to them. In some circumstances, finance and trade and industry negotiate on the items, timing, and levels of reform before the GOP CNs present their list to the IMF staff.

Historically, the Philippine Central Bank leaded toward the preservation of nationalism and clientelism. Its import control instruments isolated domestic industries from international competition and promoted the interest of corrupt and well-connected political veto players.[21] "Central Bank licensing . . . is an obstacle toward product improvement, greater completion, and the provision of supplies to export industries."[22] Its guardianship over the status quo and against unfettered liberalism extended into the battles with the Ministry of Finance and NEDA and the World Bank during the negotiations with formulation of the SAL programs, a battle that former Central Bank Governor and nationalist leading Gregorio Licaros lost. The Central Bank's import control office had two reasons for keeping its import controls close to its vest. First, like Licaros, it viewed the import control office as an agent protecting its principals, the national entrepreneurs. Second, the more control it has over imports, then the import control office would rise in power vis-à-vis other bureaucracies, and as a result, so would the Central Bank as a whole.[23]

Level IIB players also had some influence in the "accommodating" decision-making process. Often, industrial associations shielded from market forces threatened to either reduce production or shut down completely. Frequently, their interests were protected by government officials with ties to these corporations and often served as their mouthpieces. To offset their reform costs, the GOP often used "revitalization" policies, giving industries cheaper credit and technical assistance or by lowering tariffs for intermediate and capital goods.[24] The decision-making process for trade was every bit as political. During the Marcos years, trade policy was shaped in relatively closed quarters. Technocrats from the NEDA and the MTI pushed for liberalization while the Tariff Commission (a division of NEDA) held legally mandated public hearings before any tariff adjustments. These accommodating hearings served not only as the mechanism for consultation between the ministries and Level IIB actors, but also for determining an incentive priorities list.[25] Before the SAL program, these hearings were taken lightly by the industrial associations since they produced little and that Marcos typically determined many critical rates by decree.[26] At times, other tactics could be used to ensure the level of protection while at the same time appear to liberalize. During the SAL, the Tariff Commissions would reduce the tariffs on competitive products to highly protective industries. In exchange, it would also reduce tariffs to their inputs.[27] Even with opportunities to push reform, the GOP had to abandon the World Bank SAL due to balance of payments issues.

The need for import reform was critical. Development projects needed for economic growth and poverty reduction measures required financing. Typically, domestic level financing always fell short as domestic expenditures always exceeded domestic revenue, always dubious given the level of corruption in bureaucracies. This gap needed to be bridged, and international financing served as that medium. Exports generate foreign exchange and revenue, and all things being equal, would cover the gap. The revenue generated by exports varies with the type, markets, and vitality of the international economy; as a result, exports of industrial products are more valued than agricultural ones, which are more subject to market fluctuations. A country's imports can sap the country of much-needed revenue, still leaving a financing gap. As a result, the GOP had few alternatives than to seek out public and private sources. In addition to these sources, the GOP attempted to transform its inward-oriented trading policies to an export-oriented one, where imports serve as intermediately ingredients to either exports or their industries.

1984 GOP-IMF IMPORT OBSERVATION

Engulfed in distributive-type bargaining, the 1984 IMF-GOP negotiations took many detours. On the international level, IMF and the GOP preferences converged on numerous issues, such as the need to increase the country's capacity to generate exports that would bring in foreign exchange. In this area, the parties' aspiration points were closely configured, especially when it came to import reform on raw materials and intermediate goods. The parties also discussed areas where their aspiration points were far from malleable, such as import reform in areas that protected Marcos' cronies from international competition. The IMF failed sufficiency to persuade the GOP to abandon its no agreement costs. In large part, this failure can be attributed to the alignment of key international Level II players that continued to support Marcos.

The sequencing of coalition building followed an external-internal pattern. Finance Minister Cesar Virata set up the Consultative Group upon which to generate revenue for reforms. With foreign financing, both public and private, the GOP could use this leverage to initiate reforms on imports. This strategy worked at Level IIA, the cabinet level, as bureaucracies had opportunities to reform sectors. Not all parties, however, bought in. In large part, the Central Bank as well MTI resisted. For instance, the latter proposed eleven major industrial projects of questionable merit and required tariff barriers to shield them from international competition. This position stood in the way of the principles of rationalization and efficiency.[28] The MTI support for these projects was tightly held that even the World Bank gave up trying to convince them of their inefficiency. As such, when the World Bank mission arrived in the Manila, it avoided discussing these projects.[29] Likewise, highly protected import substitution industries also had to be dragged into the process.

Kicking and screaming, domestic Level IIB veto players made significant use of their access to policy-makers. Once with the high-level official, threatened industries typically threaten to reduce production levels or to leave the domestic market altogether.[30] Sometimes, access by the veto player was unnecessary as many protected industries had on their leadership chain individuals that also sat on critical government decision-making boards. For instance, Cesar A. Buenaventura served as chief executive officer at Shell Philippines LLC from 1975 to 1990, a position that overlapped with his position on the Central Bank Monetary Board (as the representative from the private sector) from 1981 to 1987. The process could also work in reverse as a government official could sit on a private company's board of directors.[31] At times, the chief executive would receive powerful veto players and use an executive order to counter a pervious order. Finally, sometime trade-offs would be made. Lower tariff rates on intermediate goods and capital would be implemented while simultaneously reducing tariff rates.[32] Ultimately, technocrats relied on Marcos to function as the "guy to get the guy (Level IIB veto players)."[33]

Import reform would serve to resolve the country's continuous structural trade imbalances. By 1983, the Philippine current account had reached unprecedented disequilibrium proportions. In 1973, the current account registered a surplus of $473 million and afterward had begun to decline steadily until it reached a peak deficit of $3.2 billion in 1982. During this period, the country's import levels also began to climb and outpace the country's export levels by $2.6 billion, a situation that was precipitated by the two oil shocks, dramatic price reduction of the country's traditional exports, and a sluggish international market.[34] The heightened trade deficit could hardly be sustained through the Philippine stock of foreign assets, as total reserves (minus gold) dropped from a high of $2.8 billion in 1980 to a 10 year low of $888 million in 1982. Foreign borrowing became necessary to continue trade flows. In 1982, the GOP secured $2.5 billion in foreign loans, but its debt service obligations surpassed disbursements by $1 billion.[35] This precarious situation continued in 1983 and was exacerbated by the August assassination of Benigno Aquino, when private foreign creditors were compelled to halt further financing to the Philippines. Afterward, the GOP announced the first of five consecutive moratoriums on the principal payments to its foreign private creditors while it continued to make payments to its official creditors. Virtually cut off from private international sources of financing, the GOP sought the assistance of the IMF as a catalyst to renew the flow of foreign loans.

During the height of the balance of payments crisis, the GOP and IMF relationship suffered considerable strains. In February 1983, the two parties agreed to the terms of an economic rescue package, but the GOP exceeded these targets several months later. The IMF abrogated the agreement and the two parties drew up a second stabilization standby package in November 1983, but one month later, the IMF uncovered "misleading" accounting practices by the Philippine Central Bank and withdrew the program from IMF executive board consideration (for more on this incident, see chapter 2). The third attempt (from March through December 1984) at an economic program was handled differently as the IMF came down hard on the GOP and forced it to meet prior-requirement targets and reforms at levels and degrees that conformed to strict IMF stabilization preferences, a distributive bargaining position to be sure. The GOP response was far from the degree or the level desired by the IMF.

International Political Economy of the 1984 Import Observation

In the context of the IMF-GOP negotiations, import reform moved intermittently, with at time inconsistent and uneven dimensions. In late 1983, the GOP reluctantly abandoned the structural adjustment course, under the SAL and ILP programs, it had adopted earlier in the decade and took a path that attempted to resolve the balance of payments crisis, a road that utilized tariff instruments and foreign exchange controls. This course, however, failed to quickly reestablish external stability and considerably exacerbated the crisis. Throughout the negotiations, the IMF held doubts on the GOP's ability to resolve the crisis, a perception based on the Marcos regime's inability to make politically difficult decisions and adhere to its commitments. As a sign toward demonstrating its resolve, the IMF adamantly insisted that the GOP initiate a series of policy moves that, among other things, include tight monetary measures and import reform policies. On politically sensitive issues to Marcos, the IMF failed to make its case clear. For the GOP, the cost of no agreement outweighed the political implications of reform. The GOP reassessed these cost calculations when the United States shifted its preferences toward the IMF target point, a critical juncture in the negotiations.

In the third quarter of 1983, the GOP addressed the country's current account problem. In October, the Philippine Central Bank depreciated the peso from P11: $1 to P14: $1, the second major depreciation in five months. One month later, the Central Bank continued to implement foreign exchange controls designed to discourage "unnecessary" imports and assure the optimum use of scarce foreign exchange reserves. Under this scheme, the GOP required domestic banks to sell all of their foreign exchange holdings to the Central Bank, where it pooled the reserves into a fund and then redistributed the foreign exchange for priority imports.[36] As a corollary, the Central Bank also set higher ceilings for payments and new import letters of credit and continued to require importers to secure import licenses to appropriate these products.

Furthermore, it also reinstituted the forward cover policy for oil companies and steel manufacturers where it promised to pay for peso depreciation differences if these companies would extend their import remittances for one year (for more on this policy, see chapter 2). Finally, the GOP increased tariffs on imports, as a means toward reducing domestic demand and increasing government revenues. In November 1983, the GOP increased the ad valorem duty on all imports and effectively raised the country's tariff nominal protection by 5 percent. The GOP subsequently increased the levy to 8 percent by April 1984 and to 10 percent by June 1984.[37] In combination, these policies were aimed to achieve two goals: secure an IMF agreement and more importantly, renew the flow of private foreign trade credits.[38]

In December 1983, the carefully nurtured and long-standing relationship between the IMF and GOP began to unravel. For several months, the GOP and the IMF conducted numerous discussions on the terms of an IMF program and were winding down negotiations on an economic program. During Minister of Finance and concurrent Prime Minister Cesar Virata's US visit, the IMF sent an examination team to Manila to investigate numerous discrepancies in Philippine Central Bank data. Later, news broke out that the Central Bank had drastically overstated its level of international reserves. Since the end of 1982, it was estimated that the gross international reserves actually had declined to United States $888 million, approximately $0.7 billion lower than earlier reported and about $0.8 billion lower than the previous year.[39] The IMF reacted defensively to the cover-up and removed the November 1983 Letter of Intent from consideration.[40] As a precondition for its return to the bargaining table, the IMF insisted that Marcos remove Central Bank Governor Jayme Laya. These calls for Laya's removal were not confined to the IMF. Foreign commercial creditors also lobbied Marcos and would not grant the Philippines another 90-day moratorium otherwise.[41] Marcos still had the military bases and the external support, but the clouds of distrust and suspicion were difficult to cut through.[42] He had to find someone to restore the government's tarnished credibility.

Negotiations on the terms of a new economic program proceeded after Marcos named Jose "Jobo" Fernandez as the new Central Bank governor, a move the IMF applauded. Together with Virata, they constituted the core of the GOP negotiation team, technocrats that understood the ins-and-outs of managing an economy. Negotiations for a new stabilization arrangement resumed in March 1984.

The IMF used had sufficient leverage to lock down key GOP players, but it lacked sufficient reach to overcome particular decision-making arrangements (to be elaborated in the following section). Virata and Fernandez shared their IMF counterparts' preferences to tighten domestic demand, an orthodox stabilization prescription.[43] In fact, Virata was driven to secure a stabilization agreement as a means to renew the flow of foreign trade credits.[44] In this respect, the key GOP players were on the same page. Despite their support, the IMF was uncertain about their ability to minimize program slippage. To minimize these occurrences, the Fund expanded its negotiations to include other key GOP players, the MTI and Budget and Management (MBM), the NEDA and other senior officials including representatives of key government corporations and the public financial institutions. In technocratic alliance terms, the IMF attempted to spread its net wide over the GOP and build coalition. Furthermore, some members of the IMF mission also met with Marcos on several occasions.[45] Marcos, however, had his preferences. Thus, the IMF strategy managed to encircle many of the critical government decision-makers, but it failed to alter the costs of no agreement perception of the central decision-maker.

The first round of bargaining found both parties in agreement over principle economic objectives. The parties agreed "to strengthen the external current and capital accounts, while allowing for the elimination of arrears and other restrictions on current payments, and on imports."[46] More specifically, the GOP proposed to reduce the external current deficit from $2.8 billion in 1983 to $1.5 billion in 1984, a reduction from 8.2 percent of GNP to 5 percent of GNP.[47] The GOP intended to meet this target through a modest 9 percent recovery in exports that complemented a 19 percent tightening of imports, which had dropped by merely 2 percent in 1983.[48] Export increases would be aided through a decline in export under-invoicing, the impact of the world recovery on commodity prices and market demand, and the maintenance of necessary imported goods, particularly for the country's nontraditional exports of electronics and garments, estimated to grow by approximately 7 percent. Imports would be subjected to considerable scrutiny to allow only priority products listed under the November 1983 foreign exchange and import control measures.[49]

Although the goal and outlined targets met with IMF approval, it found the GOP commitment to stabilization limited. Clearly, the policy instruments used to tighten international transactions could have helped mollify the crises, but in its attempt to curtail domestic demand, the GOP ignored the use of monetary tools. In fact, the GOP continued the practice of expansionary monetary and fiscal policies. The peculiar policy mix of tightening external transactions without corresponding domestic fiscal and monetary restraints largely contributed to increasing inflationary pressures and to an increased role of the black market in foreign exchange allocation. The IMF understood the reason for this policy inconsistency: "In the wake of political difficulties and a rapidly deteriorating economic situation which rendered the adoption of policy measures, at the same time, urgent and difficult."[50] The political difficulties the IMF pointed to referred to the Marcos protection of his patrimonial clients, the long-running communist insurgency, and the May 1984 Batasang Pambansa (National Assembly) elections.

Domestic political difficulties proved to be the immediate reason for an inconsistent stabilization policy; similarly, international forces provided little incentive to blend the divergent policies properly. After the Aquino assassination, $770 million in short-term loans were withdrawn, pushing the GOP on the brink of collapse. Minister of Trade and Industry Roberto Ongpin was dispatched to Brunei, Malaysia, and Singapore, with cap in hand. Brunei Sultan Hassanal Bolkiah loaned $150 million; Prime Minister Mahathir Mohamad contributed $75 million; Prime Minister Lee Kuan Yew delivered a lecture.[51] The flow of funds also came in from other sources. In October 1983, private foreign creditors ceased the flow of Philippine loans; in contrast, Japan, the United States, the Asian Development Bank, and even the World Bank continued to funnel financial assistance to the GOP and eventually released its final SAL II tranche at the end of March 1984. Not surprisingly, the GOP continued to receive additional funds from other sources as late as February 1984, even during its negotiations with the IMF. According to Economic Planning Minister Vicente B. Valdepeñas Jr., the country obtained a total of $992 million from these sources, which augment the Central Bank's foreign exchange pool. The GOP asked for bridging finance from official sources, such as the United States and multilateral institutions. This funding includes $50 million for the Asian Development Bank; $510 million from the World Bank; $217 million from US Commodity Credit Corporation; and $40 million from China.[52] The United States contributed the lion's share of the financing. The Export-Import Bank guaranteed $290 million for industrial and agricultural loans and the US government provided $900 million in aid for 1985–1989;[53] and $105 million from the US Export-Import Bank. As a result of continued access to funding, the IMF had little bargaining leverage in the early phase of negotiations.

Even when the IMF approved of a GOP measure, it often found the GOP efforts to be less than satisfactory. Since mid-1983, the GOP had devalued the Philippine peso numerous times. In June 1983, the GOP devalued the peso by 7.8 percent from P8 to P1.1 to $1. In October 1983, just before the GOP issued its moratorium, the authorities again devalued the peso to P14 to $1. At first glance, devaluation of the peso would be a proper move toward the restoration of macroeconomic stability. However, since Philippine authorities fixed this exchange rate and continued to run expansionary monetary and fiscal policies, the peso appreciated markedly and had reversed substantially the real effective depreciation that occurred.[54] As a response to this ineffectiveness, the GOP responded with another devaluation in June 1984, to P18: $1, a 28 percent decline. This move proved to be more than just a single policy change as the GOP also initiated a laundry list of stabilization measures.

Once the legislative elections were out of the way, the GOP proceeded earnestly to address the external account imbalance. The June devaluation resulted not from a government decree to an arbitrary rate, but from the GOP move to "deregulate" the import priority allocation pool by allowing domestic banks to keep 20 percent of its foreign exchange earnings. In addition, the GOP also added another two percent to the import levy and slapped a 10 percent "economic stabilization tax" on exports. With these moves, the GOP created a de facto multiple exchange rate. As the GOP slowly moved away from the regulation of external transactions, it began to shift its attention toward slowing internal transactions. By June, the GOP began to rein in monetary expansion, tighten public sector spending, and moderate wage policy.

From October 1983 to June 1984, the composition of Philippine imports sharply changed. Import controls played a significant role in this regard as it focused on the regulation of imports necessary to stimulate production of nontraditional export goods that would generate much-needed foreign exchange. In February 1984, the GOP exempted firms registered with the Export Processing Zone Authority (EPZA) from seeking Board of Investments (BOI) clearance for various raw materials, and parts and components for consumer electronic products. This move contributed to a significant change in the value of electronic exports, as the commodity averaged $85.9 thousand in the first quarter of 1984. By striking contrast, the value of exports increased to an average of $105.2 thousand in the second quarter of 1984 and $127.6 thousand in the third quarter of 1984.[55] The GOP followed this measure when it freed up raw materials targeted for the textile industry, as well as various other intermediate goods. In May, the GOP made a move that reversed the gains made to the current account, as it responded to various domestic shortages and released the ceiling limits on essential consumer imports. The 1983 import controls managed to slow consumer imports to a trickle. From October 1983 to April 1984, Philippine import levels of consumer goods averaged $15.7 million per month. For the next seven months, consumer imports averaged $21.9 million per month.[56]

Although the GOP made progress toward restoring an external balance after June, the results failed fully to satisfy the IMF. In fact, the IMF demanded that the Philippine authorities intensify their domestic oriented stabilization measures and fulfill an additional list of conditions. First, it required the GOP to dismantle its tacit multiple exchange rate, a precondition that came from IMF Managing Director Jacques de Larosière and the Executive Board. Second, the GOP needed to reduce more import restrictions. Third, the GOP needed to generate an additional P8 billion pesos in taxes. Fourth, the GOP needed to further tighten the domestic money supply. Finally, and perhaps most importantly, the GOP needed to secure commitments from a majority of the country's 483 private creditors and public bilateral lenders to fill at least 90 percent of the sought after $1.65 billion.[57] Clearly, these requirements forced the GOP to install appropriate policies that would signal to its creditors that the Philippines was on the proper road toward economic recovery. As a final incentive toward the implementation of these measures, the IMF threatened to terminate negotiations unless the GOP complied with these demands.

As the role of the country's foreign creditors in the IMF-GOP negotiations began to take greater significance, the United States also began to make its presence felt. Relatively silent during most of the negotiation phase, the United States enlarged its role as it pushed the GOP into the IMF corner. The timing of this push was significant. Some cabinet principals and high-level assistants recognized the signs and wanted Marcos out. Others, including Reagan himself, remained committed to Marcos, but wanted various degrees of political and economic reform.[58] The push started to turn into a shove in June 1984 when a US Embassy official James Nach crafted a disconcerting report that targeted Marcos as the problem. If Marcos was left to his own device, then a communist takeover would be sure to follow. "New directions from the top" were needed to right the ship.[59] Grumblings inside the Reagan administration got louder as more principals and members of Congress were convinced that Marcos had to go. In September 1984, American support for Marcos drastically eroded as the US Senate found the Marcos leadership to be "virtually bankrupt in terms of public confidence."[60] This declaration from the Senate, as well as similar positions in the House were influenced by the Nach report,[61] Filipino exiles and those concerned about developments in the Philippines.[62] The Reagan administration shared this view and believed that an upcoming World Bank agricultural loan approval would only give Marcos continued incentive to postpone economic reforms. Frustrated and unsure of the viability of Marcos, the United States voted against a World Bank loan to Manila.[63] Despite US objection, the World Bank issued a $150 million Agricultural Sector Inputs Loan that included financing for critical inputs and stipulations for reform in the agricultural sector, including changes in pricing, marketing, and institutional controls. The World Bank vote demonstrated only one manifestation of American displeasure with Marcos. Washington's loudest signal was sent in September 1984 when it announced its decision to sever official loans to Manila until the GOP capitulated on reform measures, until it had an IMF program in place.[64] The threaten reduction in foreign financing dramatically changed the Marcos' appraisal of his government's bargaining position and simultaneously increased the influence of Virata. According to Manuel Montes, this increased persuasive capacity reflects "the importance of technocratic influence varied directly with the need for foreign financing."[65] The push for reform was driven from the United States, foreign creditors, the IMF, and Virata and Fernandez. The obstruction to reform was domestically lodged.

The blockage stemmed not only from Marcos and his political preferences, but also from social veto players and within the cabinet. The principals that opposed reform benefited from the Marcos protecting their interests. Indeed, when Virata expressed to Marcos his inability to navigate the political landmines, the president replied, “Do your best in your own field, and let me know whether we can implement it politically. I can help you in that aspect.”[66] He was the “go-to-guy.” Virata had to secure the support not only from Marcos and social veto players, but he also had to move other cabinet members, other technocrats. Marcos economic lieutenants did not all subscribe to the Washington Consensus prescriptions for economic growth, some, such as Roberto Ongpin, subscribed to the Japanese strong state industrialization philosophy.[67] These conflicting networks butted heads not only on vision but also pace, investment, and the role of the state.[68] Ongpin resisted reform and closely protected his principals.

In October, under the final siege of negotiations, the GOP relented to the IMF on all issues. First, Philippine authorities dismantled the tacit multiple exchange rate and replaced it with a “free floating system.” Subsequently, the Philippine peso depreciated by 11 percent, to P19.99: $1.00. Second, the GOP removed the 10 percent ad valorem import duty and instituted numerous domestic measures to draw revenue, including an increase in the price of petroleum products. Third, the GOP allowed banks to keep 50 percent of their foreign exchange holdings, up from 20 percent. Fourth, the GOP intensified its import liberalization program through Central Bank Circular 1029, an edict that removed numerous products from the monetary authority’s scrutiny.[69] Finally, the GOP increased domestic interest rates to tighten the money supply.

The milestone Central Bank Circular continued the progress made under SALs I and II. Circular 1029 spelled out new measures that involved the regulation of import trade transactions and the liberalization of rules on monitored and regulated commodities. Key provisions of the circular focused on the new categorization scheme of import goods and the loosening of Central Bank trade scrutiny. Under this circular, Central Bank authorization for numerous imports was no longer required; however, the new procedure allowed other government bureaucracies to maintain their authority over import products. Finally, imports were classified into five categories: the freely importable, prohibited imports, banned commodities, regulated imports, and liberalized commodities.[70] Largely, Circular 1029 removed any discretionary authority that lied with the Central Bank, powers used for domestic patronage purposes. Central Bank improprieties were widely recognized by the Bretton

Woods Twins. The World Bank noted that "Central Bank licensing . . . is an obstacle toward product improvement, greater competition, and the provision of supplies to export industries."[71] Virata would be ultimately responsible for ensuring import reform.

In December 1984, the IMF executive board authorized the GOP Letter of Intent. According to the Memorandum of Economic Policy (MEP), the GOP pledged to continue its efforts to achieve and sustain substantial macroeconomic stability. From 1984 to 1985, the GOP aimed for substantial reductions in the external deficit and minimum inflation rates. The current account deficit was targeted to be reduced from 8.2 percent of GNP in 1983 to 5.0 percent of GNP ($1.5 billion) in 1984 to 4.1 percent of GNP ($1.1 billion) in 1985. These targets would be achieved through the improved competitiveness of the peso, a condition that would spur a higher level of export growth (projected at 6 percent in 1984 and 10 percent in 1985), and recovery in world demand. In addition to the growth in exports, the current account deficit would be aided by a sharp reduction on imports, projected at 23 percent in 1984 and 2 percent in 1985. Import growth would be expected to slacken in response to the peso depreciation and the tightening of domestic demand, particularly on capital goods and oil imports, a move designed to increase foreign exchange savings.[72]

Under the MEP, the GOP failed to squarely address the issue of trade liberalization, a topic that became clear several months later. In the first quarter of 1985, the GOP revised several of the original targets laid out in the MEP, especially in the area of reserve money. In addition, the authorities formally pledged to resume trade liberalization under the World Bank's SAL I and II programs. Specifically, the GOP committed itself to reduce tariffs for 47 commodities as soon as the country's creditor banks had extend a $3 billion revolving trade facility and accelerate the ILP through the deregulation of 215 other import commodities by the end of 1985. Finally, the GOP extended its commitment to trade reform through the stipulation that import duties would be lowered for all commodities with tariffs over 50 percent.[73]

The progress of trade liberalization in 1985 closely followed the ability of the GOP to secure additional sources of trade credits. In October 1984, the GOP sought a credit package valued at approximately $1 billion and had sent out feelers among its foreign creditors. The World Bank optimistically responded to the request. In contrast, Japan was more hesitant in its reply, as it waited for a go-ahead signal from other Paris Club members, a sign that was flashed with the debt restructuring agreement in December 1984. Furthermore, Japan already disbursed a $157 million commodity loan to the Philippines in 1984, but domestic players only tapped approximately $10 million of this loan.[74]

Domestic Political Economy of the 1984 Import Observation

During the 1983–1984 foreign exchange crisis, the GOP successfully overrode domestic opposition to its stabilization and structural adjustment measures. At first glance, the import controls established in late 1983 enabled the Philippine Central Bank to distribute critical foreign exchange toward the promotion of export-oriented industries that could help resolve the country's balance of payments problems. These instruments, however, were used by the Central Bank and favored banking establishments to shield Marcos cronies from the harsh economic cost associated with stabilization and structural adjustments measures. The non-Marcos cronies, increasingly concerned over the bias preferential treatment given to Marcos supporters, had begun to seek peaceful ways to remove Marcos from power. In its efforts to appease these groups, the GOP provided sufficient compensation to enable them to continue their import practices.

The Marcos government also sought to generate support for its policies by painting the use of import controls in light of its negotiations with the IMF and the national crisis, a linkage strategy that shifted the preference cost calculations for numerous veto actors. Formal and nonformal martial law decision-making arrangements revolved around Marcos. Philippine political culture is obstinately based on a patrimonial structure, where economic and political favors are disbursed to loyal supporters. In the pre-martial law period, conflict over government policy resided primarily between the executive and legislative branches, as both institutions often fought over access to distributive pork-barrel funds. Congress pursed the interest of its powerful constituency, primarily the wealthier classes of society, the landowning class, the industrialists, and commercial interests. When Marcos won his first term as president in 1964, he raised the patrimonial tradition to new heights, usurping the role of Congress and substituting himself as the patron.[75] This practice enabled Marcos to weaken the power of the legislature and eventually establish and consolidate his power base around "new Filipino elites," a group of followers who owed their allegiance to Marcos.[76] When Marcos declared martial law in 1972, he abolished the already toothless legislative branch and through chicanery, established a new constitution that relied on a strong presidency parliamentary framework. These moves did not have enough force to threaten the economic power of the dominant private actors during pre-martial law, but they "effectively shut off the majority of elite classes from political participation."[77] By 1973, Marcos had created an effective dual-structure decision-making system that enabled him to protect the crony capitalism of his political allies.[78]

The formal decision-making process also dictated the informal decision-making process that surrounded the cabinet. As the legislature was transformed into a "rubber-stamp" institution, all conflicts had been internalized within the executive. Often, cabinet members would offer conflicting proposals but arrive at no solution. Marcos became the final arbiter.[79] For the most part, Marcos often based on decisions on the impact it would have to his personal power and on the consequences to his cronies. He relied on the advice of particular confidants, such as Minister of Trade and Industry Roberto Ongpin.[80] The unbalanced power wielded by individuals such as Ongpin, who was the proverbial "guy to get the guy," tended to marginalize the effectiveness of technocrats such as Virata. Indeed, Virata was not fully aware of Marcos' hidden agenda, often finding his management efforts thwarted by the president.[81] This unending cycle of confusion is depicted by Manuel Montes:

> There was almost weekly cabinet meetings presided over by the prime minister, to thresh out issues to be brought to the President for promulgation. However these meetings didn't resolve many issues of coordination and many who presided in such meetings, including the Prime Minister, felt frustration over the fact that direct appeals (from political cronies) to the President have often overturned or suspended cabinet decisions.[82]

The decision-making style reflected the ability and necessity of Marcos to distribute economic preferences to his political base. As the chief executive, he had the institutional powers to grant, as well as withdraw, licenses, create bilateral agreements that would favor one industry over another (such as sugar with its exports to the United States), and institute, or not, export taxes. Given his closeness to highly protected sectors, Virata and Fernandez "designated" Marcos as the point-man to convince his cronies and import substitution families to surrender their rents[83] and connected the link between reform and foreign financing.[84] Under crisis conditions, Marcos played the "godfather" role. He would invite relevant private actors to the state dining room in Malacañang. He would then first have his relevant department heads view their positions on the issues. Then, he would provide private actors the opportunity to voice their concerns. Afterward, he would cloak his judgment around nationalism, sacrifice, justice, and equality, getting the private players to agree to austerity measures.[85]

Used as an instrument to distribute patronage to supporters, the Central Bank played a key role in shaping the country's trade and industrialization policies. It relied on its capacity to determine import and foreign exchange controls as primary instruments. For instance, in the 1950s, the Central Bank took direct control of import licensing, establishing a classification of each import line according to its importance to economic development or the provision of political support.[86] Central Bank control over import controls continued until the early 1980s when nationalist Central Bank governor, Gregorio Licaros, suddenly found himself "ganged up" by GOP technocrats such as Virata, Ongpin, and NEDA Minister Gerardo P. Sicat, all members of the Central Bank Monetary board.[87] Although Central Bank responsibility to control imports began to wane in the early 1980s, it still maintained a tremendous degree of control. Responsibility over imports, shifted from a centralized decision-making structure to a more diffused framework, as other agencies also had the capacity to restrict imports, such as the Ministry of Agriculture.

In a similar context with the petroleum price decision-making process, the import decision-making arrangement is also consensual in nature. Under a near-corporatists type environment, the GOP held consultations with key private actors, both veto and support, in determining the structure and levels of tariffs.[88] The institutional network was diffused, with a multitude of access points through which private actors can enter. As a matter of practice, private actors present their views to the Tariff Commission, lodged within the NEDA. In turn, the commission presents its findings to the Tariff Related Matters (TRM) standing committee, a body that functions to assist the NEDA board, composed of the president as the chair and other key economic ministers.[89] Although the Tariff Commission was the principle arena used by social actors, they can also access decision-makers through other avenues. These paths include other agencies of NEDA, MTI, the Central Bank, and the Ministry of Agriculture (MOA). Just as they can petition these particular agencies, private actors can also receive compensation through these agencies. The Tariff Commission, in conjunction with the TRM, often determines tariff rates. Another key side-payment disbursing agency is the Board Of Investments, an agency that targeted key economic sectors for investments and more importantly, granted incentives for the development of such areas. The BOI was an arm of MTI and an interagency unit that encompasses numerous heads from other departments, such as MOA, and the Ministries of Tourism (an arm of MTI), and Transportation and Communication Services (MTCS). The BOI also had close links with the import-substituting sector.[90] For the most part, those agencies that employed a consensual relationship with private actors also distributed side-payments to offset the costs of adjustment.

After the Central Bank overstatement problem, Marcos desperately needed to restore his country's credibility. Bankers heavily lobbied Marcos to dump Laya.[91] Among the list of candidates was Placido Mapa, but the banking community viewed his connection to the Philippine National Bank already tainted his reputation.[92] As a result, Virata suggested Jose "Jobo" Fernandez as the replacement. He founded and chaired the Far East Bank and Trust Company and once led the Bankers Association of the Philippines (BAP). Furthermore, he had close ties to Chemical Bank and Mitsui Bank and was involved in a joint IMF-Philippine Central Bank survey.[93] He had both the experience and the expertise to lead the Central Bank. The timing of his lofty positions also predated the rise of Marcos, meaning that he owed no loyalty to Marcos or to technocrats. This independence became a source of concern. According to Leung, "I didn't think Marcos would accept an independent, strong-willed fellow."[94] At a social gathering where both were in attendance, Virata had a lengthy and private conversation with Fernandez and discussed the aspects of the position.[95] Afterward, Virata suggested Fernandez to Minister of Trade and Industry Roberto Ongpin, who then followed through to Marcos, who had his own reservations. Ongpin explained to Marcos that he had no choice in the matter.[96] When Fernandez received the president's call, he hesitated, but when pressed, he accepted the position with the condition that he will manage the Central Bank with a free hand.[97] Marcos agreed.

Once Fernandez got in, he steered the Central Bank to addressing its foreign exchange shortage. Two private organizations, the Philippine Chamber of Commerce and Industry (PCCI) and the BAP, played key roles shaping the ongoing policies, often through the consensual decision-making process. PCCI served as the official representative of the private business sector in its dealings with the government. It represented business enterprises, commerce, manufacturing and processing, services, and business associations, among others. It generally responded to government initiatives and intervened on behalf of its members when successfully implemented policies conflicted with the interests of many PCCI members. BAP represented the interests of commercial banking firms in the Philippines and engaged in constant dialogue with the Central Bank, especially in raising the standard of banking in the country.[98] In late 1983, these two associations strongly recommended that the Central Bank implement two emergency measures. First, the Central Bank should set up a foreign exchange priority allocation system that would buy the foreign exchange holdings of all banks. Second, the Central Bank should reinstitute its forward cover policy, a mechanism that encouraged vital industries, such as steel and petroleum, to extend its import payment terms to one year. Under this arrangement, the Central Bank promised to reimburse these industries for peso devaluations. These policies, however, were the second best options. According to Sicat, the first best option would have been to let market forces make the adjustment.[99]

The best and simplest solution to the problem would have been to free the exchange rate and allow the market to solve the adjustment problem. This would have solved many things in the interim. The IMF would have welcomed this and probably as a result would have moved toward a settlement of the negotiations more quickly. But instead, the Monetary Board opted to revert to exchange controls, a less efficient method of dealing with this type of problem. The second best option was politically expedient as BAP sought to protect banks especially the small and medium sized ones.[100] To do so would require that the instruments used be more selective and discriminating so that caution would be used, and at a slower pace.[101] Exchange controls would cap the demand for the peso, but it would also provide incentives for another allocation system, one that could be controlled by the GOP.

The exchange controls led to the creation of two allocation systems. The de jure exchange controls placed tight conditions on traders. They could conduct businesses if they used payment schemes based on consignment, equity exchanges, or independent acquisition of financing received.[102] These conditions gave individuals incentives to conduct transactions through black market operations where the exchange rate was P30 per dollar whereas the official rate was P14.[103] Ironically, the GOP regulated the black market though what was referred to as the Binondo Central Bank. Roberto Ongpin ran the operations. The largest and best known Chinese black marketeers were rounded up by the National Intelligence and Security Authority and were told that it was the government's intention to organize a currency stabilization program. The Chinese black marketeers would be permitted to keep their profits from their operations provided that they cooperated with the government by buying and selling foreign exchange at government-dictated rates. In order to maintain discipline, the participants were subject to arrest if they deviated from the established rates.[104]

For this program, Onpgin set the official rate on a daily basis. He allowed the marketers to keep P0.20 per dollar, and the remaining cash would then be shipped to Hong Kong in a program that lasted for three years.[105] The existence and emerging prominence of the Binondo Central Bank proved to be a blessing in disguise. Official GOP import procedures were tedious and mired in red tape, a prospect that compelled traders to look toward the black market. With the unofficial option available and the unpredictable volatility of the Philippine peso, local industrialists also abstained from tapping the contingency pools provided by foreign donors. For non-project loans, the World Bank provided $140 million in 1984 to finance important requirements of exporters, but this fund was largely underutilized due to the lack of demand.[106] Similarly, Japan also extended a $157 million commodity loan, only to see $10 million used.[107] Thus, the black market provided the side-payment life-line that enabled non-Marcos cronies to ride-out the crisis.

In addition, the GOP also controlled the allocation of foreign exchange based upon the sector's importance to the economy. In 1983, the Philippines used 60 percent of its foreign exchange earnings for oil imports, 19 percent for food imports, 13 percent for industrial imports, and 8 percent for luxury items.[108] In 1984, the GOP adjusted this distribution pattern and concentrated on the promotion of foreign exchange generating industries. It tightened the foreign exchange allocation among various industries: $700 million for agriculture, $603 million to various manufacturing industries; $273 million for the steel industry; and the remaining $209 million for the import requirements of manufacturers.[109] In addition, the GOP also gave the export producing sector additional side-payment considerations, as it committed various credit schemes, concessional rates and financing toward increased production of agricultural and nontraditional exports. The country's garment sector also received considerable attention as the GOP attempted to increase production through maximizing the efficiency and production capacity of the country's textile sector.[110] The GOP also gave considerable financial preferences to firms with foreign partners.[111]

To support the allocation of foreign exchange, the GOP shut down the World Bank import liberalization program. In addition, the GOP also put additional taxes on imports to both discourage imports and to raise government revenues.[112] Simultaneously, the GOP expedited the production capacity of foreign exchange earning industries. In early February, the GOP exempted firms registered with the EPZA from requiring BOI clearance for the import of raw materials, parts and components of consumer electronic products. Shortly after reducing the bureaucratic oversight for export production, the GOP attempted to increase textile productivity through the liberalization of yarn, threads, and polybags imports.[113] The GOP slowly added items to its import liberalization list, with synthetic resins, resin products, and hydrogen peroxide to follow.[114] As a result of these import liberalization moves, the export-oriented industries set up under the SALs received strong side-payment considerations.

After several months under the priority allocation system, it became subject to scathing criticism with regard to its flaws and ineffectiveness. In many developing countries, a resource allocation system may be subject to corruptive practices. Such results are certain in the Philippines, where corruption and patronage practices are common occurrences. The trading and financial systems were fraught with loopholes and corruption that prevented industries not favored by the government from receiving proper critical imports. As a result of these practices, bankers, small sized industrialists, and surprisingly, PCCI called out for changes to the import procedures.[115] Furthermore, they argued that the import classification scheme poorly fit the complexities of industrial trading.[116]

Ironically, bankers, by implication BAP, and the Central Bank also received their share of indictments from industrialists. The bankers were charged with dispersing financing to selected clients. Industrialists targeted the problem to the apparent arbitrary allocation of foreign exchange among banks where export-oriented financial institutions were favored over import-oriented ones. In addition, they charged that banks had too much discretion in determining which letters of credit to open for goods that did not fall under the category of "essential producer goods."[117] As a solution to these problems,

PCCI suggested that the GOP scrap the current allocation system and institute a system that would allow all importation of raw materials, but at lower levels than in previous years.[118] Virata unequivocally rejected this suggestion.

In March, the GOP began to adhere faithfully to IMF stabilization "recommendations." The economy's excess liquidity and raising inflationary pressures necessitated the use of securities to stabilize the economy. As a result, the Central Bank marketed "Jobo" bills, treasury notes that had short maturity, large denomination, and flexible interest rates, among other things. By the third quarter of 1984, the Central Bank intensified its campaign to mop up excess liquidity as it raised the Jobo bill interest rate to 42 percent.[119] In effect, the GOP engineered a redistribution of income from the private sector to the government.[120] In addition, the higher cost of credit slammed the brakes on the economy as the volume of imports precipitately dropped, manufacturing production slowed to a trickle and unemployment increased. Meanwhile, large banks financially benefited from the rate of return of these securities.[121] Usually managed by Marcos supported cronies, large domestic banks purchased Jobo bills as a safe alternative to domestic loans, and to offset the trend toward disintermediation that accompanied the turn to high rates.[122] As expected, banks recalled many of their financing and bought the highly rewarded Jobo bills, effectively grinding the economy to a halt.

While some of the largest banks continued their support of the Marcos regime, the business community did a complete about-face. It initially supported Marcos' 1972 decision to declare martial law as a means to restore order and business confidence in the face of student riots, street protests, and disorderly conduct. Their support later eroded when Marco policies began to emphasize investments in nonperforming industries, creation of oligopolies, and the protection of cronies. The 1983 Aquino assassination fueled an impetus that fused the business community together with the Catholic Church and intellectuals forming a coalition that searched for ways to remove Marcos, preferably through the election process scheduled for 1987. Their opposition to Marcos stemmed from his authoritarian rule and manipulation of the economy on behalf of his cronies; fundamental change in the nation's social and economic system, however, was not on their agenda.[123] Business charges on the government were not isolated to Marcos alone, as they also attacked opposition party leaders who eventually reached out to business leaders to forge a coalition to rid the nation of Marcos. Business leaders, however, also managed to find other targets within the government. They attacked Virata's viability within the government and argued that his presence in the government only served to project credibility toward the international community. Such attacks were not isolated, as the Parliament itself grilled him for eight hours prior to his reelection as prime minister.[124] As domestic tensions between Philippine authorities and the business community increased, the GOP addressed what was perceived to be some of the flaws in the trading system. In large part to soothe IMF demands for reforms, as well as to address business concerns on the trading system, the GOP issued Central Bank Circular 1029.

During the negotiation process, links between the IMF-GOP negotiations and the initiation or removal of import controls were clear. In November 1983, the GOP made little effort to paint the IMF as the "bad cop." Instead, the GOP portrayed the talks as merely an attempt to restore stability to an economic situation that had spiraled out of control. For the economy to move back on track, the GOP had to mend its relationship with its private and official creditors, an outcome that would come to fruition with the procurement of an IMF stabilization agreement.[125] As negotiations progressed, the media uncovered the IMF's role in the adjustment process, a role that many domestic actors already realized. Knowledge of this link appeared to matter little. In fact, the GOP played on this information and held almost weekly consultations with various support and veto social actors. Support and veto actors eventually came to the realization that the economy's problem did not lie with the IMF, rather, the problem was domestic in origin. They fingered technocrats, in particular Virata, borrowers of foreign credit, both private and public, the Batasang Pambansa and Marcos for the deteriorating economic conditions.[126] In support of these accusations, the Makati Business Club, a non-crony private urban opposition group, released its survey of the nation's top public and private economists and 87 percent acknowledged that the current crisis was "brought on more by domestic factors." Furthermore, 73 percent of the economists surveyed responded that the government was responsible for the crisis.[127] Even with the reputation of the IMF, many sectors acknowledged that the culprit for the economic crisis resided at Malacañang Palace.

While the anti-Marcos manufacturing sector suffered economic loses from the stabilization program, the authoritarian regime's political base also incurred adjustment cost, but on more minimal levels. Within the scope of import reforms, the GOP was required to reduce the wide range of tax exemptions available to manufacturing firms through the BOI, which had been key to the development of import-substituting since the late-1960s. For the most part, IMF prescriptions that challenged these interests were outside the scope of import reforms. These stabilization and structural adjustment measures required the GOP to initiate tax reform, rehabilitate financial institutions, and dismantle public and quasi-public monopolies in the agricultural sector. The Philippines agreed to eliminate tax and duty exemptions for public corporations, although the requirement left an escape clause for the GOP to review and possibly reinstate the exemptions. For the GOP, the most controversial part of the program was the requirement to disband government marketing and price restrictions in the agricultural sector. However, while the GOP announced these reforms, it usually failed to implement them.[128]

Epilogue to the 1984 Import Observation

In 1984–1985, the GOP made successful progress toward macroeconomic recovery. The 1984 current account deficit was reduced by more than half and drastically reduced to a more manageable level of $103 million by 1985. Current account improvements were made with the help of the country's overall trade performance. The value of total exports reached $5,391 million in 1984 up from $5,005 million in 1983, a modest gain that came mainly from the increase in export prices rather than from growth in volume. The value of imports, meanwhile, declined to $6,070 million from $7,487 million in 1983, due to large reductions in import volume. Despite the reduced volume, overall imports exceeded the MEP target forecast principally because a greater volume of crude oil was imported, in part to rebuild inventories, and higher electronic component imports were required as inputs to the rapidly expanding export-based electronic industry.[129] Even with the surpassed target, the trade deficit was trimmed down from about $2.5 billion in 1983 to $697 million in 1984, and further reduced to $482 million in 1985.[130]

Trade reforms initiated by the GOP produced mixed results. For the first 10 months of 1984, the level of imports used for production registered an average of $478.3 million and dropped to an average of $396 million for the next ten months after the implementation of Central Bank Circular 1029. In large measure, the reduction of imports can be attributed to the tightening of monetary instruments. Consumer goods imports, however, illustrated a different story. For the first ten months of 1984, imports registered an average level of almost $18 million; in comparison, over the next ten months, import levels increased by an average of about $10.5 million a month.[131] The Philippine government also made modest progress in reducing import restrictions. In 1985, the BOI relinquished its control over the import of various commodities, in particular machinery and capital equipment; but the deregulation excluded products that required local content and products covered under progressive manufacturing programs. In contrast, the Philippine Central Bank continued to require prior approval for particular imports to ensure the availability of matching official financing, excluding products covered under Circular 1029. Similarly, the MTI continued to limit various commodities among a wide range of products, restrictions that included both those imposed after the release of Circular 1029 and those that existed before the circular and were to have been liberalized under SALs I and II.[132]

By the end of 1985, Philippine trade reform measures came to a sudden halt. This development came as little surprise despite the GOP intent to phase out about 9 percent of the country's total import restrictions by the end of the year.[133] Prior import reforms and stabilization measures had taken its toll on domestic industries. The footwear and garment industries experienced a massive reduction in the closing of small and medium enterprises, although the value of shipments disproportionately dropped, contributing to the reduction of firms. Similarly, the transportation industry was also hit hard, as both the number of establishments and the value of shipments plummeted. Given the recessionary conditions, small and medium enterprises put pressure on the GOP to come their aid.[134] In response, the GOP gave affected industries time to adjust and pick up the existing slack in capacity utilization.[135] This rationale carried considerable weight in the GOP calculus to suspend the ILP, a decision that was a distant second only to the snap presidential election to be held in February 1986, one year before the presidential term was scheduled to end.

Arguably, sectors that bore the cost of adjustment suffered needlessly. This assessment, however, did not arrive until the IMF sponsored economic program was already well in place. In the early stages of the program, the World Bank nodded approval.[136] Later, the World Bank adopted a complete about-face and voiced a harsh assessment on the approach adopted by the Philippine authorities and the IMF:

> In hindsight it appears that adjustment could have been both less painful and more structural if it had been done more through shifts in the composition of expenditures and output rather than exclusively through expenditure reduction. The external imbalance could have been closed by an increase in exports rather than exclusively through a decline in overall output. Instead, the real exchange rate remained stable in 1984, and it even appreciated in 1985 with the drop in economic activity. Stabilization was achieved without addressing the fundamental trade distortions that were partly responsible for the 1983 crisis. In fact, a number of stabilization policies, such as import restrictions, increased import duties and increased export taxes, were at cross purposes with the medium term adjustment requirements of the Philippine economy.[137]

ASSESSMENT OF CAUSAL MECHANISMS

The external-internal mechanism explains satisfactorily the Level IIA cabinet acceptance of the IMF prior commitment to import reform. With the World Bank SAL I and SAL II programs and financial creditors withholding credit lines, the IMF program was imperative to secure. The GOP CNs understood this sequenced process, as did Marcos. The external support for the IMF position squeezed Marcos and the trade principals, but the financing was far from dried up. Financing continued to trickle in from a critical IMF principal—the United States—and domestic sources—from the

Binondo Central Bank. With these sources, Marcos could afford to dance the line between a complete cooperative waltz and dragging his feet. Once the US Congress forced the executive to cut ties, then the external parties were unified, aligned to support the IMF. The unified stand taken by IMF principals was still met resistance at Level IIA. The Philippine Central Bank under Circular 1029 was forced to surrender its import gatekeeping capacity, but other bureaucracies still continued to put up a strong defense against import reform. A similar pattern applies at domestic Level IIA. The financial cost of a nonagreement with the IMF would be devastating to private veto players, both within and outside the Marcos circles. Limited cash would dramatically squeeze both groups, with cronies buffered by the financial impact. The tactic, however, failed to increase the costs for veto players to reform; similarly, the tactic also failed to increase the benefits for supporters of reform so that they would overwhelm the reform process with louder voices.

The deference tactic appears to explain better the connection between Level I and Levels IIA and IIB. Virata and Fernandez turn to Ongping when they need to have Marcos' ear. This tactic is reflecting in the episode where Virata and Leung went to Ongpin to float Fernandez's candidacy to the Central Bank governor position. In the case of import reform, however, the "go-to-guy" was no longer available, where competing interests would be difficult to overcome. The limited movement of the reform process at the cabinet level can be explained by the absence of a first person to contact. Ongping as the head of MTI was the obstacle. At Level IIB, Marcos is typically considered the "go-to-guy," the one that dealt with politics and used multiple means of persuasion to move private veto players. This process, however, also works both ways. Virata and Fernandez used Marcos to navigate the political landmines, but social veto players also relied on the opposite process. They relied on Marcos to navigate the bureaucratic landmines and work on their behalf. While Marcos is the first person the GOP CNs go toward when they want something done, social veto players also use Marcos as the first person they move toward. Thus, Marcos is both the first in line from the beginning of the policy implementation stage as well the first in line from the end of the policy implementation process. The limited deference opportunities explain the limited import reform.

Does the technocratic alliance strategy used by the IMF explain best the outcome? The IMF found two allies in Virata and Fernandez, but was unable to sway Ongpin and Marcos to accept the reforms. The IMF had more influence when the United States finally came around to pushing for the GOP into the IMF's arms. The tactic had no or limited application at Level IIB. As a result, the deference tactic is the best causal mechanism that promotes the interaction of international with domestic forces at all levels.

NOTES

1. Other impediments toward industrialization include the structure of exchange rate regime, the magnitude of foreign borrowing, the pattern of foreign investment, and the country's inability to respond quickly to the abrupt change in world and capital markets in the 1980s. Robert Dohner and Stephan Haggard 1994, 33–37.

2. For an excellent review of Philippine trade controls from 1948 to 1970, see Baldwin; also see Alburo, 1991; and Shepard and Alburo.

3. Sicat, 128.

4. Krugman, Alm, Collins, and Remolona, 9 and 18.

5. Virata.

6. A high EPR implies that an industry is given the protection to sell its output domestically (with limited foreign competition) as both outputs and inputs of the industry are given relatively higher tariff levels. Conversely, a low EPR means that foreign products are given relatively easier access to the domestic market. The results of Tan's study showed a structure of protection heavily favoring the manufacturing sector with an EPR of 489 as against that of agriculture equal to 100. Within the manufacturing sector, the final consumption products were heavily favored with an EPR index of 481 versus 100 for inputs to construction. The export sector is heavily penalized with an EPR index of 100 as compared to a very high EPR index of 1525 for non-exportables. Tan 1979.

7. Macapagal-Arroyo, 84.

8. Broad, 82.

9. Dohner 1988, 15.

10. Mosely, 45.

11. These items included meat and meat preparations, dairy products, cereals, fruits, and caustic soda. From the period covering January 1981 to March 1986, the ILP program covered the removal of import licensing requirements on 999 items, mostly NEC/UC and other items regulated for purposes of protecting domestic industries (e.g., part preparations, dairy products, cereals, fruits, caustic soda, cellophane, magazines, bottle containers, vinyl asbestos tiles, hydrogen peroxide, fish and fish preparations, canned sardines and mackerel). Macapagal-Arroyo, 84.

12. Sicat, 489.

13. Ibid., 489–490.

14. Mosely, 51.

15. Lallana, 37.

16. Ibid., 48.

17. For an exceptional description of "crony capitalism," see Stephan Haggard 1990 and Hutchcroft 1991.

18. Krugman, Alm, Collins, and Remolona, 18.

19. Broad, 79.

20. Sicat, 663.

21. Hutchcroft 1998, 73.

22. Quote from the World Bank, as cited by Broad, 87.

23. Broad, 88.

24. Bautista 1989, 21 and 39.
25. Alburo, 161; Montes 1991, 33.
26. Baustista, 39.
27. Lallana, 36.
28. Bautista, 41.
29. Ibid.
30. Ibid., 39.
31. Ibid., 39.
32. Bautista, 40.
33. Tandem, 2012, 28.
34. The twin oil shocks proved debilitating. In 1973, the Philippine oil import bill totaled $187 million. By 1982, oil imports registered $2,078 million and the volume decreased in physical terms. de Dios 1984, 3.
35. *World Bank Debt Tables.*
36. These priorities were (a) crude oil imports, (b) raw materials, supplies, and other inputs of export products, (c) essential grain imports, and (d) raw material of vital domestic industries.
37. Exempted from these duties were the importation of government agencies with existing obligations requiring exemptions of the diplomatic corps, personal effects of returning Filipino diplomats, and those of bonded manufacturing and bonded smelting warehouses.
38. Interview with Senior Department of Finance official, anonymity requested, April 27, 1995; and Montes 1987, 18.
39. International Monetary Fund 1984, 7.
40. Suspicions that something was foul emerged in August 1983. IMF negotiator Hubert Neiss approached Undersecretary of Finance Ernest Leung and indicated that statistics derived by several private corporations had failed to reconcile with data supplied to the IMF by the Central Bank. Leung conducted his own investigation but was unable to confirm these suspicions due to limited Central Bank access. Interview with Senior Department of Finance official, anonymity requested, April 27, 1995.
41. *Asiaweek* January 21, 1984 as cited in Yuson, 71.
42. Hutchcroft 1998, 171.
43. Orthodox packages refer to conventional forms of fiscal and monetary demand-management approaches and relatively collaboration with external creditors. In contrast, a heterodox approach places emphasis on more direct administration of prices and currency and a more confrontational approach toward external creditors. Kaufman, 64.
44. Interview with Senior Department of Finance official, anonymity requested, April 27, 1995.
45. International Monetary Fund 1984, 1.
46. Ibid., 15.
47. Ibid.
48. Ibid., 26.
49. Ibid., 27–28.
50. Ibid., 1.
51. Ongpin 2013; and Sicat, 502–503.

52. *Business Day* 1984c.
53. Gregor, 86.
54. IMF 1984, 31.
55. National Census and Statistics Office 1985.
56. National Economic Development Authority 1985.
57. *Business Day* 1984o and 1984p.
58. Karnow 1989b, 408.
59. Ibid., 406; and Blitz, 162–163.
60. *Business Day* 1984p.
61. Karnow 1989b, 406–407.
62. Sicat, 511.
63. Chand.
64. Sicat, 512.
65. Montes1991, 31.
66. Tandem 2016, 47.
67. Tiglao 2016.
68. Tandem 2016, 53–55.
69. Central Bank of the Philippines 1984b.
70. The freely importable were given due course by Authorized Agent Banks without prior Central Bank approval. Banned commodities were those which need prior Central Bank approval and consist of NEC and UC goods as well as items in tourist duty-free shops, and those used as imports in export production by firms registered with the Garments and Textile Export Board, Board of Investments and the Exports Processing Zone Authority. Regulated commodities required clearances from various government agencies although they did not need Central Bank approval and covered previously free importable items. Liberalized commodities were those that were formally banned NEC and UC items that did not require Central Bank approval.
71. Broad, 87.
72. Republic of the Philippines 1984a.
73. Philippine revision can be found in International Monetary Fund 1985, 66–75.
74. *Business Day* 1984r.
75. Doronila 1992, 23.
76. For an excellent review of the pillages that occurred under the Marcos regime, see Dohner and Intal, Chapter 2.
77. Montes 1991, 42 and 44.
78. Haggard, 224.
79. Interview with Senior National Economic and Development Authority official, anonymity requested, June 27, 1995.
80. Interview Senior Department of Finance official, anonymity requested, May 29, 1995.
81. Ibid.
82. Montes 1991, 32.
83. Tandam 2014, 351.
84. Haggard, 244.

85. Sicat, 240–242.
86. Shepard and Alburo, 176.
87. Broad, 87–90.
88. Technically, a corporatist's framework refers to the coordinated efforts of the government, business, and labor to set economic growth target levels. The absence of labor in the Philippine decision-making structure precludes, at least with regards to imports, labeling this situation as corporatists.
89. In addition to the president, the board also includes the NEDA secretariat as vice-chairman, the executive secretary and the secretaries of finance, trade and industry, agriculture, environment and natural resources, public works and highways, budget and management, labor and employment, and local government as members.
90. Haggard 1990, 225.
91. Joaquin, 293.
92. Ibid.
93. Hutchcroft 1998, 171.
94. Yuson, 60.
95. Sicat, 509–510.
96. Joaquin, 293.
97. Yuson, 63.
98. Sicat, 704.
99. Ibid., 514.
100. Hutchcroft 1998, 173–174.
101. Sicat, 514.
102. Dohner and Intal, 525.
103. Ibid., 525–527.
104. Ongpin 1986, 6.
105. Ibid., 2011.
106. International Monetary Fund 1985, 56.
107. *Business Day* 1984r.
108. Ibid., 1984b.
109. At the time, the preliminary $603 million budget for the manufacturing sector of consisted of $394 million for export industries, subdivided into imports of textiles ($1.54 million); chemicals ($168 million); pulp and paper ($54 million); and semiconductors ($16 million). Later the GOP expanded this list to include foreign exchange allocation for the transport and energy industries, for transportation spare parts and non-oil energy and energy-conserving capital goods. *Business Day* 1984f.
110. In the early 1980s, the textile sector received considerable World Bank attention. During the crisis, the GOP also paid careful attention to the textile sector as it allowed rebates of up to 40 percent of their electric bills if their power cost reached 20 percent of their production cost, a condition that occurred often as most textile millers were inefficient. Despite this and other tax breaks, textile manufacturers operated at only 50 to 60 percent due to the economic crisis. *Business Day* 1984s.
111. Dohner and Haggard, 48.
112. Dohner and Intal, 527.
113. Central Bank of the Philippines 1984a.

114. For instance, in the garment industry, local millers can offer tax and duty-free textiles to garment exporters with bonded manufacturing warehouses. The Board of Investments would then issue local millers with tax credit certificates equivalent to the tax and duty garment exporters would have paid had they bought imported raw materials. Thus, the scheme allowed local textiles to be priced competitively with imported textiles.

115. *Business Day* 1984e.

116. Ibid., 1984c.

117. The allocation of foreign exchange corresponds with the previous level of foreign exchange a bank possessed. Banks generate dollars and remit them to the Central Bank, which in turn, allocates the remitted dollars to different banks based on their previous dollar remittance volume. *Business Day* 1984g.

118. *Business Day* 1984e.

119. "Jobo" refers to the nickname pinned on Central Bank Governor Jose Fernandez. Ironically, this nickname referred to his penchant for distributing patronage jobs.

120. Montes 1989, 34.

121. Dohner and Haggard, 51.

122. Ibid., 51.

123. Reid and Guerrero, 21–22.

124. Makati Business Club 1984b.

125. *Business Day* 1984a, 1984b, and 1984c.

126. Makati Business Club 1984a and 1984b.

127. Ibid., 1984c.

128. Dohner and Haggard, 53.

129. International Monetary Fund 1985, 12.

130. Central Bank of the Philippines 1984c, 1985, and 1986.

131. National Economic and Development Authority 1985 and 1986.

132. Ibid.

133. These imports included iron and steel products, fabrics and textiles, cotton and synthetic yams and fibers, and paper and paperboard products. IMF 1985, 10.

134. *Philippine Daily Inquirer* 1985a.

135. Ibid., 1985a and 1985b.

136. Mosely, 53.

137. World Bank 1986a, x.

Chapter 8

1986 GOP-IMF Import Observation

The restoration of democracy in early 1986 caused a heightened sense of political hope and a shroud of economic uncertainty. The February 1986 election and the famed "EDSA Revolution" led to the ouster of dictator Ferdinand Marcos after almost twenty years in power.[1] In his place as president, Corazon Aquino made wide-sweeping political changes on both national and local levels in an attempt to clean up the "Marcos mess." As Aquino pushed out traces of the old regime, she brought in associates who had pledged their support for her during the presidential campaign. Often, these supporters were given high-ranking cabinet positions as political largess, forming a cabinet that yielded a relatively broad middle-left to solid right political and economic orientation. Often, individual members would often blurt out inconsistent or contradictory statements that sent out mixed signals regarding the administration's economic policy direction. This inconsistency characterized the nature of the Aquino government in both its dealings with international and domestic economic actors.

Without a clear understanding of the economic policy direction the GOP intended to take, the IMF walked the thin line between the paths of cooperation and optimism and hesitancy and skepticism. The first path developed from the IMF endorsement of the Baker Plan and the ouster of Marcos, whose political liability was well known in the international community. The second path derived from the make-shift coalition that supported the Aquino candidacy, as she was beholden to numerous political and economic groups and had yet to consolidate her hold on power. The IMF suspected that the Aquino administration might deviate from market oriented reforms as a means to draw upon popular support. To hedge against this possibility, the IMF engaged in a strategy of cross-conditionality with the World Bank, an approach that tied together the policies of the Bretton Woods twins before they disbursed funds to the Philippines. This effectiveness of this strategy was mixed, as the Philippines dragged its feet on numerous occasions.

The IMF tried to push the Philippines toward implementing import reforms. It was marginally effective as the IMF lacked clear support from the IMF Executive Board. In opposition to the IMF's position on import reform, a staunch cabinet opposition group slowed the pace of reform. Later in the negotiations, the IMF received critical information that allowed them to shift the GOP's perception of the costs, information that derived from international Level IIB actors who were attempting to push forward GATT negotiations and stimulate an international economy in recession, goals that prompted the IMF to take a distributive stance on the issue of import liberalization.

Just as trade reform had proved to be a sticking point in the negotiations, it also demonstrated to be a point of contention in domestic affairs as well. Aquino's cabinet "rainbow coalition" represented a mixture of domestic anti-Marcos interest, an environment that made the task of formulating a consistent economic policy difficult. The IMF adamantly pushed for continued implementation of the suspended Import Liberalization Program (ILP), initially drafted under the auspices of the World Bank in the early 1980s. Similarly, particular GOP departments supported this move, but the two years of economic hardship made this decision politically unpalatable. The constraints toward trade reform were further reinforced by the selection of a close ally of domestic industries as minister of Trade and Industry. Faced with both public and private sector oppositions to trade liberalization, domestic Level IIA actors continued to engage the veto actors under the auspices of the consensual and diffused institutional arrangement that structured the decision-making process and drew out the pace of import reform.

INTERNATIONAL POLITICAL ECONOMY OF THE 1986 IMPORT OBSERVATION

The issue of credibility was the focal point of the 1986 IMF-GOP negotiations. After the removal of Marcos, the new Aquino administration attempted to revive a sluggish economy through a strategy of government oriented growth policies in heavy combination with private participation. This strategy was often undermined by dissenting cabinet members as they constantly sent mixed signals to both private and public, international and domestic, financial and investment actors over the approach the GOP would take toward the economy. Given these contradictions, the Bretton Woods twins and industrialized countries had reservations about the GOP capacity to carry out economic reforms.[2] This suspicion was insufficient to prevent the IMF from proceeding with negotiations, but it did compel the IMF to engage in a cross-conditional strategy with the World Bank, a tactic that required the GOP to meet the conditionality requirements of both financial institutions. Unwavering on its demands for import reforms, the IMF insisted that the GOP continue with the trade reform policies promised under the Marcos regime. The GOP took marginal steps toward import reform, but the acceptance of liberalization initially fell on deaf ears, especially Aquino's. The IMF eventually succeeded in altering GOP preferences when IMF Managing Director de Larosière revealed the international nature of the IMF insistence on import liberalization.

GOP financial developments in the 1986 presidential elections proved to be both predictable and serendipitous. Faced with the prospect of having to own up to numerous policy blunders, Marcos was surrounded with political problems and was forced to call for elections in February 1986, one year sooner than expected. Although political opposition had been largely splintered, Marcos still initiated numerous schemes to seal his victory: increase wages for public employees, increase government borrowing from the Central Bank, lower prices of petroleum products, postpone the implementation of the ILP, an extension granted by the IMF, and purchase votes.[3] Marcos' extensive use of an expansive fiscal policy, a norm in Philippine politics, began to unravel the 1984–1985 stabilization and structural adjustment gains previously achieved by the GOP. During the elections, it was unclear how much high-spending and free-wheeling Marcos had conducted. Afterward, the IMF learned that the state-owned Philippine National Bank (PNB) suffered massive public withdraws of approximately Pl.9 billion, an indication that the money supply had in all likelihood surpassed the targets established in 1984.[4] This development proved unfortunate for the Philippines as the IMF intended to release its final 1984 Special Drawing Rights (SDR) tranche to the GOP, an installment that would have triggered another $350 million in new loans under an agreement made with the country's foreign commercial lenders.[5] After numerous discussions with the IMF in March, the GOP laid down the options of either requesting an extension to the terms of the 1984 agreement or drawing up another stabilization program.[6] The GOP chose the latter alternative. In line with the Baker Plan, the IMF agreed to allow the GOP to draft a program that put more stress on economic growth than stabilization.[7]

The next step placed was to draw up an agenda for Philippine economic recovery. To expedite matters, the IMF presented the GOP with an economic package proposal that increased the tax component of petroleum products and emphasized the ability of the GOP to meet its debt servicing obligations.[8] The prospects of putting additional restraints on the Philippine economy proved unappealing; thus, the GOP had refused to consider the IMF proposal and proceeded to draft its own economic program, a plan that incorporated an agenda posited by a team of economists from the University of the Philippines, a study referred to as the "Yellow Book."[9] This proposal recommended that the GOP engage in pump-priming policies to kick-start the economy, policies that among other things, involved a realignment of the country's trade and industrial policies to make the trade and production incentive structures as neutral as possible. Transparency would be achieved through a significant reduction of trade taxes, the complete elimination of quantitative restrictions, and a flexible exchange rate policy.[10] Ironically, the "Yellow Book" position on the removal of discrimination between tradable and non-tradable goods closely mirrored the IMF concern.[11]

The GOP desire to develop and implement its program proved to be a microcosm in the larger scheme of the negotiations as uncertainty about the capacity of the Aquino government to implement its promises became a critical issue. The new GOP cabinet was made-up as a result of a desire to reward the loyalty of people who had helped Aquino during the campaign rather than a desire to advance any particular class interest groups or economic philosophy.[12] Aquino's first cabinet appointee was Jaime V. Ongpin, an outspoken critic of Marcos who espoused the virtues of removing government from the economy. In the area of trade reform, his position was quite clear: "Philosophically, I don't believe in protection, period."[13] As Aquino's economic manager and finance minister, Ongpin fought to have "Jobo" Fernandez, who was considered "untainted" under the Marcos regime, retained as Central Bank governor.[14] Ongpin also supported the candidacy of Solita Monsod for economic planning minister and later, secretary general of the NEDA. An academic who helped draft the "Yellow Book," Monsod supported the removal of trade barriers as a mechanism to relieve pressure from the consumer who "has been ripped off for years."[15] Together, these individuals formed the core "free trade bloc" during the early years of the Aquino government.

Despite the exalted position of the "free trade bloc," this alliance met considerable resistance. Aquino appointed Jose Concepcion as minister of Industry and Trade. This move was criticized in the press as a payoff for his leadership of National Movement for Free Election (NAMFREL) and its apparent support of Aquino during the campaign. Concepcion had close links to Cardinal Jamie Sin, who was instrumental in organizing electoral opposition to Marcos.[16] In addition to his political leanings, Concepcion strongly reflected the sentiments of the import substitution sector, where his twin brother Raul was a major player. With his close links to the import substitution sector, Jose Concepcion's preferences on trade were quite clear. He strongly opposed moves to resume trade reforms, as a senior official in the department characterized: "He is a protectionist. He never pretended to be anything else."[17] In principle, he accepted and advocated the argument posited by local manufactures, a position that stressed the dire consequences that would befall local industries due to the lack of domestic demand and the GOP's inability to curtail smuggling. With respect to the IMF and structural adjustment, Concepcion acknowledged that these policies would put the economy on track as it had in the past. However, the social and political repercussions, in his view, were unacceptable and hence, he preferred to draw out trade liberalization programs in the long run.[18] In adopting this stance, Concepcion projected the notion that MTI's only "constituency was the industrialists, he didn't consider the trading sector and as a consequence, didn't look at the bigger picture [of long-term economic growth]."[19] Isolated, Concepcion would not have been able single-handedly to oppose the cabinet "free-trade" coalition; however, Concepcion secured the support of both Aquino and Minister of Labor Augusto Sanchez. With the president's backing, Concepcion managed to delay or make exceptions to the GOP trade reform position, a stand that attempted to continue a protectionist line.[20]

The democratic nature of the cabinet decision-making process revolved in large part from the designs of Ongpin, and to a smaller measure, in the Aquino decision-making style. In IMF-GOP negotiations, Ongpin sought formally to include the input of other government agencies, in an effort to derive a common position against the IMF and to reach an agreement that had the support of all interested government agencies, a necessary step for sustaining the economic programs. Although he may not have realized it, Ongpin gave factions within the cabinet, such as the protectionist bloc, opportunities to compound, delay, and complicate negotiations with the IMF. The decision-making framework was compounded by the Aquino managerial approach. Unlike Marcos, who was the final arbiter of disputes among cabinet members, Aquino initially preferred to maintain her distance from the conflict and wait for the cabinet to reach a consensus. These two factors created a daunting obstacle for technocrats, as they often met staunch opposition to IMF type economic programs and had limited capacity or opportunity to shift cost perceptions. In cases of a stalemate, technocrats needed to reach the ear of Aquino, who often shied away from such discourse. Under these conditions, the strategies for building coalitions had limitations.

The IMF had limited reach in its capacity to persuade the GOP live up to its economic commitments underscored the IFI stand on negotiations. To address this concern, the IMF joined forces with the World Bank. The Bretton Woods twins used a bargaining strategy that relied on, among other things, "a clear division of labor (in monitoring the country's compliance with the programs agreed with the Bank and the Fund)."[21] The sequence was clear. The IMF would go first, then the World Bank would follow. In essence, the cabinet's lack of clear economic focus and the inability of the IMF to extend its preferences beyond the GOP's "free trade" cabinet coalition prodded the IMF and the World Bank to institute a strategy of cross-conditionality, a tactic designed to boost the bargaining power of the Bretton Woods Twins in relation to the recipient country. The leverage of this strategy received considerable boost when other international public donors linked their financing with the IMF "seal of approval."[22] As a result of linking the IMF with donor aid, it appeared that the IMF was again in a position of bargaining strength.

Despite the increased bargaining strength of the IMF, Ongpin and Fernandez, who agreed with the IMF principle on free trade, were unable to change the preferences of the protectionist bloc that surrounded Aquino. From the beginning of the negotiations, the IMF mission moved away from its emphasis on stabilization measures and focused on structural adjustment reform policies, primarily regarding sugar and coconut monopolies, government financial institutions, and ILP.[23] With its interest clearly on the bargaining table, the World Bank strategically placed its representative in the IMF-World Bank negotiations, an arrangement that continued.[24] Throughout the negotiations, the IMF insisted that the GOP continue with the import liberalization of over 1,000 commodities, a reform scheduled to take place by May 1986. In contrast, the protectionist bloc headed by Concepcion pushed to defer the process until domestic industries had sufficient time to recover from the economic recession suffered under the Marcos years and the authorities had installed adequate safeguards to protect these industries against smuggling and unscrupulous invoicing.[25] As a demonstration of its commitment to the principle of import liberalization, the GOP deregulated 802 import items, albeit in piecemeal fashion.[26]

In moves that coincided with negotiation rounds, the GOP begrudgingly accepted to IMF demands. The GOP understood the urgency attributed to securing an IMF program and the importance of keeping the IMF at the bargaining table, as it proved to be the catalyst toward private foreign financing. At first, the GOP released from regulation 143 items days before a May deadline.[27] After the IMF continued to insist on full ILP implementation, the GOP released an additional 81 items and still later, 400 more items from regulation. Shortly after the third liberalization round, the IMF and GOP came to terms on the major targets of the economic program, with the ILP, among other things, remaining to be later worked out. To close the deal, the GOP issued feelers on a possible compromise over the remaining 383 "sensitive" items; the GOP offered to surrender some import items to deregulation, in exchange, the IMF would extend the phase-down period of the ILP.[28] The IMF flatly rejected this offer. In early September, the GOP announced that it would release eighty additional products for liberalization after it conducted additional public hearings before the affected industries, an offer that enabled the GOP to further delay import deregulation.

COG involvement in the negotiations managed to help bring the talks to conclusion. In September 1986, Aquino made her first official visit to Washington, DC where she met with American officials and dignitaries and had attempted to solicit increased pledges for economic aid. Attempting to capitalize on her "Good Will," Aquino linked American economic aid with the impending leases on the American military installations at Clark Field and Subic Bay. In return, the United States pledged to increase its donation. Aquino concluded negotiations with the IMF when she met with de Larosière and World Bank President Barber Conable to discuss the status of the standby agreement and reassure them of the GOP intent to adhere to its economic reform commitments. De Larosière was reluctant to compromise for numerous reasons. First, the GOP needed to resolve a contentious issue with the World Bank before he could endorse the program. During its Economic Recovery Loan talks with the GOP, the World Bank insisted that the GOP take measures to reform the Philippine Development Bank, an issue that had to be first worked out.[29] Second, de Larosière received considerable pressure from industrialized countries to do something about the tariff and nontariff barriers to trade that existed in developing countries.[30] Third, the Uruguay Round was launched in September 1986, attempting to remove government interference in agricultural and tertiary products. In these discussions, the IMF played only an information and technical role, but it emphasized the need for unhindered trade as a means toward securing balance of payments equilibrium.[31] The timing of the Aquino, de Larosière and Conable meeting and the Uruguay Round was not coincidental. Finally, de Larosière was scheduled to step down at the end of 1986, and he wanted to tie up loose ends.[32] Highly constrained by these factors, de Larosière had no room to compromise on the issue of import liberalization. Aquino reluctantly accepted this prerequisite and authorized the initiation of the program, with some caveats. Although de Larosière approved the negotiation standby agreement, he had to await the outcome of the ongoing World Bank-GOP Economic Recovery Loan negotiations. Clearly, de Larosière's constraints, especially from the industrialized countries, played a significant role in determining the outcome of the negotiations.

Under the 1986 IMF-GOP Standby Agreement, the GOP promised to continue the progress made under SAL I and II. The GOP announced a schedule for the liberalization of all items remaining from the April 30, 1986, list; another schedule for the liberalization of items previously scheduled to be liberalized on December 31, 1986, and June 30, 1987; and new tariffs for items that have been liberalized. Furthermore, the GOP promised that these liberalization schedules would not extend beyond April 30, 1988, and tariff adjustments would not result in a level that exceeds the maximum rate of 50 percent.[33]

The GOP made similar trade reform promises to the World Bank. Under the 1986 Economic Recovery Loan program, the GOP attempted to undo the damage done to the reform program during the 1983–1985 crisis. Of the

25 percent of total imports covered by prohibition or by a requirement of "prior approval" from government agencies, about half, or 936 items, mainly consumer goods and raw materials, had already been scheduled for liberalization by the Aquino government when negotiations with the Bank began. As a result, the Bank focused its position on intermediate products, and asked for an additional 292 restrictions to be removed by April 1988. In addition it asked for a new program of liberalization and a review of the tariff structure to be presented before release of the second tranche.[34] Unlike the broad scope of the 1986 Standby Agreement, the World Bank confined its conditionality to a few key policy issues, which noticeably focused on allocation rather than the redistribution aspect of the "Yellow Book." These issues formed part of the agenda of the first two SAL loans: trade liberalization, tax reform public investment program, and the reform of the Government financial institutions.[35]

DOMESTIC POLITICAL ECONOMY OF THE 1986 IMPORT OBSERVATION

The Aquino government opened up the process considerably by allowing for consultation through the trade and industry ministry[36] and other bureaucracies. The process continued to resemble the "substitutive" Marcos process as new Trade and Industry Minister Jose Concepcion Jr., interjected his protectionist preferences.[37] Concepcion and his twin brother Raul, who himself was a powerful player in resisting petroleum price increases, had interests in the flour and appliance manufacturing industries. While technically divesting himself from these interest and turning over management to his family,[38] he acknowledged the need to reform, but he insisted that safety nets be in place to cushion the exposure to competition to domestic industries. The reforms were instituted piecemeal, a little here and a little there. This process was the result of the consultation of Level IIB veto players. In 1987, when Congress was restored, it joined the fray as it gave itself the power to delay action on tariff changes and conduct public hearings.[39]

During the early years of the Aquino administration, the GOP made significant progress in reviving the suspended ILP of 1981. After nearly three years of stagnant economic performance, beneficiaries of long-standing import substitution policies feared that cheaper foreign competition would finish many off local industries. In response to these prospects, manufacturers who were previously excluded from the political process under martial law took advantage of the revived democratic atmosphere and petitioned the GOP (principally through MTI in general, and Concepcion in particular) to postpone ILP implementation. The GOP listened and responded to these cries by attempting to forestall trade reform until the proper safety nets were in place. These cries were more than window dressing; they were real as unemployment and underemployment were reached devastating levels. Invariably, these intentions were overshadowed by the GOP need for external financing. To push through with the ILP, the GOP proceeded to issue commensurate side-payments among the various effected sectors and link import reforms with the current IMF-GOP negotiations.

As much as cabinet institutional changes were critical in 1986, personalities also became instrumental in determining policy preferences. Clearly, Concepcion and Sanchez played critical roles in shaping the nature Philippine economic policy. Along a similar vein, executive secretary Joker Arroyo was instrumental in shaping Philippine policy on human rights abuses and the informal decision-making process. As a human rights lawyer and a long-time confidant of Benigno Aquino, he was dedicated to bringing human rights abusers under the martial law regime to justice. Furthermore, he also proved to be a chief obstacle to the informal cabinet decision-making process. Also a long-time friend of Corazon Aquino, Arroyo sought to insulate her from making difficult decisions by regulating issues that came up for consideration, as he often screened memos and allowed them to pile up on his desk. His interference, leftist oriented economic leanings, calls for limitations on foreign business penetration and selective repudiation of foreign debt, and administrative ineptitude often left many members of the business community calling for his resignation. Arroyo resigned his position after the August 1987 military coup attempt against the government.[40] His departure, along with others in the cabinet, marked two significant changes under the Aquino administration. First, it shifted cabinet political and economic preferences toward the right. Second, it also forced Aquino to take a more personal, visible role in managing her government.[41]

The often acrimonious and factious process at the cabinet level reflected the divergent interest of social actors on import reform. During the IMF-GOP negotiation process, numerous domestic interest groups favored liberalization of the country's trading system.[42] Under the 1984 IMF-GOP Stabilization program, various domestic manufacturing sectors experienced severe under-capacity problems that slackened employment and production. Real GNP dropped 6.4 percent and 4.4 percent during the 1984–1985 years, respectively. Likewise, the unemployment rate hit an average of 10.8 percent and just as critically, underemployment registered an average of 34.7 percent. These developments had to be balanced with a need to reform the trading sector where distortions were plentiful. According to the "Yellow Book," and its subsequent follow up the "Medium Term Philippine Development Plan of 1987–1992," reform was geared to increase industrial competition. To meet this end, the GOP wanted to rationalize the tariff structure and use tariffs as the principle instrument while reducing or eliminating nontariff barriers to trade. Furthermore, it would attempt to level the playing field for domestic industries against unfair trade practices such as smuggling.[43]

In general, some domestic industries applauded the GOP's efforts to restart the ILP and deregulate inputs to domestic industry.[44] Phase I of the program, scheduled to begin on May 1, 1986, would have lifted import controls on items mostly used as inputs to domestic industries, such as the garment, agricultural, packaging, handicraft, and construction sectors. Although some sectors looked forward to continued implementation of the ILP, others were less than enthusiastic about the prospect of foreign competition. In principle, sectors long protected by import controls agreed with the notion of import liberalization; however, they also called for the GOP to pave out a "more level domestic playing field." For instance, the long-time shielded textile industry, led by the Textile Mills Association of the Philippines (TMAP), supported the GOP intention to release import controls provided that the authorities managed to set up the necessary safety nets and preconditions (i.e., anti-smuggling programs and compensation packages). These demands, however, were both extreme and difficult to implement in the short term, as among other things, the TMAP argued that the industry must reach 85 percent capacity and be given a grace period of five years before it would agree to the ILP.[45] Similarly, PCCI also conditionally agreed to the restart of the ILP. These concessions centered on a two-year ILP extension, implementation of safeguards against smuggling, dumping, undervaluation, and mis-declaration of import practices, reduction of the interest rate and revitalization of infrastructure.[46] In contrast to these positions, the National Economic Protectionism Association (NEPA), an organization of domestic manufactures that promoted a "Filipino First" economic policy, unequivocally opposed liberalization as they said it would reverse decades of efforts to promote the economic advancement of Filipinos, generate greater unemployment, and import dependence. Clearly, these veto actors were not ready to endorse and resume the ILP immediately.

Without question, opposition to import liberalization sought time to recover from the economic quagmire suffered under the martial law years, and to this end, the Aquino administration made a temporary pact with labor. Sanchez was her point man, but his relationship with labor was mixed. The KMU viewed him as a supporter of its causes; the TUCP viewed him as an obstacle to its interests.[47] Regardless, Sanchez bargained with these interests. Upon his insistence, labor leaders of the four top unions agreed to a six-month moratorium on strikes. In exchange, the GOP agreed to drastic changes in union organization rules. It allowed security workers and government employees to unionize. Furthermore, the pact lowered the number of votes needed to form a union, from a two-third requirement to a simple majority. Despite this pact, strikes actually increased from 371 in 1985 to 571 in 1986.[48] Sanchez was subsequently dismissed from the cabinet in November 1986 for being too far to the left.

Its populist image notwithstanding, the GOP kept closer ties with industrial management than with labor. Before the May 1, 1986, ILP start date, the Aquino government regularly consulted with PCCI, Employers Confederation of the Philippines (ECOP), and various other industry sectorial task forces on the issue of trade reform. The GOP used its diffused and consensual institutional arrangement to solicit veto and support actors' preferences. Through the MTI, NEDA, the Tariff Commission, and Malacañang Palace itself, the GOP provided a multitude of platforms. These consultations occurred more frequently when negotiations began between the IMF and the GOP.[49] These consultations invariably linked IMF talks with ILP reform, where Concepcion strongly favored a staggered ILP phase-down.[50] The diffused nature of the institutional arrangements allowed the veto actors multiple access points to the decision-making process. Ironically, NEDA held open consultations while MTI conducted closed meetings.[51] Malacañang held its doors open to a broad spectrum of opposition as the league of governors and city mayor petitioned Aquino to suspend further implementation of the ILP. Aquino accepted this request only until after public hearings determined the ILP impact on business and the population.[52] Although veto actors strongly voiced their opposition to import liberalization, their voices were in some respect muted by the presence of support actors, especially those that relied on cheaper imported intermediate goods and the need for trade credits, a resource that could only be provided with the successful conclusion of an IMF program. Commercial banks applauded the need for import reform, especially when they could open letters of credit to actors without securing prior government clearances.[53] Other support groups were certain industrial sectors, such as the agro-industries, the packaging industry, the handicraft, and construction industries. With their participation in the decision-making process, support actors eroded the political power of veto actors.

The GOP issued side-payments to offset the cost of adjustment. At each import liberalization round, the GOP exchanged measures of compensation for opposition consent to liberalization. It used both indirect and direct means to compensate veto players. In early May 1986, the GOP agreed to push for the implementation of the Comprehensive Import Supervision Scheme (CISS), a measure that improved the GOP ability to record the value of imports. In 1987, the GOP signed a ten-year contract with the Swiss-based Societe Generale de Surveillance (SGS), a private firm that carefully monitored the value of imports. These measures enabled the GOP to restrict smuggling, a constant problem due to the country's high tariff rates. Perhaps more importantly, the GOP also exchanged the delisting of regulated items for increases in tariff rates. Under the second, third, and fourth liberalization stages, the GOP agreed to increase temporarily tariffs of equal protective effect, by almost 20 percent in some cases.[54] As an added side-payment, the GOP extended a new package of BOI incentives for scarcely occupied, export-oriented, and labor-intensive industries, incentives that accounted for only a small percentage of the country's national product.[55]

The IMF-GOP negotiations concluded in September 1986, but controversy over the ILP continued. The GOP agreed to proceed with trade liberalization, but had to tread carefully in the fragile political environment. As a result, in its Letter of Intent, the GOP made it clear that it would proceed only after studies, public hearings, and consultations were completed.[56] In October, Concepcion tried to buy more time by asking the IMF to delay the implementation pending more consultations and more in-depth studies. The IMF clearly rejected this proposal.[57] By the end of 1986, 936 items had been liberalized, with the rest scheduled for liberalization by April 1988. This still left approximately 500 items subject to import restriction. Overall, the GOP had been more successful in eliminating restrictions on consumer goods imports than it had on industrial intermediate goods dominated by larger domestic producers with high sunk costs.[58]

EPILOGUE TO THE 1986 IMPORT OBSERVATION

In the two ensuing years that followed the 1986 IMF-GOP agreement, the Aquino administration made significant political and economic gains. Aquino's popularity, which reached an approval rating of over 70 percent, helped pave the way toward cleaning up the Marcos mess.[59] With this support, Aquino formed a constitutional convention to draft a new government charter; a document based almost exclusively on the country's 1935 constitution, and oversaw a plebiscite and relatively peaceful elections for congressional and local offices. Like the pre-martial law years, Senate members were elected on an at large-basis, a procedure that maintained their independence from the executive but also perpetuated its indebtedness to the traditional ruling class. The lower chamber, by contrast, was elected on a single-member district formula. Election results provided overwhelming support for Aquino, as her party won twenty-two of the twenty-four available seats in the Senate and over 70 percent of the seats in the house. Re-installment of the legislature had two results. One, it pushed the democratic process forward. Two, it ended Aquino's ability to initiate law by decree. Under the new constitutional arrangement, the House of Representatives was given the power to determine tariff rates and the Senate the authority to propose amendments or concur.[60] The addition of another layer of decision-makers failed to provide a significant obstacle toward increased economic prosperity and macroeconomic stability in 1987 and 1988. In 1986, the Philippine economy grew by a scant 2 percent of GNP over the previous year, a performance that was followed by growth of 5.7 percent in 1987 and an unexpected growth rate of 6.7 percent of in 1988. As a manifestation of the improved economic conditions, the overall balance of payments remained on the credit side. In contrast, the current account balance, $996 million in 1986, recorded deficits in 1987 and 1988, of $444 and $373 million respectively. The country's trade balance proved largely responsible for the current account deficit as the level of imports outpaced the level of exports for 1987 and 1988, by $1,017 billion and $1,018 billion, respectively.

Domestic demand for imports reached unprecedented levels, but with a change of emphasis. In 1988, raw materials and intermediate goods imports accounted for almost 51 percent of the nation's imports and far outpaced the share of imports in other categories. Cotton imports increased by 50 percent, iron and steel by almost 40 percent and textiles by over 16 percent.[61] To its credit, the GOP liberalized imports according to the schedule presented to the IMF. Under the final phase, the GOP intended to liberalize import controls on 157 items: 94 items grouped under List A (e.g., brand new trucks and engines, machineries and equipment and spare parts) and 63 items under List B (e.g., cement clinker, radiation-emitting apparatus). Products that are on list "A" were available for immediate liberalization. Products on list "B" were subject to further consideration before they can be liberalized.[62] Obviously, determining which list a product falls under was critical.

Although the GOP implemented the ILP on schedule, the pace of liberalization began to slow after April 1988. For the most part, trade liberalization consisted mainly of consumer goods, whereas the controls that represented a real obstacle to export competitiveness—cotton fabrics, cement, and steel being politically sensitive commodities, remained in force. Beginning in 1988, NEDA announced that 114 products were nonnegotiable for reasons of national security and 104 more would certainly be liberalized. After this stage, raw materials and intermediate goods became more difficult to liberalize.[63]

ASSESSMENT OF THE CAUSAL MECHANISMS

Like the 1986 Petroleum Observation, the deference tactic at domestic Level IIA had limits securing its target veto player, Aquino. The mechanism was limited by the consensus institutional arrangement established under the new government. For Ongpin, he needed to secure first and foremost, the support of Concepcion, as Aquino and others deferred to his position. Concepcion, whose strong ties to domestic import-substituting industries, agreed with the principles of free trade but had strong consideration on the timing and required safety nets to catch industries that lacked the capacity to compete against foreign goods. He also had held issues toward providing a level playing field for domestic industries, promoting an anti-smuggling campaign. The consensus institutional arrangement made Concepcion a key veto player. The causal path toward Aquino's endorsement traveled as followed: Ongpin-> Concepcion->Aquino. The causal path also traveled from the back door to Aquino as she met with other COG (IMF Managing Director de Larosière and World Bank President Conable) to discuss snags and concerns. Once she was persuaded, Aquino met the IMF's conditions.

Deference at Level IIB was more complex. Marcos had direct access and familiarity with many principals, and as a result, he could attempt to sway their preferences using many of the instruments at his disposal, not to mention his charm. Aquino, on the other hand, lacked the regular pattern of interaction with the import-substituting industry, and as a result, for her it was more difficult for her to persuade them. Instead of an arrangement of familiarity between agent and principal, the social veto players relied on bureaucrats, panels, and commissions to press their advantage. They had multiple access points to halt the process: MTI, NEDA, the Tariff Commission, and Malacañang Palace. In contrast, to the petroleum observations, the decision-making panels and the compensation halls were one and the same. This arrangement allowed the principals to keep the discretion of Concepcion on a tight leash. The deference path took multiple steps and routes to get to the agent, allowing him to put the breaks on reform, or at least serve it piecemeal, because of the costs to the principals appear daunting and inundating as they come from many bureaucratic avenues.

The external-internal tactic slowly nudged along the prior-conditions piecemeal. As the delegated agent in pursuit of economic reform and stability for international Level IIB actors and its sister institution, the IMF staff laid down the need to renew fully the World Bank structural adjustment programs. Failure to do so would prolong the round of negotiations and heighten the costs of no adjustment as the public and private financial community tied all aid to the IMF program. The IMF principals were all behind their agent. Drying up of financial credit lines would bring the economy to a halt, increasing the cost associated to the democratic transition. For the GOP, the cost was fixed; import reform, whether a little or a lot, yielded either an IMF program with replenished financing or nothing at all. There was no middle ground, nothing in-between, a threshold for which Concepcion searched. The GOP CNs pursued the reforms; they just had MTI bureaucratic obstacles.

At domestic Level IIB, the tactic did not only increased the costs of no agreement, but also the benefits of an agreement. Without question, the IMF role as a catalyst for continued financing was imperative, but it was paralleled by the continuing round of the GATT negotiations, including the nascent Uruguay Round talks and regional free trade negations with the country's ASEAN partners. Global trade was heading more toward liberalism, integrated financial and production markets were just starting to open. The benefits of being inside with the leaders of trade were more advantageous than being on the outside. Furthermore, Reagan, a vicious proponent of free trade, argued for more trade opportunities, and as a result to drive more US exports for the economy. The international tide seemed too good to ignore.

Can there be another explanation for the outcome? From the IMF side, it managed to secure the support of two interlocutors, Ongpin and Fernandez. Its reach, however, seemed limited as it had limitations convincing Concepcion. The backdoor, however, allowed an opportunity for the Bretton Woods Twin's COGs to convince Aquino of the requirement for reform. The front door and the back door approaches cornered Concepcion to agree to the reforms. The tactic provides a compelling explanation for the implementation, but it provides a less than compelling argument for the timing, why import reform occurred in stages. At Level IIB, the technocratic alliance tactic is limited to explain the reform, with the exception that the prevailing international environment encouraged supporters of reform from export-oriented industries, overwhelming opponents of reform. These voices for the status quo, however, were not run over.

The GOP treated import reform with great care and delicateness. The lessons from the Marcos years helped to aid the first year of democratic transition. The Aquino government was certainly concerned about legitimacy and had to maintain as best as it could the face of reform, which also meant transparency for all groups, which later excluded the left. As a result, access to halls of decision-making was wide open, more studies were conducted, and pass-through of prices was capped. A critical addition to these changes were tariffication schemes. The rules and norms of governing changed, and the Aquino administration received a passing grade, at least for the first year.

NOTES

1. EDSA is the acronym for Epifania de los Santos Avenue, the road that connected the two military bases where anti-Marcos military leaders held up after a failed attempted coup.

2. Interview with Senior National Economic and Development Authority official, anonymity requested, May 23, 1985.

3. Both opposition leaders Salvador Laurel and Corazon Aquino threw their hats into the presidential ring. Archbishop Cardinal Jaime Sin brought together the two

sides to form a "shot-gun wedding," an arrangement that placed Aquino as the presidential candidate and Laurel as the vice-presidential candidate. Communist factions refused to participate in this make-shift coalition.

4. Banal III.
5. *Business Day* 1986d.
6. *Bulletin Times* 1986b.
7. *Manila Bulletin* 1986b.
8. Ibid., 1986c.
9. Alburo, et al.
10. Ibid., 179–195.
11. Interview with Senior National Economic and Development Authority official, anonymity requested, May 23, 1995. The recommendations had little to do with the restoration of Philippine democracy. The authors drafted the proposal and had planned to submit it to the GOP as a policy recommendation for economic development; regardless of who the head of the state was at the time. This intent was evident as the preface failed to directly address either Marcos or Aquino. Interview with Senior Budget and Management official, anonymity requested, May 23, 1995.
12. Mosely, 54.
13. Gigot 1986a.
14. Charges of graft and impropriety later surfaced to tarnish the Fernandez image.
15. *Manila Bulletin* 1986j.
16. NAMFREL is a private, Catholic supported agency dedicated to monitoring elections to ensure fairness.
17. Interview with Senior Department of Trade and Industry official, anonymity requested, August 7, 1995.
18. World Bank 1986a, 1.
19. Interview with Senior National Economic and Development Authority official, anonymity requested, May 23, 1995; and interview with Senior Department of Trade and Industry official, anonymity requested, August 7, 1995.
20. Interview with Senior National Economic and Development Authority official, anonymity requested, May 23, 1995.
21. Ibid.
22. Mosely, 56.
23. *Manila Bulletin* 1986h.
24. Interview with Senior NEDA official, anonymity requested, July 28, 1995.
25. Ironically, the World Bank was sympathetic to the GOP position. *Manila Bulletin* 1986f.
26. General Agreements on Tariffs and Trade, 13.
27. Central Bank of the Philippines 1986a. Central Bank Circular 1100 allowed the import of various commodities without the approval of numerous government agencies, in particular, the BOI. The liberalized products were primarily: hydrogen peroxide; paper waste; basic iron and steel; synthetic resins and resin products; synthetic yams, fibers, and threads; paper and paperboard products; fish and fish preparation; tires; and textiles.
28. *Manila Bulletin* 1986l and 1986n.

29. *Manila Chronicle* 1986c.

30. Interview with Senior Bangko Sentral ng Pilipinas, anonymity requested, June 15, 1995; and interview with Senior Department of Finance official, anonymity requested, April 27, 1995. Ironically, in early 1986, outgoing World Bank President A. W. Clausen attacked the United States and other industrialized countries for their import restrictions of farm products and textiles from developing countries.

31. Boughton, 1007.

32. Interview with Senior Department of finance official, anonymity requested, April 27, 1995; and Kahler 1993, 387.

33. Republic of the Philippines 1986d and 1987.

34. Mosely, 57.

35. Ibid., 56.

36. Alburo 1991, 161.

37. Interview with Senior Department of Trade and Industry official, anonymity requested, August 7, 1995.

38. Thompson 1992, 44–45.

39. Alburo 1991, 163.

40. Reid and Guerrero, 124.

41. Ibid., 143.

42. According to a Makati Business Club survey, 69 percent of respondents were supportive of import liberation policy. Caution should be used about relying on the accuracy of this statistic, as only one-fifth of the club's membership responded to the survey. Of the eighty-four respondents, the manufacturing sector represented forty-six companies, services, seventeen, banking and finance thirteen, agribusiness and food twelve, trade ten and commercial/retail five. Makati Business Club 1986.

43. Menardo, 3.

44. For Instance, waste paper, iron and steel products, synthetic resins, polyester staple fiber and filament yam, paper and paper products, fabrics and tissues, ceramics, tires and pottery.

45. *Manila Bulletin*, 1986i.

46. Ibid., 1986a.

47. West, 114.

48. National Statistical Coordination Board. 1987. Strike figures for 1986 were preliminary.

49. *Manila Bulletin* 1986g.

50. Ibid., 1986d.

51. Ibid.,1986j.

52. Ibid.,1986i.

53. Ibid.,1986g.

54. Dohner and Haggard, 73.

55. Dohner and Intal, 443.

56. Republic of the Philippines 1986d.

57. *Manila Bulletin* 1986p.

58. *Dohner and Haggard*, 73.

59. *Business International* 1989.

60. See the 1987 Constitution of the Republic of the Philippines. Article VI. Sec. 24.

61. Data taken from National Economic Development Authority 1989.

62. Macapagal–Arroyo, 85.

63. Mosely, 60.

Chapter 9

1989 GOP-IMF Import Observations

Philippine economic performances in 1987 and 1988 gave the GOP plenty of reasons for exultation; yet, for the most part, the next series of IMF-GOP negotiations reflected the uncertainty regarding the pace of the country's economic growth and its debilitating effects. The domestic side had an optimistic appraisal of Philippine economic performance in the medium-term, 1987–1988. In contrast, the IMF viewed the 1989–1991 prospects of the country's economic performance with caution. Although much acrimony surrounded this aspect of the November–December 1989 negotiations, both parties agreed on the need to continue Philippine efforts to liberalize the trading system. Despite the animosity that surrounded the talks, both sides agreed on terms of a Memorandum of Economic Policy in early 1989. The 1989 Import Observation stands out from the other import observations in one significant respect; the GOP later submitted an addendum to the original MEP. As a result of this addition, the observation can be broken down into two distinct phases. The first stage (late 1988–early 1989) focused on crafting the initial MEP agreement, where import reforms received a broad consensus from both the IMF and the GOP. On domestic Level IIB, public veto actors had a hand in negotiations, but little social veto actor or support actor involvement transpired. This is referred to as the 1989a Import Observation. The second stage (end of 1989 to late 1990) revolved around the MEP addendum. Under this stage, the GOP decided to accelerate the pace of import reforms and it pushed through an ambitious new tariff reform program, a program that received heavy cannon fire from social veto actors. This is referred to as the 1989b Import Observation.

The 1989a and 1989b Import Observations also differed significantly from the previous two observations in two other respects. First, the institutional decision-making structure was more diffuse, a condition that allowed more players into the process. Under the 1984 and 1986 Import Observations, the decision-making process remained exclusively within the domain of the executive branch. In contrast, the restoration of an effective and at times adversarial legislature played a more debilitating role in the policy process as the preferences of lawmakers had to be taken under consideration. While this institutional structure was existent under the 1989a and 1989b observations, it failed to play a significant role in the IMF-GOP negotiation process until the 1991 and 1994 observations. The August 1987 military coup attempt dramatically changed the shape of the cabinet policy process. First, it led to the ouster of Ongpin, and more importantly to the process, to the resignation of Arroyo, whose departure paved the way for pro-business oriented proposals to reach Aquino. Second, it also changed the route of the policy process. Rather than leave proposals to the discretion of the executive secretary, proposals were now routed through the cabinet's security cluster before they were permitted to be presented to the full cabinet. Departments that failed to abide by this procedure incurred the wrath of Aquino and were required to remit the proposal and receive approval from the security cluster.[1] This procedure had minimal impact on the deliberation of import reform, but played a critical role in the formulation of petroleum pricing policy.

INTERNATIONAL POLITICAL ECONOMY OF THE 1989A IMPORT OBSERVATION

In the first phase of the 1989 Import Observation, holdover Central Bank Governor Fernandez and Finance Minister Vincente Jayme agreed with the IMF on the need for continued structural adjustment. Both sides already shared similar preferences. In comparison to the state of economic despair that existed under the later Marcos years, the Philippine economy made significant gains during the first three years under Aquino. These accomplishments, however, were achieved through transitory means rather than through substantial and sustainable measures. Imports increased, but it did little to inject added value to its exports nor were the backward and forward linkages strongly forged. The economy grew, but it did little to help the plight of those most in need. Indeed, more work needed to be done. According to a World Bank report, despite the measured success achieved, the Philippines continued to exhibit glaring institutional and structural flaws, some of which have plagued the system for decades. Given these problems, the GOP needed to focus its efforts to "increase the overall efficiency of the economy, in agriculture, industry and the public sectors, while ensuring that the benefits from growth are distributed more equitably with the population and between the regions." In general, the GOP had to reform incentives and distortions away from protected industries and cities and toward agriculture and labor intensive export industries. To address these issues, the GOP needed to reform numerous areas: the exchange rate and trading structures, investment and savings incentives, infrastructure and energy problems, land reform and public investment, among others.[2] The World Bank report was just one problem that concerned the GOP.

Clearly, the more pressing immediate concern involved the GOP fiscal shortfall. To resolve this issue, the GOP attempted to obtain access to addition sources of funding, primarily through external sources. The United States was one source, as it was in the midst of negotiations with the GOP over the lease value of the American military installations. A second source was the IMF, as GOP qualitative performances under the 1986 agreement availed it to tap the IMF Extended Fund Facilities (EFF), a growth oriented program, not a stabilization one. Finally, official foreign creditors initiated the Philippine Assistance Program (PAP), formerly referred to as the Multilateral Assistance Initiative (MAI). Although the PAP promised to provide more lucrative funding, access to its coffers was conditional on receiving an IMF "seal of good housekeeping."

On the issue of import reform, IMF and GOP preferences converged before the proceedings of negotiations so an external-internal external-internal external-internal pattern was present. The GOP presented its structural reform program to the IMF. Unlike the fiscal and monetary reform targets, structural adjustment talks proceeded with relatively little controversy, especially in the area of import liberalization, an area that both parties wanted continued. According to a Central Bank official, the GOP was bent on continuing the program, whether the IMF favored the program or not.[3] The IMF viewed this intent with much optimism, but it also cautioned about the projected results that may occur from the move. In its Updated Philippine Development Plan, the authorities assumed that import and export targets would reach annual growth rates of 12.5 percent and 13.5 percent, respectively.[4] Given the slow pace of the global economy, the IMF found this appraisal to be extremely optimistic. According to a GOP official, "The IMF found these targets unrealistic, and saw the need to bring more realism into the targets."[5] Given this insistence, the GOP reassessed the country's import and export targets to 9 percent and 10 percent, respectively.

Like the 1986 negotiations, the Department of Trade and Industry (DTI, formerly the MTI) made its presence felt during the talks. Although the GOP reiterated its commitment toward swift import liberalization, DTI insisted that particular industries be given sufficient time to adjust their efficiency levels to prepare itself for competition, without losing the flexibility required for bilateral or multilateral trade negotiations.[6] The Uruguay Rounds and ASEAN Free Trade Area (AFTA) negotiations were underway.

The GOP insistence was accepted by the IMF, which allowed it to continue its tariffication practices of setting up tariff barriers in exchange for reducing nontariff barriers to trade. In addition, the DTI insisted on the establishment of a mechanism that would provide "an effective and efficient administrative machinery that can prevent mis-declaration and ensure customs duty collection." In this regard, importers would be required to submit import documents to the DTI so that it could detect possible mis-declarations in quantity, value, or description. The DTI, in turn, would forward its findings to the Bureau of Customs for investigation.

In June 1989, the IMF Executive Board approved the Philippine economic package. Growth rates under the EFF agreement were targeted for 6.5 percent of GNP until 1992, as insisted by NEDA head Solita Monsod. In exchange for the higher growth targets, the GOP agreed to tighten significantly its fiscal and monetary policy, and continue its structural adjustment efforts, in particular, import liberalization. The MEP outlined the dates of the planned deregulation phases. Before June 30, 1989, the GOP promised to release a minimum of ninety items from List B, where products were given consideration for liberalization; however, Congress would be able to introduce tariffs on these items by the end of September 1989. The second and third significant reform measures would be taken up no later than December 31, 1989, and by June 30, 1990, respectively. Under each round, at least forty-five additional items were to be selected from List B, for which tariffs would also replace the protection measures by the end of the economic program, mid-1992. Finally, for all List B items not covered by under the previous release phases, tariffs would replace quantitative restrictions no later than the end of 1994.[7]

DOMESTIC POLITICAL ECONOMY OF THE 1989A IMPORT OBSERVATION

With continuation of the ILP (reduction of import quotas), long protected industries sought safety mechanisms to ease the cost of adjustment. In early 1988, Congress passed a measure to increase tariff rates on various products from 0 to 50 percent to a higher floor of 5 percent and a higher ceiling of 75 percent. These adjustments were viewed as necessary to protect industries that had suffered from the effects of the ILP. Ultimately, Congress failed to satisfy the concerns of all domestic constituents. As one senator stated: "We tried to make it as fair and reasonable as possible. It was a very difficult task because when we favor one group, another group complains."[8] Aquino signed the legislation, but vetoed the 75 percent tariff ceiling given to sixty products.[9]

Even with the passed tariff legislation, the tariff rates continued to remain flexible. Sectors subject to import liberalization were allowed to petition the Tariff Commission, a standing body of DTI, to increase automatically their tariff rates through a demonstration of potential injury. In such instances, the Tariff Commission would consider numerous variables: the state of readiness of the industry to competition and external factors, infrastructure, and interest rates.[10] DTI also called for higher tariff rates as it continued to push for a slower pace of trade liberalization, and promoted the use of a 0 to 100 percent tariff system. Just as tariffs were subject to increases, they were also subject to decreases. During 1988, the Tariff Commission received numerous requests from several industries seeking a reduction on import tariffs of approximately 310 commodities ranging from consumer items to industrial products such as fuel and palm oils, packaging material, and soybean. Although the GOP responded favorably to these requests, it also emphasized that these reductions were temporary.[11]

ASSESSMENT OF THE CAUSAL MECHANISMS

The CNs acquired the deference consent of DTI and the Tariff Commission, addressing concerns of domestic constituents. Much like the 1986 Import Observation, this process takes time to process as multiple veto players attend numerous outlets to voice their concerns. Without question, the need to strengthen legitimacy requires that the GOP proceed through the process methodologically and with significant input from groups learning to operate a more democratic political process, especially one that has opened up the system with a new legislative branch. In contrast, the external-internal tactic has limited capacity to explain the slow, incremental steps taken by the GOP—using tariffication instruments to provide economic and political cover for itself. Given the international economic climate regarding the Uruguay Round and AFTA, it is logical as a first option economic choice to decrease import restrictions to set up the country for negotiations. The GOP, however, went with the second option political choice, provide protective tariff cover first, then bring down rates. Again, the political logic is justified under the external-internal tactic.

Is there an alternative explanation for the effect? The outcome was defined as "qualified success," where reform was staged incrementally. The technocratic alliance can explain the link as capturing the DTI principal (Jose Concepcion) and getting him to accept the virtues of free trade. While agreeing to the notion of the Washington Consensus, DTI also had to balance these ideas with its domestic principals, where reform without safety nets would lead to devastation.

1989B IMPORT OBSERVATION

In contrast to the 1989a Import Observation, the second Import Observation was conducted under completely different circumstances. Wary of signs that the economy was heating up, the GOP submitted an addendum to the MEP in late 1989. The addendum pointed out the significant progress made by the GOP with regard to imports, and as a result, the government asserted its continued commitment to such reforms. More importantly, the addendum called for tightening of fiscal and monetary instruments to combat inflationary pressures that became manifested in the economy (see chapter 4 for more details). One of the focal points of fiscal tightening was petroleum price hike, a measure that (in)directly led to the 1989 military coup attempt. The GOP's legitimacy, not to mention its survival, was on the line.

International Political Economy of the 1989b Import Observation

The December 1989 coup had visibly shaken the Aquino administration, and it turned toward a road designed to strengthen its vulnerable political base through the implementation of a market-oriented restructuring program. In February 1990, Aquino administration cabinet members held a secret meeting where they painfully acknowledged the perpetuation of poverty in the metropolitan and rural areas of the country, the structural impediments that handcuffed economic growth, and the inability of authorities to address adequately these issues, among numerous others.[12] In this soul-searching meeting, tempers flared and the Aquino cabinet agreed on the need to speed up growth measures through greater reliance on market-forces.[13] The New Economics Program (NEP), as the plan was commonly referred to, attempted to shake not only the country by instituting wide-ranging reforms, but also to break out of the traditional government practices used to resolve short-term problems. As stated by new Department of Finance Minister Jesus Estanislao, architect of the reform package:

> In the past, we have been over-relying on monetary tools. When you have a problem, you tighten up the market. For the past 20-25 years, we have been trying to solve our problems using this solution. What we want to do is play (the new) economics on the Central Bank and the [International] Monetary Fund . . . an economy run by interest rates cannot continue. Aside from using monetary tools of the monetary authorities, we want to expand the number of tools.[14]

Among the tools suggested by Estanislao was the use of a "flexible fiscal policy," along with the implementation of full trade liberalization. Without a doubt, the NEP had its supporters. The IMF, in particular, jumped on board in support of the program. According to Estanislao, "The IMF interposes no objection to the Philippine economic reform package. It took notes of the determination of the Republic of the Philippines to address its long standing structural problem."[15] Indeed, the IMF was pushed out of the spotlight early in the initial development stages of the program, as Estanislao explained: "The MEP has a lot of IMF (influence) in it. This (New Economics) is a Philippine program. We're formulating it on our own."[16] In this context, the GOP secured Level IIB actors first, then its Level I counterparts. The pattern resembled an inside-outside pattern. While the IMF agreed on board with the program, it expressed misgivings about the ability of the GOP to implement the plan as outlined.[17] Indeed, Level IIB actors were not persuaded. To make the program easier for them to swallow, Estanislao went to great lengths to present the program as a "Philippine Program," and present it from out of the IMF shadow. This case was particularly true when it came to tariff reform measures.[18]

Tariff liberalization became the focal point of the NEP. In general terms, the NEP package pledged to "fundamentally reform the tariff, foreign trade and investment regimes, and introduce measures to strengthen external competitiveness and to make the economy more open to increase export growth and inflows of foreign investments." Regarding trade, the proposal promised to rationalize further and simplify the country's tariff structure with fewer and lower tariff rates.[19] Later, the GOP attempted to reform dramatically the country's tariff system, but domestic opposition gained momentum and managed to force Aquino to declare the reform in abeyance. The tariff reform system of the NEP was officially abandoned in October 1990, under the weight of the 1990–1991 economic crisis that engulfed the Philippines.

Domestic Political Economy of the 1989b Import Observation

Since 1986, the GOP maintained ongoing consultations with veto actors on trade liberalization. These discussions centered on the scope of the ILP and the proper mechanisms that would protect domestic industries that resulted in the use of tariffication, of tariff protection mechanisms that replaced quantitative import measures. For the most part, this mechanism allowed the ILP process to continue relatively unabated, although criticism continued to sound from veto actors over the pace and direction of ILP. Under the 1989b Import Observation, the mainstay ingredients for minimizing distributive conflicts, the instruments of consultation and side-payments were used sparingly. Furthermore, the negotiation process followed an internal-external-internal external-internal external-internal pattern, where the latter would be a negotiation process with Level IIB veto actors. The results proved disastrous.

In 1989, the IMF-GOP MEP program began to slip away from the grasp of the Philippine authorities. Economic activity dipped to 5.6 percent of real GNP in contrast to 6.7 percent in the previous year. This disappointing development translated into other unfavorable trends. Import levels outpaced export levels by $2.5 billion. The money supply jumped from $59 billion in 1988 to $78.5 billion in 1989. Inflation crept up to over 10 percent. To a large extent, the GOP had been a victim of self-infliction as it heightened domestic demand through expansionary fiscal policies, private sector minimum wage increases, and price support for oil consumption. As a demonstration of its resolve to contain these debilitating effects, the GOP submitted to the IMF an MEP addendum that reaffirmed its commitment toward tighter fiscal and monetary restraint.[20] According to an official account, the promises made in the addendum put "us on a very tight leash."[21] Concerning trade, the GOP pledged to continue lifting import restraints on the remaining items to be liberalized. The GOP had accomplished much, but much more work was left. Between April 1986 and September 1989, the GOP liberalized 88 percent of its imports and intended to liberalize 370 remaining items before the end of 1992.[22] The efforts of the GOP to stabilize the unwieldy economy had devastating results. The GOP found the most debilitating response to be from a rebel military force that initiated another military coup attempt, the most devastating among the numerous attempts incurred under the Aquino administration.

The December 1989 military coup attempt prompted the implementation of emergence trade liberalization measures in early 1990. The transportation and food sectors were left with shortages of the highest magnitude. The GOP addressed these shortages by instituting emergency measures that temporarily reduced tariffs in these areas.[23] Meanwhile, the GOP also took this opportunity to reduce tariff rates on imports that would enhance the efficiency of the manufacturing, energy, textile, and banking industries. Although the GOP addressed the needs of these sectors, it continued virtually to ignore the import needs of agriculture sector, an area where large distortions persisted. During the tariff reduction design, the Department of Agriculture submitted an extensive list 224 of goods for tariff reduction consideration; however, its suggestions were largely ignored to ensure protection against certain domestically produced agricultural machinery. The agro-machine traders had accumulated approximately six months of inventory; as a result, imports would have affected their cash-flow positions and certainly aggravated the current glut of agro-machineries in the market. In fact, prices of agricultural equipment had been dipping due to excessive supply and not tariff reduction.[24]

The repercussions of these emergency measures were soon manifested. Tariff liberalization in these areas provided substantial relief for these ailing sectors, but it also cut into GOP coffers, as statistics for the first quarter of 1990 revealed a dramatic shortfall in government revenues. In conjunction with unexpected government expenditures derived from residuals of the 1989 coup attempt, lost revenues from the tariff reduction exacerbated the GOP fiscal position, a shortfall that the Central Bank was forced to cover by increasing the money supply. Congress, however, refused to assist the GOP by increasing "sin" taxes.[25] As an alternative, Estanislao proposed the implementation of a 5 percent across-the-board levy for some 3,000 imported items, a measure aimed to generate approximately P5 billion in revenues. The Aquino cabinet postponed its decision on this measure as it would trigger commodity price increases and could be used as a rallying point for "rightist" elements against the government.[26] Without a doubt, the 1989 military coup attempt had left a considerable mark on the Aquino cabinet.

In February 1990, Philippine decision-makers were at a cross-road. In a secret meeting, GOP sequestered cabinet members discussed possible solutions toward restoring life in a stagnant economy. Estanislao, who had another ally in Cayetano Paderanga Jr. who took over Estanislao's vacated position as chief of NEDA, argued for an outward oriented economy, which would include, as a critical component, tariff reform. Estanislao clashed with Concepcion on this issue without a resolution. Other cabinet meetings were held and the two continued their sparing. Aquino did not attend these meetings as she distained discord.[27] This process had its virtues. Toward the end of the administration, the decision-making process became more cooperative. As Paderanga recalled: "If anything, under the Aquino administration, we learned how to govern. The government grew from the Marcos [decision-making] style, moved to the Aquino style of the early years and finally, we [the departments] learned how to cooperate. Dissenting departments began to compromise on many issues and reach an agreement before we submitted a proposal to the full cabinet."[28]

While the cabinet had failed to reach a consensus on a general economic orientation, plans were developed by NEDA to reform the tariff structure, a critical component of an outward looking economy. NEDA, along with its agency the Tariff Commission, designed a system of tariff rates that focused on simplification of the tariff rates and the elasticity of supply products in the economy. The program was designed to phase-down import tariffs within a three-year period. The tariff reform was essentially the brainchild of NEDA, but a dynamic spokesperson was sought to facilitate and overcome domestic, public, and private opposition to the proposal; hence, the program was given to Estanislao. As the point-man for the reform program, Estanislao held continuous negotiations with other government agencies, most notably, DTI. Although the original NEDA proposal called for a three-year tariff phase-down period, Estanislao, with the support of newly appointed Central Bank

Governor Jose Cuisia, pushed for a simplified three-tier tariff system with a bold one-year phase-down period.[29] In contrast, DTI Secretary Concepcion wanted a five-tier tariff system with higher rates to protect local industries and a phase-down period of five years.[30] In exchange for its consent, DTI, especially its investment arm the BOI, pushed for the reduction of tariffs to 10 percent on agricultural and plant equipment and spare parts. Furthermore, BOI insisted on the continuation of a limited set of incentives for export-oriented industries, pioneering major industries and selected small and medium industries. Estanislao agreed to these demands.[31]

By June, the cabinet still failed to reach a consensus on the economy. Aquino forced Estanislao and Concepcion to hash out their differences before an arbiter, Science and Technology Secretary Ceferino Follosco. Estanislao garnered the support of Follosco, and subsequently Aquino accepted the export-oriented economy proposal. This new program was termed New Economics Program (NEP) with Estanislao as its flag bearer. The NEP aimed to drive the economy toward higher growth rates at levels comparable to the Philippine's highly successful Southeast Asian neighbors. The premise of this program focused on the need to break away from "traditional" economics, away from the protectionist measures that favored business, and shift domestic incentives toward the Filipino consumer.[32] Among the most discussed and highly controversial topics was the acceleration of trade liberalization measures, namely tariff reform.

Despite the optimism that surrounded the NEP, several groups chose to approach the package with much skepticism. Academicians from the University of the Philippines School of Economics seriously questioned the adequacy and scope of the NEP to achieve its lofty goals.[33] The manufacturing groups threw its support for the program under the condition that the authorities succeed in the control of smuggling.[34] Large, medium, and small scale industries were less inclined to support the package as they were destined to lose their tax exemptions and investment credits under the NEP.[35]

In July 1990, during a congressional recess, Aquino signed Executive Order 413. The measure would go into effect thirty days after it was published in two major newspapers. EO 413 generated significant domestic controversy regarding the lack of industrial consultation. The Makati Business Club criticized the administration for not actively pursuing consultations with relevant industries.[36] These accusations were inaccurate, as the GOP had conducted numerous consultations with the private industry during the design stages, discussions that were given considerable media attention. Such input, however, had been labeled as "transitory and meaningless" where opportunities to express the private sector's viewpoints were limited to five minutes.[37] Furthermore, consultations were mired in obfuscation as hearings for the NEP were also used as discussions for the Uruguay Round tariff reduction deliberations.[38] Perhaps the most surprising aspect of the consultations regarded the duration of the process. The private sector expected consultations to continue and eventually include congressional participation; but the GOP suddenly halted the process and unveiled EO 413.[39]

Like the NEP, EO 413 carried a significantly broad scope of detractors. The Buy Philippine Made Movement Inc., headed by industrialist Raul Concepcion, brother of DTI Secretary Jose Concepcion, asserted that it would have no objections to the NEP tariff reforms if the GOP can prove "beyond a reasonable doubt that smuggling [could] be contained."[40] PCCI also criticized the tariff reform package by pointing out that reforms would put many domestic industries at a disadvantage in comparison to their foreign competitors given the handicaps that domestic firms faced, such as high interest rates, poor infrastructure, and the high cost of power, among others.[41] Finally, the sugar interest argued that the classification of particular raw materials would injure the industry and lead to the displacement of approximately 4 million workers.[42] For the most part, these arguments rang true. The level of imports had increased in 1990 from the previous year. Employment in critical sectors, such as textiles, footwear and wearing apparel, furniture and fixtures and machinery, had dropped. Finally, the costs of borrowing money also increased as interest rates crept up from 7.9 percent in January 1990 to 11 percent in July 1990.[43]

The GOP extended side-payments as part of the effort to win over both public and private domestic supports for EO 413. In May 1990, the Department of Finance (DOF) and the BOI agreed to continue the special fiscal incentives for a limited set of investment priorities that focused on export industries, pioneering major industries with significant technology transfer, and selected small and medium scale industries.[44] In 1990, the Incentives Investment Program (IPP), run by the BOI, distributed a total of P103 billion in incentives to various sectors as manufacturing received over P41.9 billion, energy-related projects P23 billion, tourism-oriented projects Pl2.8 billion, and infrastructure/industrial service facilities with P7.2 billion, as the major recipients of incentives. These side-payments, however, proved insufficient to offset the cost of EO 413.

Private interest made every attempt to arrest the tariff initiative. On one side, the private sector sought congressional assistance. During the design stages of the program, congressional input was conspicuously absent, an intentional oversight on the part of the GOP. This exclusion had three purposes: First, it shielded Congress from domestic pressure. Second, it maintained the traditional tacit understanding between the branches that the executive should deal with tariffs.[45] According to the 1987 Philippine Constitution, the legislature was given responsibility with determining tariff rates. However, the executive and the legislative branches realized that the former would be better suited to this task. Finally, it saved Congress from public humiliation. According to a senior NEDA official, "It would seem like a slap in the face to Congress if we drew up the plan and presented it [to them]."[46] After the GOP unveiled EO 413, small and medium scale industries pressured the legislature to enter the controversy.[47] With congressional assistance, the opposition aimed to forestall the publication of EO 413 in newspapers, a strategy that would also delay implementation of the measure.[48] In addition, the media provided the vehicle for the public dissent, as it published commentaries and interviews from economists and business leaders.[49] The second front went through the halls of the GOP itself. While the small and medium scale industries relied on formal procedures, the larger players rushed the GOP through informal channels. Criticism through family and personal avenues came up through the DTI and Malacañang Palace.[50] After the constant barrage of domestic pressure, in July 1990, Aquino issued a memorandum that directed the DOF, NEDA, and other concerned agencies to hold the implementation of EO 413 in abeyance. The executive order was later abandoned in October, when the economic crisis of 1990–1991 became more pronounced and after the Iraqi invasion of Kuwait.

This attempt at reform fell before it even started. The absent variable from this reform attempt was the specter of IFIs. Manuel Montes notes:

> What is notable about the 1990 tariff reform program and Secretary Estanislao's overall reform package is that there was a conscious, and largely successful, effort to design a "Philippine package," that is, a set of reforms independently designed and implemented by the country in its own interest. In particular, at the time it was proposed, the tariff reform program went beyond the items that were under discussion with the World Bank and the IMF.[51]

Epilogue to the 1989b Import Observation

In the 1989b Import Observation, the GOP found its chief battle to be under the second phase and against its domestic constituents. The first phase of the Observation had little elements of distributive conflict between the IMF and GOP, save for issues regarding growth rates and deficit levels. On the issue of imports, the preferences of the IMF and GOP converged. For the second phase of the Observation, negotiation battles were confined to the domestic level. Economic conditions, on both the macro and microlevel, had constrained opportunities for business actors, especially those in the non-tradable goods sector. These actors negotiated with the GOP through the consensual and diffused institutional framework, only to find doors previously opened to be locked shut. Furthermore, their access to compensation was severely limited. In effect, the institutional mechanisms that mitigated the distributive conflict failed to provide adequate protection. However, in the mid-1990s, EO 413 became mute as other developments took center stage; yet, the GOP had learned valuable lessons from this observation.

ASSESSMENT OF THE CAUSAL MECHANISMS

The 1989b Import Observation is distinct from the other observations. In the other cases, the IMF was front and center in the observations. In this case, IMF was in the wings, well away from the audience. Its presence was still felt as the economy started heating up and the GOP needed to hit its targets to continue receiving IMF and foreign financing. The external-internal tactic at Level IIA was implicit, as both Estanislao and Concepcion understood what was at stake with the creditors. This tactic was also implicit at Level IIB as the GOP shined away the light from the IMF but it continued to shine it on the Uruguay Round negotiations, which bled with the EO 413 proposals. Furthermore, the benefits of reform were viewed as a given and the costs were emphasized, primarily by the social veto actors. This explanation is sound.

The deference tactic provides a better explanation on the outcome. Clearly, Estanislao needed to secure the endorsement of Concepcion, providing another Estanislao -> Concepcion -> Aquino arrow diagram process. The problem rested in the link between Estanislao and Concepcion, where the two continued to miss on points. Aquino subsequently intervened when she anointed Follosco as the arbiter. Having both principals agree on the program, the hardest part was to sell it to Level IIB players. There was no deference player as Aquino and the Congress stayed away from the fray. Instead, Estanislao and the DTI simultaneously pursued the support of numerous industrial veto players, dangling the promise of tariffcaion in an otherwise fast paced consultation process. The GOP failed to win over critics such as Buy Philippine Made Movement Inc., PCCI, the Makati Business Club, and sugar interests. In retrospect, getting these associations, or at least a few of them, on board might have convinced others to also support EO 413. Opposition to EO 413 was fierce. The veto players used multiple formal and informal channels to prevent the publication of the order and to rescind it. Groups attempted to sway the media, but to no avail. Finally, opposition groups used the media to voice their opposition and apply pressure to relevant bureaucracies, the Congress, and even to Malacañang itself. While the deference tactic was used at a minimum, if at all, at the beginning of the process, it was used extensively at the back end.

The GOP applied new rules, norms, and processes to secure EO 413, trying to shift the equilibrium. The process as in other circumstances would rely on the success of previous efforts to craft an explicit or implicit ratification. The highlight in this episode was the removal of the IMF from the process.

In "traditional" IMF-GOP negotiations, the media highlights the IMF specter and the conditionalities, a shadow that was removed and in its stead was an attempt to demonstrate its nationalist tendencies by showcasing the program as its own. This disconnect may have been the problem because there were little benefits with which the GOP could connect and bring out supporters of reform. This "sovereignty moment" could also be a response to the 1989 December coup, with which Aquino made appeals for nationalist unity. Perhaps, this same call was assumed to be able to produce a similar effect. In addition, the GOP rolled out EO 413 with due speed, and as some argued, with haste. It consulted with multiple veto players, but it did so with a short duration, hoping to avoid extensive criticism. The agent attempted to shift the preferences of its principals, but in doing so, did the opposite, it reinforced it. The principals pulled in the slack on the reins.

NOTES

1. Interview with Senior National Economic and Development Authority official, anonymity requested, June 27, 1995.
2. The report was issued in October 1988. The concluding portion of the report was reproduced by the *Manila Chronicle* 1988b, 1988c, 1988d, 1988e, and 1988f.
3. *Manila Chronicle* 1988a.
4. Republic of the Philippines 1990a.
5. Interview with Senior Department of Trade and Industry official, anonymity requested, August 7, 1995.
6. Ibid.
7. Republic of the Philippines, Memorandum of Economic Policy, as published in the *Manila Bulletin* 1989b and 1989c.
8. *Manila Chronicle* 1987.
9. Republic of the Philippines 1988b.
10. Interview with Senior Department of Trade and Industry official, anonymity requested, August 7, 1995.
11. *Manila Bulletin* 1989f.
12. *Manila Chronicle* 1990c.
13. Interview with Senior National Economic and Development Authority official, anonymity requested, June 27, 1995.
14. *Manila Chronicle* 1990g.
15. Ibid., 1990k.
16. Ibid.,1990d.
17. Interview with Senior National Economic and Development Authority official, anonymity requested, June 27, 1995.
18. Interview with Secretary of Finance Jesus Estanislao, April 20, 1995.
19. Macapagal-Arroyo, 85.
20. Republic of the Philippines. Addendum to the Letter of Intent, as published in the *Manila Chronicle* 1990a and 1990b.

21. Interview with Senior National Economic and Development Authority official, anonymity requested, July 28, 1995.
22. Republic of the Philippines. Addendum to the Letter of Intent. As published in the *Manila Chronicle* 1990a and 1990b.
23. Republic of the Philippines 1990.
24. *Manila Bulletin* 1990b.
25. Ibid., 1990c.
26. Ibid.
27. Lallana, 39.
28. Interview with Senior National Economic and Development Authority official, anonymity requested, June 27, 1995.
29. Interview with Senior National Economic and Development Authority official, anonymity requested, June 27, 1995.
30. *Manila Bulletin* 1990g.
31. *Manila Chronicle* 1990i.
32. Ibid., 1990g, 1990h, and 1990j.
33. Alonzo, et al.
34. *Manila Chronicle* 1990g.
35. Ibid., 1990h.
36. Makati Business Club 1990.
37. Interview with Senior Chemical Industries Association of the Philippines official, anonymity requested, August 9 and 10, 1995.
38. Of the twenty-two products recommended for tariff binding, seven were tropical products and most were not currently produced locally.
39. Interview with Senior National Economic and Development Authority official, anonymity requested, June 27, 1995.
40. *Manila Bulletin* 1990f.
41. *Philippine Star* 1991.
42. *Manila Bulletin* 1991c.
43. Central Bank of the Philippines 1992.
44. *Manila Bulletin* 1990d.
45. Interview with Senior Department of Trade and Industry official, anonymity requested, August 7, 1985.
46. Interview with Senior National Economic and Development Authority official, anonymity requested, June 27, 1995.
47. Small and medium scale sectors provided the most vocal dissent, but this was ironic because the reform package was not meant to hurt the small and medium scale industry and effective, because large domestic oligopolies bad been identified as having the most to lose from liberalization, and the only objectors to the program; Montes 1992, 19.
48. *Manila Bulletin* 1990f.
49. Montes 1992, 19.
50. Ibid., 18.
51. Manuel Montes as cited by Lallana, 40.

Chapter 10

1991 GOP-IMF Import Observation

Trade liberalization under the 1989b GOP-IMF stabilization program came to a sudden halt. In mid-1990, the GOP unveiled EO 413, an attempt to make drastic changes to tariff reforms, a move staunchly supported by the IMF but opposed by defiant domestic industrialists. After the GOP attempted to reconcile its differences with domestic manufacturers, it found itself caught in the grip of another economic crisis. Like the 1983 balance of payments crisis, the GOP attempted to meet the challenge, in part, through the introduction of an import levy. As a result of the implementation of this mechanism, as well as through tighter monetary and fiscal instruments and foreign exchange devaluation, the Philippine economy slowed to a trickle. In mid-1991, the GOP attempted to initiate another tariff reform system, an effort that met with a larger measure of success.

Under this observation, the strategies used shift adjustment costs had mixed results. A banker and technocrat, Philippine negotiator Estanislao used his institutional capacity as the country's economic manager and chief negotiator to access critical IMF executive board members and solicit their support for his solution to the Philippine balance of payments crisis and the GOP budget deficit. In principle, Estanislao agreed with the need for tariff liberalization and expressed the intent of the GOP to forge through on reforms of this nature later. On the domestic side, Estanislao followed a three pronged strategy to receive veto actor endorsements. First, he managed to get Concepcion on board early, the "guy to get the guy." Second, he linked domestic negotiations with the GOP commitment to tariff reform, although in a more subtle manner. Finally, he recognized the need to compensate, perhaps excessively, veto actors for the costs of adjustment. Under the 1991 Import Observation, the Philippines continued to take steps toward tariff reform, just at a slower pace.

INTERNATIONAL POLITICAL ECONOMY OF THE 1991 IMPORT OBSERVATION

Under the 1991 IMF-GOP Import Observation, the GOP made modest strides in tariff reform. The path, however, was not a straight one. Developments, such as numerous natural disasters, the 1990–1991 Persian Gulf crisis, the US military installation leases, and an international financial consortium, combined to force the GOP to take a slight detour. It initiated an across the board import levy as a solution toward resolving the country's macroeconomic disequilibrium. Initially, the IMF had strong reservations about this solution, but these misgivings were swept aside when Estanislao managed to generate significant support from international Level IIB actors, a strategy borrowed from the pages of the IMF.

In 1990, the GOP was overwhelmed with numerous misfortunes. The 1989 coup attempt and various natural disasters forced the GOP to expand its fiscal and monetary operations, pushing these levels near the point of surpassing the targets promulgated in the 1989 MEP addendum. The Persian Gulf War proved to be the final catalyst that drove the GOP over the targets. By the end of September 1990, the rising international cost of crude oil and oil products, rising interest rates, a burgeoning fiscal deficit, and a fragile political environment forced the GOP into dire financial straits. Given the present course of the economic situation and with little hope for recovery in sight, another balance of payments crisis was on the horizon.

To relieve GOP from its financial pressures, the BAP intervened. It consulted with Central Bank officials, and afterward, it reinstituted an import control program similar to the mechanism first used in 1983. Under the system, the BAP placed barriers to trade by refusing to issue letters of credit for imports that failed to meet its criteria. By taking steps to restrict imports, the banking community relieved the GOP from the predicament of having to break its IMF and World Bank trade liberalization commitments.[1]

By the third quarter of 1990, the GOP made a run to restore macroeconomic stability. It allowed the peso to devalue precipitously from P25.00 to P28 to the $1. It increased the price of petroleum and petroleum products and instituted additional taxes. Finally, it stepped up open market operations to rein in an expanding money supply. Yet, by the fourth quarter of 1990, economic conditions spiraled out of control, and the GOP had failed to reach the targets set out under the 1989 MEP addendum. Given the GOP inability to restore fiscal and monetary stability and regain equilibrium to the current account, the IMF had abrogated the 1989 MEP, and began negotiations with the GOP to draft a new economic program.

Both parties sought a quick resolution to the country's economic plight. The difference between the two, however, lied in the means to achieve this goal. Under the context of the situation, the IMF viewed trade liberalization as nothing more than "part of the decorations on a Christmas tree."[2] Given the gravity of the situation, Estanislao proposed that the Philippines reverse the progress made on trade liberalization and institute a temporary 5 percent across the board import levy to raise revenue to curtail imports and reduce domestic demand. Estanislao suggested this course of action after Congress had again failed to pass several critical tax generating measures.[3] Reluctantly, the IMF agreed to this measure as the only alternative remaining to narrow the government's revenue shortfall. By the end of November, the IMF tentatively agreed to terms on an economic program provided that the GOP tap additional revenue sources.

The GOP effort to generate revenues resulted in political disaster. By initiating the import levy, the GOP was compelled to pass along the increased cost of imported petroleum products and oil to consumers as failure to adjust prices only added to the government's fiscal troubles. The GOP again adjusted domestic petroleum prices, but dissent from members of the cabinet and various interest groups forced the GOP to reduce these adjustments partially.

Unable to generate the necessary revenue, the GOP looked toward other revenue measures and Estanislao proposed that the GOP increase the import levy by an additional 4 percent. Originally the IMF resisted this measure. It acquiesced when Estanislao mentioned that unless the IMF agree to this measure and authorize a standby agreement. The GOP would be ineligible to secure PAP funds, and a balance of payments crisis would be inevitable.[4] Furthermore, Estanislao had secured the support of Japan, which agreed to pressure the IMF to understand the plight of the Philippines.[5] Faced with the prospect of constituency pressure and an apparently severe Philippine balance of payments crisis on the horizon, the IMF agreed to this measure as long as the levy did not distort incentives against various products.[6] In other words, as long as the levy was imposed across the board, it was acceptable.

The GOP and IMF agreed to terms on an economic program in February 1991. Under the context of the NEP, the GOP used fiscal instruments as the centerpiece toward the restoration of macroeconomic stability. The GOP promised to bring down the overall fiscal deficit through measures to raise revenue, a reduction in expenditure appropriations, and a gradual elimination of the oil subsidy deficit. The import levy remained a critical mechanism toward a tighter fiscal position, a measure that lasted until July 1991 and was replaced, for the most part, with tariff reforms as a mean to strengthen the country's economic growth. The new tariff program reduced rates at a slower pace than the initial tariff schedule for most products with rates of 0 to 50 percent to 10 to 30 percent, and lowered rates to around 10 percent for raw materials, around 20 percent for intermediate goods, and around 30 percent for luxury consumer goods, with a temporarily higher tariff rate as well as a 3 percent rate for basic raw materials. Initially, the GOP set tariffs at a significantly higher level and brought them down afterward. Simultaneously, the GOP also instituted the release of quantitative restrictions with the new tariff schedule. Exceptions to the phase-out of quantitative restrictions were maintained for those restrictions imposed for health and national security reasons, and for products covered under approved industrial rationalization programs.[7]

DOMESTIC POLITICAL ECONOMY OF THE 1991 IMPORT OBSERVATION

In contrast to the previous Import Observations, domestic negotiations between the GOP and its constituencies, both public and private, extended to unprecedented dimensions. Learning from the lessons of the 1989b Import Observation, the GOP returned to the decision-making process it relied on in the past, a consensual framework that allowed for a multitude of public and private veto actors to participate in shaping and framing the tariff system. This arrangement managed to contain the distributive conflict that normally resides in the discussion of an issue such as tariff reform. Typically silent in this type of negotiations, Congress took a larger role in the articulation and moderation of veto player interests. Private veto associations also had a large role in this process, in particular groups that had switched positions, from its preferences for import protection to tariff liberalization. The GOP also emphasized its international commitments to the GATT and its IMF commitments, which overlapped to a large extent. Instrumental to these domestic negotiations was the capacity of the GOP to compensate losers to import reform. Although it used numerous compensation mechanisms, the GOP heavily relied on tariffication, an instrument that fulfilled two objectives: the distribution of side-payments and the generation of revenues for the GOP.

A traditional concern, the government deficit problem shaped the nature of GOP responses in the 1991 Import Observation. In the third quarter of 1990, GOP efforts to address the concerns regarding EO 413 became moot as the Philippine government felt the tightening of its purse. The GOP was forced to revise the course of tariff reduction due to economically devastating international and economic events, developments that put considerable strains on the GOP's budget. Congress could have come to its aid with revenue generating legislation, but it was reluctant to move into the troubled spotlight and tax its principals. As a last resort, the GOP installed a temporary 5 percent across-the-board import levy, a move that increased the landed cost of crude, and as a consequence, forced the GOP to increase the domestic price of petroleum products. Days later, the GOP succumbed to political pressure and reduced the adjusted prices. This policy disaster again forced the hand of the GOP as it increased the import levy by another 4 percent.

The levy measure drew criticism from within the cabinet. NEDA provided tacit resistance to the implementation of the measure, as it would slow the pace of the economy, but it eventually threw in its hat with the plan.[8] Bitter criticism came from the powerful Department of Budget and Management (DBM, formerly the MBM) as it opposed the fashion in which the levy was introduced: as a temporary measure. As a senior Budget and Management official remarked, "What good does the levy do if you announce that it is only temporary? All the public sector has to do is wait several months until the levy is repealed and then import."[9] To be sure, some dissent was present regarding this course of action. Regardless, the degree of insulation, access to the COG, and the particular power afforded to the country's chief economic manager allowed Estanislao to push through the import levy. Throughout this course of events, Estanislao continuously reminded his detractors that the country would run into more daunting long-term economic dangers unless it could tap into PAP coffers, a source that would be closed to the GOP without an IMF program.[10] This strategy to link the two domains followed an external-internal path.

The import levy proposal also received mixed response from private sources. For instance, former NEDA Chief Solita Monsod publicly argued against the levy as it would place considerable burdens on the average consumer and the poor. As an alternative, she revived her calls to go into temporary arrears to foreign commercial creditors, among other prescriptions.[11] Unswayed by Monsod's appeals, business leaders tacitly supported Estanislao's plan. The crisis and a sense of nationalism, more than anything, motivated their position. As Estanislao put it: "They decided to bite the bullet given the nature and condition of the economic problem. It was a national emergency. They wanted to do their part to help."[12]

After the economic crisis waned, the GOP put plans forward to reinstate the tariff restructuring program. As outlined in the 1991 MEP, the GOP began its efforts to reduce the 9 percent across the board levy and simultaneously reinstitute the tariff restructuring and import liberalization program. During the June 1991 legislative recess, Aquino issued Executive Order (EO) 470, a mandate that restructured tariff rates.[13] Although both EO 413 and EO 470 were issued during legislative recesses, the latter was formulated under rules and processes that rested more on consensus and an expanded negotiation process.

Broad and deep based, the GOP tariff reform decision-making process encouraged the participation of numerous groups. Unlike the previous observation, in the 1991 Import Observation, the GOP expanded the number of agencies involved in the formulation of the tariff system. An interagency committee created and placed under the direct control of the NEDA board, the Tariff Commission became the focal point for tariff negotiations.[14] As an added voice for social actors, the Congress also took an active role in the negotiation process, often holding joint legislative-executive hearings with the Tariff Commission and the DTI. The diffused institutional arrangement allowed a myriad of intermediate actors to access the decision-making process and present their aspiration points on the issue.

Trade reform brought in a wide-range of social actors. For the most part, PCCI and the Makati Business Club led the charge for the non-tradable goods sector. Using arguments familiar to the GOP, these associations contended that economic conditions, infrastructure, and energy inefficiency, the Philippine experience with numerous brown outs, and smuggling problems would put domestic industries at a distinct disadvantage when asked to compete with foreign goods. Other industries, primarily those that produced raw materials, also presented similar arguments. The sugar industry asserted that GOP intentions to reduce the tariffs on sugar and saccharin would cause severe dislocation problems to the countryside. However, these industries had to share their political space. Cognizant of the direction that the international economy in general and Uruguay Round in particular were taking, the long-time protected industry also joined discussions, but in support of tariff reforms. Speaking for the trade industry, the Textile Mills Association of the Philippines (TMAP) had realized that it needed to use better quality imports if it were to compete on a global level and pushed for a favorable and simpler tariff structure. It strongly supported this target position that it went so far as to offer to subside local textile millers in exchange for lower tariff rates.[15] In the absence of tariff reforms, TMAP would not be able to maintain its status quo position for long. Linkage to the Uruguay as well as the IMF targets outlined in the 1991 MEP, contributed greatly to shifting the cost perceptions of veto players.

The GOP made it clear to link international and domestic factors. The move toward more a more neoliberal economic world was gaining momentum. The European Union, the Uruguay Round, the Association of Southeast Asian Nations (ASEAN) Free Trade Area (AFTA), and North American Free Trade Area (NAFTA) negotiations were proceeding at a seemingly unstoppable pace. If the Philippines were to benefit, not to mention survive, from such conclusions, then it would have to make its industries more competitive and ultimately more efficient. To move it this direction, the GOP used the foreseeable economic landscape as leverage against domestic veto players. According to Emmanuel C. Lallana, the GOP played the international card to nudge reforms as well as prevent policy reversals. Paderanga admits that the reformists used the Uruguay Round negotiations as a pretext to get trade reform going. They also employed the Brussels UR negotiations round to justify passing the new tariff code before Congress opens. Paderanga also used international commitments, like membership in AFTA, as a means to prevent regression on trade reforms. He describes their action as akin to an "individual using whatever firewood he can get to throw in the fire."[16]

International leverage pushed the Philippine economy toward a more neoliberal posture, but it also had to be aided with domestic side-payments. At the initial stages of the negotiations, the GOP suggested using the release of tax credits to extend subsides to export oriented industries. A DOF instrument, tax credits had a time lag, a delay that could have hampered the competitiveness of export oriented industries.[17] For these industries, tariff reduction would provide a quicker instrument. The problem remained with veto actor support, however. Tariffication proved to be an integral instrument, used by the Tariff Commission, to gain support from veto actors. In May 1991, the GOP approved the 1991 ILP, which included 95 areas of economic activity compared to 261 in the 1990 ILP, but of these 95 areas, 92 were carried over from the previous plan and 3 were new inclusions. Although EO 470 reduced tariff rates within five years, it initially raised rates before bringing them down to more manageable levels. In some instances, tariff rates that were originally set at 30 percent were raised to 50 percent under the program, and eventually reduced to 30 percent again. In addition, the GOP agreed to hold back the tariff cut on capital equipment for one year, exempt some tariffs from reform by keeping imports under particular industries to 50 percent, and secure approximately 5 percent more in revenues.[18]

With EO 470, the GOP established a gradual and longer tariff reduction package. Similar to its predecessor, EO 470 retained the four-tier system, where the fundamental differences hinged on the average nominal tariff rate, currently at 25 percent, duration of the phase down, and the tariff levels that commodities would be levied at the beginning of the program. EO 413 was designed to drop the nominal tariff rate to 19 percent within a three-year period. In contrast, EO 470 dropped the current average nominal tariff rate to 20 percent within a five-year period, and was designed to correct for long-existing distortions that existed between agriculture and manufacturing sectors by making greater tariff cuts in the latter sector.[19] Efficient resource allocation had its price as the GOP stood to lose approximately Pl6.55 billion in revenue over the five years.[20]

Examination of the restructured tariff rate promulgated under EO 470 highlights the industries that received strong consideration during the consultation rounds. Consumer goods, in particular footwear, clothing, and tiles, retained a 50 percent tariff level during the full five-year period. In contrast, tariffs for various kinds of finished garments and wood products remained at 50 percent at the beginning of the phase, but were due to tail down to 30 percent during the last year. During the first year of completion, tariffs for textiles were set at an average of 30 to 20 percent and remained at that level for the duration of the program.

EPILOGUE TO THE 1991 IMPORT OBSERVATION

In 1991, the GOP managed to restore an overall macroeconomic equilibrium. It turned its 1990 balance of payments troubles around from a deficit of $185 million to a 1991 surplus of $1.4 billion. This remarkable turnaround was aided by a marked improvement in the Philippine current account as the merchandise trade deficit had been reduced by $8 million through an increase in the value of exports by $653 million and a decrease in the value of imports by $155 million. This development was significantly aided by a drop in the value of the Philippine peso in relation to the American dollar, from an average of P24.30: $1 in 1990 to P27.48: $1 in 1991. The successful turnaround, however, derived primarily from increased non-merchandise trade by almost twice the level from 1990, $1.3 billion versus $611 million.[21] Similarly, the country's capital account also improved by $1.3 billion in 1991. Finally, increases in the level of long-term loans and short-term capital also helped to rectify the balance of payments problem.

For the most part, the GOP's ILP deeds matched its promises to the IMF. In early 1991, the GOP selectively engaged in a program that attempted to deal with the country's tariff problems through the deregulation of imports for various sectors. This trend continued throughout 1991 and culminated in the launching of EO 470. From April 1986 to November 1991, the GOP deregulated a total of 1,438 commodities. The total value of these items amounted to $2,162 million during the first eleven months of 1991 represented around one-fifth of total imports during the same 1990 period. Raw materials (54.4 percent) accounted for a majority of the imports, as consumer goods (31.5 percent) and capital goods (13.8 percent) constituted the remainder. By the end of December, the GOP retained 419 items that remained to be liberalized and 303 were scheduled to be liberalized in phases that did not last past end-1994, except for a few items that required a longer implementation period to allow affected domestic industries time to adjust. One hundred and sixteen items remained restricted due to security and health reasons.[22]

By mid-1991, the brunt of the economic crisis had waned. Domestic interest groups increasingly voiced their displeasure with cost of imports and their production levels, and called for dismantling of the 9 percent levy. Importers contended with a double hex, the devaluation of the peso and the levy, although a study later showed that the former impacted more heavily on imports than the latter.[23] The 9 percent import levy, initially scheduled to be phased out in June 1992, was reduced by 4 percent in August 1991 for all products except for crude oil.[24] To continue efforts to raise revenue, the GOP increased the import levy on petroleum products to P0.95 per liter on imported crude and Pl.00 per liter on imported oil products, a tax that was commonly referred to as the "Estanislao levy."[25] Interestingly, the levy adjustments were made in conjunction with the introduction of EO 470.

ASSESSMENT OF THE CAUSAL MECHANISMS

Deference stated early. After the Persian Gulf crisis, the imperative for foreign financing dramatically grew. The reduction of exports, increase in oil prices, peso devaluation, and reduction of OCW remittances exacerbated the macroeconomic deficit. Without a doubt, this sea of bad fortune, in addition to domestic self-inflicted wounds, pushed the GOP to seek remedies from its foreign creditors, starting with Japan, which was the leader of the PAP. With the deadline looming and without the IMF program in place, the GOP would face another debt crisis. The international financial shortfall was significant, but the IMF insisted that domestic revenues helped cover the expenditure-revenue gap. The Congress was the first group Estanislao approached. This request was rejected, forcing the GOP to pursue revenue generating measures to which only the executive had access. In addition to these measures, import reform played a critical piece to ensuring that industrialized exports and not finished imports resolve financial woes. This time, Estanislao linked NEDA and its Tariff Commission with DTI, getting a wider range of voices to promote import reform. These actors gave Estanislao a broader network and vision with which to work. NEDA broadly focuses on development, whereas DTI by itself has principals that emphasize industry and trade. At Level IIB, the move to Congress also served well. While it was reluctant to increase taxes under the electoral cycle, it enthusiastically participated in the negotiation of EO 470 to provide services to its principals, also in the context of the electoral cycle. This can be viewed as taking a "negative" function and turning it into a "positive" one. Deference served Estanislao well.

External-internal explains well the observation. By getting outside parties critical to the IMF-GOP negotiations, the CNs were able to apply significant potential costs of a nonagreement on the laps of the IMF. A missed PAP pledging session would make a worse situation worst. External players were also critical as everything was laid out in the context of the Uruguay Round and AFTA negotiations, compelling Level IIB opponents to EO 470 and an assured international economic landscape to accept the terms as well as the side-payments. For the proponents of trade, the benefits of the impending landscape was too much to pass up. They chose to line up now, express their preferences, and be out the inside now instead of being on the outside later.

The technocratic alliance has limits explaining this observation. The impetus for EO 470 rested on the domestic side of the equation, not the international. The two sides already agreed on import reform with the IMF urging caution and the GOP being ambitious. The IMF had some marginal leverage with the CNs as EO 413 was a template for the new program and DTI and NEDA were already onboard, to which Aquino already deferred. This condition yields two assessments. One, if we examine the alliance tactic from the beginning of the process where preference already matched, then we could conclude that the tactic is overdetermined. The IMF had a marginal effect. Two, if we examine the alliance tactic at the end of the process where IMF and GOP Level IIA preferences synced, then the tactic proved effective. At Level IIB, the evidence of an IMF technocratic snare is unclear and as a result, we must extrapolate from the evidence at hand. During the 1991 Petroleum Price Observation, the IMF sent staff members to convince domestic principals on the necessity and urgency of the stabilization and structural adjustment measures. We can presume that if IMF staffers attempted to persuade Level IIB veto actors on one issue, with a high level of probability, they also tried to do the same for import reform. Much like we derive two assessments from the tactic at Level IIA, we can derive two at Level IIB. One, the GOP, NEDA, the TC, and the legislature were already extensively involved in the negotiations process. If the IMF did have an effect on social veto players, there is a strong chance it was marginal. Again, the underlying issue is how much supporters of import reform already existed or were convinced before the IMF arrival. Two, the end result speaks for itself. EO470 was implicitly ratified and as a result, we can conclude that the tactic served as the mechanism that linked international with domestic politics. This assessment, however, may also be overdetermined as it marginalizes the existing structures and processes transpiring at Level IIB.

The 1989b and 1991 Import Observations are an excellent example of learning. In these examples, we have the same initial variable, CN Estanislao, steering both observations. Under EO 413, the process started with a narrow player portfolios, finance and trade and industry. In contrast, EO 470 stated out with broader player portfolios, finance and economic development and the Tariff Commission. NEDA's emphasis on economic development broadens the focus of negotiations with a wider set of preferences than with a narrower one. Furthermore, going to the Congress also assisted the process. It served as a consultation mechanism both for Level IIA and Level IIB actors, steering both their agents and listening to their principals. Just as importantly, it closed the "backdoor" channels. Congress could no longer be available to opponents of reform after the announcement of EO 470.

NOTES

1. *Manila Chronicle* 1990m.
2. Interview with Secretary of Finance Jesus Estanislao, April 20, 1995.

3. Republic of the Philippines 1990d.
4. Interview with Secretary of Finance Jesus Estanislao, April 20, 1995.
5. Ibid.
6. Interview with Senior National Economic and Development Authority official, anonymity requested, May 23, 1995.
7. International Monetary Fund 1991. Attached is the Republic of the Philippines 1991 Memorandum of Economic Policy.
8. Interview with Senior National Economic and Development Authority official, anonymity requested, June 27, 1995.
9. Interview with Senior Budget and Management official, anonymity requested, May 23, 1995.
10. Interview with Secretary of Finance Jesus Estanislao, April 20, 1995.
11. Collas-Monsod 1991.
12. Interview with Secretary of Finance Jesus Estanislao, April 20, 1995.
13. Republic of the Philippines 1991. Executive Order No. 470.
14. The NEDA Board is composed of the president as chair, the director general of NEDA secretariat as vice-chair, and the executive secretary, the secretaries of finance, trade and industry, agriculture, environment and natural resources, public works and highways, budget and management, labor and employment, and local government as members.
15. *Manila Bulletin* 1991b.
16. Lallana, 40.
17. *Manila Bulletin* 1991a.
18. Ibid., 1991d.
19. Among the industries affected by reductions in the EPRs for manufacturing are textiles and footwear, chemicals and chemical products, nonmetallic mineral products, basic metals and metal products, and machinery, including machinery, including electrical and transport equipment, particularly capital equipment.
20. *Manila Bulletin* 1991e.
21. Overseas Contract Workers (OCW) accounted for a significant portion of this increase. OCW receipts rose by 16.2 percent from higher earnings, (up by $318 million) as the number of deployed workers increased by an astonishing 37.9 percent to 615,019. Central Bank of the Philippines 1992.
22. Ibid.
23. Antonio 1992.
24. Republic of the Philippines 1991b.
25. Republic of the Philippines 1991c.

Chapter 11

1994 GOP-IMF Import Observation

The environment under which the 1994 GOP-IMF negotiations transpired was dramatically different from the previous observations. On the international side, the Cold War had ended and the sense of necessity embodied in the US-Philippine relationship had also drawn to a close. The 1991 US military withdrawal ended ninety-four years of American military presence in the Philippines.[1] Coinciding with these developments, free trade organizations, such as the GATT and later the World Trade Organization (WTO), the European Union (EU), the North American Free Trade Area (NAFTA), and the Association of Southeast Asian Nations Free Trade Area (AFTA) gained international acceptance and prominence. As a result of the emergence of the liberal international trading order, IMF preferences at Level I became superfluous. The GOP had acknowledged the need for free trade, especially when protectionism meant being left behind by its international trading partners. The GOP negotiating team held the same principles on international trade as their IMF counterparts. Similarly, the domestic side also experienced dramatic changes. In 1992, the Philippines experienced its first relatively peaceful transfer of leadership since 1964, as Corazon Aquino gave way to Fidel Ramos. In addition, the GOP took great strides to reform the economy, to move it toward a liberal economy. Despite these historic developments, two things remained the same: a sluggish economic growth rate and a tenuous political landscape.

The tactics used GOP generated satisfactory results; their limits, however, also were quite clear. The IMF technocratic alliance tactic encompassed key players within the GOP negotiation team, but the coalition failed to generate sufficient weight to modify substantially COG and other public actor preferences. GOP negotiators already agreed, to a large extent, with the IMF. Regarding import reforms, the lack of dissension was the result of similar economic preferences on the part of the negotiators and the ongoing international trade negotiations currently underway. Furthermore, the external-internal pattern continued to pressure domestic veto players. In comparison with international negotiations, domestic bargaining between the GOP and social import veto players was far from sanguine. Domestic negotiations hit many bumps and holes, but were never thrown from the road. Engulfed in a consensual negotiating framework, the GOP and veto actors discussed the issue at hand, especially in relation to the upcoming international talks. The negotiations provided the GOP with added support from Level IIB support actors, who were able to diminish the power of veto actors. In addition, the GOP also disbursed side-payments, in the form of direct and indirect instruments to gain veto actor support. Despite their grumbles and loud bantering, the veto actors accepted continued reform of the country's tariff system.

INTERNATIONAL POLITICAL ECONOMY OF THE 1994 IMPORT OBSERVATION

Negotiations between the GOP and IMF over the terms of the country's economic program centered again on the resolution of the GOP fiscal picture. The IMF focused its attention on the GOP capacity, or incapacity, to generate the necessary levels of domestic revenues to finance its expansionary fiscal policy, and in turn, establish the appropriate infrastructure, transportation, and energy needs for private sector led growth (see chapter 6 for an elaborate account of the GOP fiscal problems). Deficiencies in the tax base and collection procedures required the GOP to turn its attention to alternative sources, primarily international and domestic loans. Indeed, the IMF was concerned with reducing the current account deficit by controlling the government deficit, an issue that resonated throughout the negotiations and had a significant role in stretching out the negotiation process from March 1993 to June 1994. Embedded within this issue, both parties realized the importance of growth-oriented structural adjustment measures, especially through trade liberalization, a reform that would lead to lower input costs, increased global competitiveness of the country's exports, increased exports, and increased foreign exchange earnings. Despite the consensus on trade liberalization between the IMF and the GOP, the latter wavered on its position, a stand made necessary by powerful domestic forces. Although its tariff reform program was inconsistent in some areas, the GOP still managed to take steps toward reforming its tariff system.

After his 1992 electoral victory, Ramos faced daunting challenges. He inherited a country mired in political, economic, and social despair, as Communist insurgency, pervasive poverty, and crime and civil disorder permeated throughout the nation. Ramos pledged to attack these problems by improving the country's economic growth performance, a promise that placed the GOP in a position to prime-pump the economy and "stimulate growth by focusing public investments on the highest productive and economic returns," and initiate market-oriented reforms to provide the private sector with an environment conducive business.[2] Ramos referred to this massive effort as "Philippines 2000," the drive to push the country into Newly Industrialized Country (NIC) status by the turn of the century.[3] These initiatives, however, required tremendous levels of capital and the GOP was hard pressed for cash, as the GOP operating expenditures, capital outlays, and net lending increased over the years. Domestic sources had been virtually tapped out and the GOP turned outward for funds to fuel its projects. Foreign official and private donors and lenders, however, were reluctant to issue monies or trade credits to the GOP without an IMF endorsement.

The decision to turn toward the IMF was not made without considerable deliberation. For the GOP, the problem was flexibility. It feared that it would tie itself to an agreement and be unable to alter its policies to fit its particular needs. Despite this caveat, the Ramos administration wanted to continue the outward oriented policies pursued by Aquino. This would include fiscal reform, monetary discipline, and further structural reforms such as trade and investment liberalization and deregulation, with or without an IMF agreement.[4] After considerable deliberation, the GOP found that the cost of a "no-agreement" outweighed the benefits of trudging through the country's development program without an IMF agreement.[5] As another incentive, the GOP viewed the prospective agreement as an "exit program," a package designed to get the Philippines out from under the tutelage of the IMF once and for all. Again, the IMF and the GOP set up negotiations to draft another economic program.

Unlike the 1986 GOP-IMF Import Observation under the Aquino administration, the GOP bargaining position on imports, as well as on additional targets, was relatively coordinated and uniform. This position was made possible through two factors. First, the GOP used the bargaining format, initiated in 1990, which allowed various agencies the opportunity to negotiate on a multiagency basis with the IMF. Previously, the IMF met with GOP agencies on a one to one basis. This new format required the GOP to bargain among themselves, and reach a consensus on various positions before negotiating with the IMF. Second, GOP consensus on bargaining positions was also derived from the preferences of the Ramos cabinet members. While its decision-making process was similar to the later Aquino years, the Ramos cabinet also consisted of individuals who shared similar preferences for market-oriented economic reform. According to acting Finance Minister Ramon del Rosario, "It wouldn't make any sense to form a cabinet that wasn't on the same page."[6]

The early IMF-GOP bargaining rounds brought out both the similarities and the differences between the chief negotiators. Beginning in early 1993, the GOP and IMF negotiation teams readily agreed on the principles of the economic program, where fiscal and monetary reform would be the guiding foundation that was complemented with structural reforms. Disagreements between the two parties, however, revolved around the technical aspects of the program, the specifics of the economic targets, and perhaps most importantly where trade was concerned, with the timing.[7] The GOP representatives, del Rosario Jr., and Central Bank Governor Jose Cuisia Jr., were not protectionist oriented by any stretch of the imagination, but they were concerned about the pace of liberalization, the effect it would have on the manufacturing sector, and its domestic, political, and economic consequences.[8] Del Rosario wanted: "[Trade] liberalization to move more quickly in some areas and slower in other areas. While I agree with the principle of trade reforms, it would be good for some companies and unfair for others." Given this case, del Rosario added, "We preferred to move [trade liberalization] slower."[9] In contrast, IMF mission chief Bruce Smith pushed for trade reform that followed a faster pace. The IMF felt that without trade and other structural reform measures in place, the Philippines would reach a balance of payments crisis every time real economic growth hit above 3 percent, and would continue to be trapped in the same "boom or bust" cycle that the Ramos administration was trying desperately to escape.[10] Much like the 1984 Observation, the IMF wanted GOP prior commitments in place before it agreed to the terms of an economic program.[11]

The Ramos administration commitment to import reform was spelled out long before the start of IMF-GOP bargaining rounds. In July 1992, the GOP issued a new tariff reduction schedule that allowed the GOP to continue its efforts to deregulate imports through the ILP, and implemented measures to deregulate the foreign exchange system. In early 1993, the GOP followed up trade reforms as it reduced tariffs for packaging materials and intermediate goods.[12] This progress hit a snag as the GOP later deferred the tariff schedule on agricultural products, in particular corn. The Bretton Woods Twins responded to this move with considerable criticism. World Bank Director for East Asia and the Pacific Calixto Madavo assailed the GOP for its failure to ease trade controls on agricultural production. The IMF took a similar stance, and indicated that the GOP had sent the wrong signals to the domestic and international communities regarding its commitments to implement structural reform policies. "Wrong signals . . . wrong signals. Why are you doing this? It is not towards the right direction," Undersecretary of Finance Ernest Leung paraphrased the IMF when it assessed the GOP move. The Philippine technical team responded that the country's agricultural sector, which accounts for 30 percent of the gross national product and 45 percent of the labor force, was a crucial segment of the economy that deserved consideration. "It (the agricultural sector) is therefore of a higher sensitivity," said Leung.[13]

Without question, the IMF was extremely concerned about the signals the GOP would send to domestic and international private actors. In past negotiations with the GOP, the IMF had always been wary of US efforts to push for a relaxation of conditionalities or quicken the pace of negotiations. American concern for its client rested on its interest in the military installations. With the removal of US bases, the pressure to speed up negotiations or loosen the terms abated; yet, the IMF continued to feel pressure from international actors. As recounted by del Rosario, the IMF desperately needed a success case to point toward, to demonstrate that its policies did more than just stabilize, but also to aid in sustained economic growth. The Philippines was to be that case.[14]

International scrutiny aside, the negotiators failed to see eye to eye on the pace of the talks. Bruce Smith continuously doubted the credibility of the GOP to live up to its commitments and questioned every move made by the GOP. Smith had doubts about his counterparts, in his mind, "We were always trying to pull a fast-one."[15] Christopher Browne, the IMF Resident Representative to the Philippines, showed his frustrations over the pace of negotiations and prior-conditions implementation as he publicly made very scathing remarks directed toward the Philippine authorities. These remarks did not go unanswered as Ramos called for the removal of the IMF negotiation team. In June 1994, Kadhim Al-Eyd replaced Smith and Howard Handy took over for Browne. The switch coincided with a GOP change of its own. Del Rosario and Cuisia stepped down, due to purely domestic reasons, and were replaced by Leung and Gabriel Singson, a long-time Central Bank deputy governor, respectively.

With Leung leading the way, the GOP placed the negotiations on hold and intensified its commitment toward instituting the appropriate structural policies. In July 1993, the GOP increased the import levy on oil and petroleum products by Pl.00, a move that eventually became known as the "Leung levy," and that forced the GOP to increase the price of petroleum products in early 1994 (see chapter 6).[16] Furthermore, the GOP placed a moratorium on tax exemptions for projects registered with the BOI, an annual drain of approximately P9 billion to Pl0 billion per year. In exchange for the moratorium, the GOP lowered the duty on capital equipment. Privatization efforts also intensified. The IMF later warned the GOP that this move would only generate revenues in the short-term, and should not be relied on as an indicator for projections of future government revenues.

During the IMF-GOP talks, trade liberalization became a more prominent international issue. On a regional scale, in early 1993, the ASEAN member nations agreed to form a free trade area (AFTA), a cooperative that would reduce intra-trade tariffs to 0 to 5 percent by the year 2003. In congruence with AFTA, the GOP also participated in the Uruguay Round negotiations, where issues of trade liberalization were certain to have a reverberating effect on the Philippine economy; this would especially be the case since the Philippines had joined an association of agricultural exporting developing countries, referred to as the Cairnes Group. In line with these multilateral institutions, the Philippines also linked itself with the nascent Asian-Pacific Economic Cooperation (APEC). Thus, while the IMF played a role in GOP trade liberalization, its importance paled in comparison to the more pronounced international dynamics that sounded the calls for free trade. The GOP import policy revolved around its commitments to these international organizations and moved away from the IMF.

The GOP chief negotiators, as well as domestic Level IIA players, such as the COG, were already more or less on the same page as their IMF counterparts. Timing remained the only difference between the two parties. In addition to this condition, the urgency for IMF recommendations on structural adjustment policies waned as international trade organizations became more prominent. In striking contrast to the 1984 and 1986 Import Observations, the 1994 Import Observation can be characterized for its harmony rather than discord.

By mid-1994, the parties reached an agreement. After months of deliberation, Congress passed various revenue generating measures, included the controversial expanded value added tax. Ramos approved the bill, and within hours, Malacañang Palace and the IMF agreed on a new Memorandum of Economic and Financial Policy (MEFP). Under the terms of the program, the GOP targeted economic growth to increase from 2.3 percent of GNP in 1993 to between 4.5 and 6 percent for every year thereafter until 1997. Despite the increase in growth, the GOP also promised to hold down inflation, which ran at 10 percent by early 1994 to about 4 percent, the average rate within the Southeast Asian region. The GOP backed these targets with promises of oil deregulation, bank liberalization, removal of import restrictions, and completion of its program to privatize state-owned companies.[17]

DOMESTIC POLITICAL ECONOMY OF THE 1994 IMPORT OBSERVATION

During the first half of the Ramos administration, trade reform vacillated between continued progress and subtle reversals. The administration set a course to implement import reforms, a path that fell in line with the liberalization of the global economy. To be sure, the administration would face a stiff challenge from domestic veto actors. In overcoming this opposition, the administration used linkage and more importantly, side-payment strategies. Taken as a whole, the Ramos administration achieved a large measure of success in its efforts to initiate import reform, a signal that showed private investors that the government was serious about structural adjustment. Disaggregated, however, these measures revealed that Philippine politics continued to be tightly veiled in patron-client ties.

The 1992 national elections had the makings of a potentially divisive political landscape. In the presidential campaign, Fidel Ramos was viewed as nothing more than a dark-horse candidate, even with the support of outgoing President Corazon Aquino, and the military and business communities.[18] He seemed to lack appeal, charisma, and focus, especially in comparison to more colorful candidates; however, his access to GOP coffers and its organizational machinery, and the presence of Imelda Marcos in the election, contributed to give him a sufficient 23.4 percent of the popular vote, enough to carry the election.[19] Despite his victory, Ramos' coat-tails failed to translate into congressional seats for his party candidates, the Lakas-National Union of Christian Democrats (NUCD), an outcome that left control of the institution in the hands of the opposition Laban ng Demokratikong Pilipino (LDP) party. The divided government made the prospects for effective governance remote. A LDP leader predicted: "There will, of course, be a honeymoon period [with the new president]. After that it will be quarrel-some."[20] This prognostication might have been actualized had the prevailing conditions remained, but Ramos and his party offered lucrative patronage incentives to the National People's Coalition (NPC) to form a coalition with the Lakas-NUCD party and LDP members to defect from their party. Between the May 1992 general elections and June 1993, seventy-one members of the House switched parties to join the Lakas-NUCD party.[21] In contrast, the Senate remained firmly under LDP control. Party affiliation, however, remains a tenuous indicator of legislative behavior in a culture overwhelmingly dominated by patronage politics.

The election results placed Ramos in a precarious situation. Uncertain of his political base, he took cautious and conservative approaches toward reform. Much like Ongping's argument against the wall of favoritism that Marcos had built. Ramos and his presidential security adviser Jose Almonte pushed for reforms to strengthen the state.[22] Their close bond started in 1956 when Ramos and Almonte served as a captain and second lieutenant during a military skirmish with Philippine rebels, the underling asked, "Why are they rebelling?" "Why are we killing our own people?"[23] Almonte, a close confidant of Ramos, concluded that Philippine political, economic, and social institutions created inequality, which could be rectified, in simple terms, by creating a "level playing field."[24] The Ramos government would have to strengthen state autonomy and insulate it from patronage politics to pursue Washington Consensus policies.[25] At the financing level, Ramos put Ramon del Rosario as his Ministry of Finance. His family owned Philippine Investment Management (PHINMA), a holding company that included cement, banks, and minerals. Holdover Jose Cuisia continued his role as Governor of the Central Bank. Lila Bautista, and later Rizalino S. Navarro served as DTI heads. Ramos extended an offer to Rizalino "Roy" Navarro, a friend of del Rosario, to serve as Ramos' DTI principal. He strived for consensus and eventually became a close adviser to Ramos. [26] With this cabinet in place, the economic policies were decidedly liberal and export-oriented, but the delicate political landscape required slow treading.

The reform process, as covered under the petroleum observation, was anything but straight. In a sense, Ramos continued to behave more like a candidate, compromising on many positions, trying to be popular, and reaching out to all sectors, rather than a president.[27] From another perspective, Ramos had great respect for the democratic process and wanted a "constructive consensus" to put in its imprint for its softer version of the East Asian Industrialization model.[28] Even steps that his administration had taken to resuscitate the stagnant economy were clouded with caution, an approach that had given the administration its share of victories. The successes were clear. The GOP temporarily eliminated tariffs on power generating equipment to resolve the country's power shortage and on cement and cement clinkers to stimulate growth in the construction industry.[29] It also took another step toward trade liberalization as it lifted all but one minor restriction on the flow of foreign exchange.[30] This trend continued as the GOP deregulated import controls that ranged from consumer durables to agricultural products.[31]

Another clear-cut case of reform accomplishment was the implementation of Executive Order 8, the tariff reform initiative.[32] This program was developed through the joint-agency Committee on Tariff Related Matters, which consulted with various sectors of the economy. The committee generally followed a rule of thumb of simply doubling the then existing tariff schedule during the first year of the executive order, to be gradually brought down back to its original level after five years.[33] A comparison between E0 470 and EO 8 proves quite revealing. Under the five-year phase out program EO 470, consumer durable goods tariffs (e.g., refrigerators, freezing equipment, and air conditioners) were scheduled to be reduced between 20 and 45 percent.[34] In contrast, under EO 8, tariffs for these products were initially increased to 100 percent and gradually reduced to rates comparable to those spelled out under EO 470; however, EO 470 dropped the duty to 30 percent by the end of 1995 while EO 8 dropped the tariff to 30 percent by the end of 1996. Similarly, the GOP also issued side-payments to powerful actors in the agricultural sector, such as producers of crustaceans, corn, various meat products, and poultry. These industries received added protection as tariffs for their products increased between 60 to 100 percent. In short, the Ramos administration continued to provide substantial protection to industries with political clout.

For the Ramos government, the external-internal linkage strategy generated effective results. The GOP shifted the issue of import reform away from the narrow band of local-oriented domestic producers and toward broader issues of industrialization, especially through the "Philippines 2000" program, and globalization, through the international forums of GATT, WTO, AFTA, and APEC.[35] Linkage of domestic Level II negotiations with a larger agenda, "Philippines 2000," and with numerous international actors produced the intended effect. First, veto actors were no longer fighting with the GOP alone on import issues; rather, the GOP had become one of many opponents. Veto actors had to divide their attention between a multitude of forums and could no longer focus on the GOP itself. Second, the linkage forced domestic veto actors to adjust their perceptions of costs, as they realized that they could no longer doggedly adhere to their protectionist positions as their competitors in similar situated industries in other developing nations might become more efficient, leaving the Philippine industries behind in its wake.[36] Finally, by presenting negotiations on the import reform program within a larger agenda of economic growth and international competition, the GOP managed to draw into the decision-making process support players, actors that could apply pressure on veto actors to alter their perceptions of costs. The Confederation of Philippine Exporters and Philippine Exporters Foundation were two of the more prominent export-oriented organizations that made its presence felt when it spoke with a unified voice when it combined in 1994 into the Philippine Exporters Confederation.[37] While the association amplified its voice, it also operated in a more institutional friendly environment under Ramos as it was accredited by the government.[38] By adopting a larger set of issues, the GOP was able to use its institutional framework to mitigate the distributive conflict that existed between veto and support players.

With the presence of numerous international forums, it is doubtful that a GOP external-internal linkage strategy with the IMF would have shifted cost calculations of veto actors. First, the GOP fiscal situation was far from catastrophic, as in previous IMF-GOP Observations. Second, the Fund continued to press the GOP for import reforms, but on a dimension less than that experienced under the 1986 Import Observation. Indeed, the international climate and the emergence of trading regimes moved into the spotlight and gently nudged the IMF off the stage. Third, the IMF image began to change dramatically since the days of Marcos. The media portrayed it more fairly, a perspective it earned through stellar work from Ramos, improved economic conditions, and the work ethic of IMF negotiator Al-eyed.[39] The relationship broke traditional boundaries as the IMF staff even shared their computer diskettes with GOP negotiation teams.[40]

Not surprisingly, the country's slow economic recovery prompted a different response from other sectors of the economy. Frustrated by long experience with economic stagnation, much of the business community, as well as many in the labor movement, initially indicated a receptiveness to new strategies.[41] The electronic and garment industries continued to push for reduced tariffs to lower production cost, increase efficiency, and increase global competitiveness. Similarly, exporters complained about the effects of the foreign exchange system as they often incurred inflated cost from domestic banks and an exchange rate that often trimmed the international competitiveness of their products.[42] The GOP responded to these calls as it funneled government support through BOI tax incentives, and lower capital cost.

Although the emergence of support actors helped to minimize the power that veto actors wielded, the GOP continued to keep the lines of communication open and maintain the consensus that Ramos wanted. The GOP permitted affected sectors opportunities to express their grievances to the Tariff Commission; a dubious gesture early in the process, as many domestic sectors had little information regarding the tariff levels of comparable products in other countries. The GOP soon corrected this oversight.[43] Often, opposition to reform used this forum also to apply its linkage strategy. For instance, it argued that the pace of EO 470 and EO 8 tariff reductions were more stringent than reductions required under AFI'A guidelines, and that these schedules exacted debilitating tolls on many domestic producers as they made high tariff cuts on finished products without making corresponding reductions on raw materials. Furthermore, opposition argued that the GOP tariff reduction programs made unilateral cuts in tariffs at rates faster than the country's ASEAN neighbors were scheduled to do so under AFI'A, and this asymmetry undermined Philippine bargaining power at the AFI'A negotiation table since the GOP had unilaterally drop its tariffs at a faster pace and at lower rates.[44]

Under this observation, veto actors also relied heavily on the legislature as an instrument of interest articulation. As promulgated in the 1987 Philippine Constitution, the legislative branch was responsible for determining tariff rates, a right that was passed along to the executive branch when Congress was in recess. The problem with the legislature taking a lead role in determining tariff levels is that it lacks insulation from well-organized constituents and is often the object of political pressure, especially in terms of patronage politics. As a result, Congress often remains sensitive to the calls of protectionism, in particular from the agricultural and import-substituting sectors. As it learned lessons from the 1989b Import Observation, the Ramos administration instituted a new tariff policy framework that both embraced and distanced the legislative branch. In conjunction with the Tariff Commission, the legislative branch held consultations with veto and support players over the content and pace of import reform programs. These initiatives, however, would emanate from the executive branch, as it would announce them when Congress was in recess. This process effectively bypassed the legislative decision-making process, a process that provided veto actors with numerous opportunities to water down tariff reform programs.

The GOP managed to bypass effectively constitutional constraints that favored legislative preferences. This process, however, is not absolute. Before she left office, Aquino signed Republic Act No. 7607, the Magna Carta for Small Farmers.[45] This law codified government commitment toward the protection of the agricultural sector, especially from the ravages of foreign competition. As a means toward this goal, the Ramos administration was legally obligated to increase tariff protections for corn, bovine and its products and poultry, measures that were supported by DTI, DOA, and NEDA. Ironically, small farmers had little hand in tariff increases as large producers of these products benefited from these increases.[46] A 1992 World Bank study found that six agribusiness conglomerates control roughly 80 percent of the commercial poultry market, as well as 60 percent of the commercial feed market and 40 percent of the commercial pig market.[47] In addition, the GOP suspended another executive order under the insistence of DTI, which argued that 20 to 50 percent tariff reductions created considerable distortions for the manufacturing sector.[48]

Indeed, the agricultural sector continues to remain a powerful force in Philippine politics. Prior to the Uruguay Round, the Philippines effectively prohibited the importation of pork and poultry.[49] When the GOP agreed to the Uruguay Round Agreements, it was required to implement import reforms in the importation of various meat products by July 1, 1995. The GOP finally allowed the importation of pork and poultry products in October 1996, as it issued import licensing to domestic importers "who have no known interest in importing."

After the GOP added additional agricultural protection, it proceeded slowly to inch its way forward toward the deregulation of trade controls. With continued pressure mounting against the GOP, Philippine authorities found that it could not implement these measures unilaterally. Indeed, the GOP held year-round public hearing reviews that addressed the determination of the tariff band and tiers, the tariffication of the remaining quantitative restrictions, and a possible reimposition of special duties in place of ad valorem duties on certain items and the proper timing of tariff adjustments.[50]

The battles were not confined to the executive and legislature; they subsequently moved to the judicial system. This new arena shortly became prominent in the new Ramos administration. Early on during the administration, the GOP sold the state-owned Manila Hotel to a consortium of Malaysian and US investors. Before the sale was finalized, local investors that also bid on the property petitioned the Supreme Court to stay the sale and instead grant it to local investors. This remedy would be based on the Philippine Constitution that requires giving Filipinos preferences in ownership. The court agreed. In another case, DTI chief Navarro just concluded with 124 other countries the Marrakesh Final Act, the provision that closed the Uruguay Round and thrust the WTO into operation. After constant battles on the Senate floor and behind closed doors, the Philippines ratified the provision. One day before the Philippines was to be deemed a founding member of the WTO, three Philippine senators issued a stay on the treaty on the grounds that it violates the Philippine Constitution. Again, the battleground was in the judiciary. Three years later, the court denied the petition.[51]

In June 1994, Malacañang Palace and the IMF agreed on a new MEFP. Under the terms of the program, the Ramos administration laid out a series of structural adjustment commitments to the IMF. These promises included deregulation of the oil industry, liberalization of the banking sector, removal of import restrictions, and completion of its plan to sell state-owned companies to the private sector.[52] In addition, the government also aimed to boost growth from 2.3 percent in 1993 to 4.5 percent continuing upward every year to about 6 percent by 1997.

Almost immediately, the GOP delivered on its trade reform promise. In July, the GOP issued EO 189, a measure that reduced tariffs on capital equipment to bring down the cost of producing goods and dampen the effect of the E-VAT on commodity prices.[53] The Executive Order was used as Congress was not in session. According to the BOI, tariff reduction meant foregone revenues of P 540.9 million to the government, and increased foreign exchange outflow of $144 million. In September, the GOP reformed tariffs on imported raw inputs used by garments, plastics, and chemical industries. It also made other liberalization moves by December. Although these moves coincided with the signing of the MEFP, the GOP intensified its move in preparation for the international trade environment that the GATT agreement would create. These intentions, however, failed to sit well with domestic manufacturers of non-tradable goods as the Federation of Philippine Industries (FPI) strongly opposed the GOP plan to liberalize trade protection without proper consultations with the private sector.[54] The Chemical Industrial Corporation of the Philippines, Inc. cautioned that this could have far reaching effects on industries, in terms of plant closures and mass layoffs, and reduced incentives for foreign competition to invest in the Philippines.[55]

EPILOGUE TO THE 1994 IMPORT OBSERVATION

The GOP committed itself to meet high levels of economic growth as a percent of GNP, a formidable challenge to be sure. Export promotion through efficient use of resources was the key. To meet this challenge, the GOP drastically reduced duties across the board for various capital equipment, components, and spare parts. These measures aimed to improve the domestic energy industry, the transportation sector and revitalize an impoverished infrastructure, moves that were crucial to the GOP drive to reach NIC status by the year 2000. Not to neglect its foreign exchange generating capacity, the GOP also lifted trade barriers on the importation of raw materials for the country's twelve export winners, groomed by the GOP to become leaders in export production.[56]

Tariff protection on local agricultural products, however, continued to plague the economy. In October 1994, the GOP imposed bans on the imports of garlic, onions, potatoes, and cabbage, a policy that was instigated at the insistence of political pressure from the northern provinces. Chicken products have historically been highly protected because of the relatively strong lobby of the many large producers in the industry.[57] This measure, however, drew concerns from the country's leading industries. According to a Makati Business Club official: "The government should make its policy clear so that the private sector would know where it stands and where it will go. Business decisions are being made on the assumption that the environment will be conducive and that there would be stability of policies over time."[58] As evident under this example, political patronage continues to play a key role in determining economic policy.

ASSESSMENT OF THE CAUSAL MECHANISMS

The external-internal tactic yielded a positive causal link between the international and domestic arenas. With the context of the US military bases pushed off to the side, the GOP relied on the dynamics of the international economic landscape of liberalization. The Uruguay Round and AFTA environments required the harmonization of foreign economic policies with domestic import and trade reform, and the Ramos administration intended to ride these waves. Given the preferences for Washington Consensus type policies, both sides fielded a relatively narrow gap between them. The sticking points were how to get there and the obstacles faced. Level IIA principals knew what was at stake. The same theme ran through Level IIB. GOP first turned outward with the free trade negotiations and then turned inward making arguments about commitments and costs, information that initially remained concealed until the Ramos administration revealed them. Level IIB veto actors need tariffication and other side-payments to ride out the cost of reforms.

The deference tactic had a similar effect at Levels IIA and IIB. At the former, cabinet principals were already on the same page and variations among them were at the margins. In the formal arena, the GOP used issues of industrialization and approached export groups in what later came to be known as the government sanctioned Philippine Exporter Confederation. Its support, in addition to the use of side-payments, assuaged the backlash of highly sensitive industries, such as agriculture. Indeed, DTI and the chemical industries were on the outside looking in, protesting the effects of EO 189.

A technocratic alliance can explain the ratification at Level IIA, but not at Level IIB. Like the deference tactic, the IMF used an alliance net that ensnared individuals whose preferences nearly matched their own. Furthermore, the IMF under Al-Eyd received glowing praise from both the GOP and its bureaucracies and with the Philippine media. This broad-based net ensured that the preferences on design closely matched. The net had its limits in implementation, as there is little information to show that the IMF networked Level IIB veto actors. Again, taking from the 1994 petroleum price episode, we can surmise that the IMF also attempted to generate support for import reform. Again, in most cases, we can also surmise that the IMF preached to the choir.

The GOP learned from the previous IMF-GOP import processes. After the failure of EO 413, the GOP continued to learn and keep the Congress away from import reform by rolling out programs when it was in recess. Furthermore, it also continued the practice of pushing tariff reform through the Tariff Commission. It also received pushback on some of its policies, but it could use side-payments to offset the costs to injured domestic players.

NOTES

1. The deadline for renewal of the leases on the military establishments was September 16, 1991. Negotiations on the terms of the leases began on May 14, 1990. However, the Mt. Pinatubo volcanic disaster forced the United States to abandon Clark Air Base. The United States and GOP agreed on a ten year, $230 million, lease extension for the Subic Bay Naval Base. The Philippine Senate rejected ratification of the treaty by a vote of twelve to eleven. The United States handed over to the Philippines Clark Air Base over in November 1991 and Subic Bay Naval Base on November 23, 1992.
2. State-of-the-Nation address of President Fidel V. Ramos, as reprinted in the *Manila Chronicle* 1992b.
3. The “Philippine 2000” concept was developed during the Aquino administration. Ramos brought the program to greater prominence.
4. Thompson 2011, 137–142.
5. del Rosario Jr., 46.
6. Interview with Secretary of Finance Ramon del Rosario Jr., May 11, 1995.

7. Interview with Senior National Economic and Development Agency, official, anonymity requested, July 28, 1995.
8. Coincidentally, both del Rosario and Cuisia were former bankers.
9. Interview with Secretary of Finance Ramon del Rosario Jr., May 11, 1995.
10. *Manila Bulletin* 1993a.
11. Interview with Secretary of Finance Ramon del Rosario Jr., May 11, 1995.
12. Republic of the Philippines 1992f.
13. *Manila Bulletin* 1993d.
14. Interview with Secretary of Finance Ramon del Rosario Jr., May 11, 1995.
15. Ibid.
16. Republic of the Philippines 1993.
17. *Philippine Daily Inquirer* 1994c.
18. Thompson, 178.
19. Voters that still supported the Marcos regime split their votes between Imelda Marcos and Eduardo Cojuangco, a Marcos crony. Reid and Guerrero, 240.
20. Tiglao and Tasker.
21. Riedinger, 140.
22. de Dios and Hutchcroft, 55.
23. Almonte.
24. Bernardo and Tang, fn. 7.
25. Ibid, 14.
26. Alcuaz.
27. Brillantes, 225.
28. Thompson 2011, 144–145.
29. Republic of the Philippines. 1992c; and Republic of the Philippines 1992d.
30. Central Bank of the Philippines 1992c.
31. Republic of the Philippines 1992b and 1992d.
32. Republic of the Philippines 1992f.
33. Virata and Associates, 58.
34. Industrialist Raul T. Concepcion headed the coalition of domestic manufacturers for nontradable goods.
35. These issues where highlighted during Ramos' 1992 inauguration speech and the Philippine Medium-Term Development Plan. The speech was reprinted in the *Manila Chronicle* 1992a.
36. Hutchcroft 1996, 5.
37. Lallana, 50.
38. de Dios and Hutchcroft, 52.
39. Interview with Senior International Monetary Fund official, anonymity requested, July 11, 1995.
40. Interview with Senior National Economic and Development Authority official, anonymity requested, July 28, 1995.
41. Hutchcroft 1996, 7.
42. Tiglao 1992.
43. *Manila Bulletin* 1993b.
44. Ibid., 1992.

45. Republic of the Philippines 1992b.
46. Ibid.
47. Riedinger, 143.
48. *Manila Bulletin* 1993c.
49. *Business World* 1997.
50. *Manila Bulletin* 1994j.
51. Thompson 2011, 247–249.
52. *Philippine Daily Inquirer* 1994c.
53. *Manila Bulletin* 1994l.
54. Ibid., 1994k. In the early 1990s, PCCI began to move away from its emphasis on manufacturing and concentrate more on the industries of commerce and services. This move prompted the manufacturing coalition of PCCI to break-away and establish its own association, the Federation of Philippine Industries. Raul T. Concepcion eventually became the president for this organization.
55. In 1995, controversy arose over the tariff schedule of petro-chemical imports, a clear case of a Philippine infant industry. Japanese investors became hesitant to invest heavily in the Philippine petro-chemical industries with minimal tariff levels. Domestic manufacturers argued that tariffs needed to be raised. Interview with Senior Chemical Industries Association of the Philippines official, anonymity requested, August 9 and 10, 1995.
56. *Manila Bulletin* 1994k.
57. Ibid., 1994n.
58. Ibid., 1994m.

Chapter 12

Conclusion

This book applied the case study as the method toward advancing our knowledge and tools of analysis. It applied the case study as a hybrid deviant case and hypothesis generating analysis. The IMF-GOP negotiations provided a multifinality pattern on outcomes. When we disaggregate the outcomes by Levels IIA and IIB, we derive twenty-fours outcomes (see Table I.2). When we compress the outcomes to reflect the inclusiveness of Levels IIA and IIB and focus only on the Level IIB outcomes, we generate five different outcomes: success, limited success, qualified success, reverse limited success, reversed failure, and no attempt. "Success" shows that the policies were implemented and sustained. "Limited success" demonstrates the policies were implemented, but insufficient to produce the desired outcome. "Qualified success" shows a phased in process or implemented as a last resort, a deal breaker. "Reverse limited success" shows policy implementation but some element was reversed. "Reversed failure" demonstrates policies that were implemented but reversed in its entirety. Finally, "no attempt" is evident. Several of the similar outcomes, however, have different causes, reflecting an equifinality pattern at work. Tables 12.1–12.4 show that when we disaggregate the effectiveness of both the deference and external-internal tactics, we have up to thirty-five different combinations of patterns as some observations may share the same outcomes, but they vary in how they were achieved. The multifinality and equifinality patterns and the design of the case study allows me to treat each thirty-five differentiated outcome as a deviant case. According to Arend Lijphart, such cases "uncover relevant additional variables that were not considered previously, or to refine the (operational) definitions of some or all of the variables."[1] On the hypothesis generating side, this work seeks to assist the research agenda on the Two-Level approach in

IMF-recipient country studies regarding negotiated outcomes. For both of these points, I posit the following propositions.

I present two major conditions and numerous minor conditions to frame the typology. The first major condition focuses on the CN's economic and political leverage. If the costs and benefits leverage from the IMF and its principals can be clearly mapped out, then the CN has strong leverage to garner support of either the deference player or the COG. With unclear leverage, the CN has qualified or limited success. In the 1984 observations, foreign financing continued during the negotiations process, limiting the IMF's credibility to act as a catalyst for foreign financing. Furthermore, the US congress and executive branches may also send mixed signals (1984 and 1986 observations), obfuscating the signals sent from the IMF. In contrast, a clear "backward mapping" from the IMF on the costs and benefits via a clear signal can provide the CN with leverage against both Level IIA hurdles as well as against international objections. In the 1989 observation, the guaranteed financing from PAP drove a clear leverage path for the CN.

The second major condition derives from the Helmke and Levitsky framework. We can apply the formal and informal interplay of institutions. When informal institutions function in "complimentary" form, the formal rules dominate and produce equilibrium outcomes that all parties accept lest they be subject to punishments for their violation. The informal rules help facilitate the equilibrium outcomes. Informal institutions serve an "accommodating" function when the formal and informal rules produce a suboptimal outcome that players accept in letter but their behavior violates the spirit. An example is the US executive practice of the impoundment of appropriated legislative funds. When formal rules fail to produce an outcome and informal rules provide a structure that allows for a derived outcome, they "substitute" for the formal rules. Finally, when formal institutions are ineffective and the procedures are not systematically enforced, thereby allowing actors to ignore or bypass the rules, then the informal rules produce a "competing" effect. This typology is consistent with observation made by Randall W. Stone, informal procedural institutions produce an "accommodative" effect that undermines IMF preferences for prudent liberal oriented economic policy, and Shepsle, multiple principals and variation of preferences both among and within principals can produce significant discordance between principal and agent.[2]

The presence or absence of these cleavages can be used to either undermine or strengthen the position of the CN. She uses these signals to derive concessions with her Level IIA and Level IIB constituents. One of the clearest examples lies in the 1984 GOP-IMF context. The Reagan administration supported the Marcos government and continued to provide funding while commercial banks pivoted their position around an IMF program. Furthermore, the republican administration was at odds with democratic lawmakers on the viability of Marcos to reform or defend US interests. The US preference ambiguity gave Marcos a lifeline to postpose reforms. Key international actors were clear, an IMF program must be in place before foreign financing can continue, setting up the IMF as a catalyst. But the spirit of this position was violated by United States, which continued to funnel money to Marcos. Furthermore, the United States could not be punished for its transgression. As long as the accommodative informal rules are in play, the CN had no leverage with Level IIA players. Once cleavages among principals are narrowed, then CNs can use the international environment to shift Level IIA preferences. "Agency slippage" where the IMF and its principals have drastic differences in preferences can also work in favor of the CN. She can also lobby the principals and the World Bank to set up a funding pledge session that can speed up the timing of negotiations, forcing the IMF to accept terms that it might otherwise not, as signified in the 1989 and 1991 observations. Again, the accommodation interplay is at work as the IMF might again serve as the catalyst for foreign financing, but the behavior of its principals undermines the spirit of its leadership. GOP CNs Jayme and Estansialo took advantage of the PAP timing with Japan and used it as leverage to shift IMF preferences especially regarding timing. The IMF authorized programs much sooner that it perhaps preferred. These points bring us to the first proposition.

> Proposition 1: The CN has clear leverage to move Level IIA players if the IMF and its principals send consistent signals. This is a necessary, but not sufficient condition, for reform success.

The second major condition is the Level II institutional process through which the CN must proceed. At Level IIA, the CN must secure the COG through the formal and informal institutional interplay. Under Marcos, the CN would proceed through a "substitutive" arrangement to either secure or lose Marcos' support. The rule breaker here was Marcos, who could await a cabinet decision then could turn around and void it, undermining the capacity of the cabinet to develop and adhere to a decision-making process and ultimately weakening its credibility. In the 1986 and after observations, CNs faced in most cases a complementary and sometimes substitutive informal network and made deals to secure the endorsement of cabinet subgroups and the trade minister before the COG would come on board. The COG defers to the cabinet subgroups and trade ministers. Attempts to break the rules and submit proposal before proper vetting by subgroups were punished by admonishment from Aquino. She, however, as the COG could stick to the rules of the cabinet but could undermine its intent to secure a viable outcome with the IMF.

The second condition also examines the interaction at Level IIB where the institutional arrangements vary by issue. On petroleum pricing, the price increase institutional design process divides actors into two categories: those receiving the benefits of the price increase, not including the GOP, and those that incur the effects. The first set is the petroleum companies, which expect that once they begin the formal petition procedure before the ERB, then they receive adequate cost recovery. They abide by complementary rules where its violations are legally enforced by the GOP. The groups incurring the effects are divided into three categories. The first is the group that participate in the cost adjustment meetings through the tripartite/multisectoral or NWPC, which meet before the price increases or shortly afterward. Sanctioned by the government to participate in these conventional participation meetings, they provide input and assessments regarding their economic postures and the level of adjustment, if any, required to maintain their livelihood. They expect these meetings to produce viable outcomes that all accept and violation of these rules are subject to punishment. This grouping falls under the category of complimentary informal rules. The second grouping enters the process typically at the ERB hearings. Typically comprised of labor, transportation, and consumer organizations, this grouping typically loses on the petroleum price decision and as a result, declares its intention to engage in unconventional participation practices such as strikes and demonstrations. This behavior is neither government sanctioned nor punished. Simultaneously, they also seek remedies through their respective bureaucratic agents in the form of sanctioned pass-through (i.e., fare increases). This grouping can be labeled under the accommodating informal institution in that its behavior both produces suboptimal outcomes and violates the spirit of the process. The third grouping involves those that may participate in the ERB hearings, disapprove of the decision, and then participate in protest, strikes, and demonstrations, which cannot be punished unless a violence threshold is met. Unlike the previous grouping, this group lacks the agents to represent them in pursuing their interest or passing their increased costs to other actors; as a result, their demands are typically unmet. They have no incentives, short of membership mobilization limitations, to cease this behavior. They expect unfair treatment from the rules and have no incentives, short of membership mobilization limitations, to cease this behavior and every incentive to sustain their levels of unconventional participation for as long as possible. This group falls under the substitutive category. In addition, the more bureaucracies and legislators that act as agents for the principals that become active late in the process, the less the chance the government has of sustaining price increases.

Import reform institutional players can be broken down into two primary categories, import substituting industries and export industries. Through complementary informal rules, the design of the process is heavily back ended. At the front end of the process, these players abide by the rules and the procedure. As covered early in chapter 7, these industries participate in executive branch hearings (e.g., Tariff Commission, Board of Investments, NEDA, etc.) and voice their grievances, in some cases being heard, in other times, ignored, leading the conclusion that these hearings are meaningless. This process typically occurs before the declaration or issue of import reform. After the new rules are revealed, veto players that lost attempt to restore their gains or minimize their loses by entering backroom channels through members of the legislature, the executive branch, or the president, or in some cases, mobilize public support. The design of the process promotes accommodating informal behavior. These two categories lead to the following main proposition.

> Proposition 2: The type of informal institutional system determines the capacity of the CN to use deference and external-internal sequencing to secure both the COG support and social veto player support.

Pivoting on Proposition 1, I draw out three propositions for CN behavior.

> Proposition 3: If the informal institutional procedure is arranged in a "complementary" position, where the principals (e.g., United States, Japan, commercial banks, etc.) and the agent (IMF) have narrow cleavages then the CN can use the external-internal tactic to shift Level IIA preferences, including the COG. This condition is necessary, but not sufficient for reform.

In the GOP cases, the CN had the maximum probability of credibility to use the external-internal tactic as the IMF was leveraged as the catalyst for continued foreign credit. Without such funding, the GOP economy would grind to a halt. Furthermore, in the context of the Uruguay Round and AFTA, the Philippines would be left out of the benefits engaged in free trade. See Table 12.1 for the Maximum Probability outcome type.

> Proposition 4: If the informal institutions have an "accommodative" function, where multiple principals and the agent have significant preference distances from each other, then the CN lacks the capacity to use the external-internal tactic.

In 1984, foreign creditors delegated the IMF to ensure the repayments of debt and its servicing. It served as a catalyst for extended foreign financing. This delegation, however, transmitted an unclear signal as the United States continued to transfer aid. It made attempts to apply the tactic as less than credible. Furthermore, the United States also had divisions within the cabinet and partisan cleavages both within and between the legislature and the executive. If the CN attempted to link external costs and benefits to internal cost and benefits, then he would lack credibility.

> Proposition 5: If the preferential distance between the principals and the agent is wide, then the CN can exploit the leverage and use it against the IMF.

In cases where the principals were ready to finance the GOP via the PAP, but the IMF stood in its way, then the divide can be leveraged by the CN to get the

Table 12.1 External-Internal Typology at Level IIA

Outcome Type	*Outcome Result*	*International Level IIB Leverage*	*Domestic Institutions*	*Primary International Actors*	*Level IIA Domestic Veto Players*	*Evidence*	*Notes*
Diffused Leverage	Success	Clear	Complimentary	Uruguay Round, and AFTA	Trade Ministry	1994 I	Strong commitment to emerging international/ regional free trade environment.
Maximum Probability	Success	Clear	Complimentary	IMF Principals	Security and trade bureaucracies	1989a I, 1989b P, 1989b I, 1989c P, 1991 P, 1991 I, and 1994 P	Principals, World Bank import requirement reforms, and PAP funding link with IMF program.
Mixed Signals	Qualified Success	Clear	Complimentary	IMF Principals	Rainbow cabinet	1986 P	Leverage limited due to both optimism of the new Aquino government, caution by private creditors, and declining oil prices. Also, Consensus cabinet rules.
Principal Uncertainty	Qualified Success	Unclear	Substitutive	IMF Principals and World Bank	Trade Ministers	1984 I and 1986 I	Principals preferences initially unclarified and changed. Reforms added at last minute.

Modest Credibility	Qualified Success	Unclear	Substitutive	IMF	COG (Marcos)	1984 P	Reform implemented in stages, but timing was important.
Principal Usurpation	No Attempt	Clear	Accommodating	Japan/PAP, Camdessus and European Principals	Security Cluster	1989a P	Japan/PAP timing favored GOP resistance.

IMF to concede on numerous sticking points. In the absence of this concession, then the GOP would miss the financing window. In the 1989 and 1991 observations, the IMF agreement had to precede the PAP, without which, the GOP would have suffered from balance of payments problems. Clearly, the IMF would have contributed to a problem it was purported to resolve. Using Proposition 2, I derive the following propositions.

> Proposition 6: If the CN faces accommodating informal functions, then he can use the deference tactic with the COG to secure the social veto player ratification.

Under the 1984 and 1989c, petroleum price observations, the CN used the deference tactic. The tripartite secured mainstream social veto players (i.e., labor) ratification via wage increases, who subsequently accepted the price increases. Other groups outside the tripartite process secured their own side-payments via their respective bureaucracies. The excluded veto players, numbers that turned out to be small in comparison to turnouts in other years, subsequently marched the streets. In the 1984 petroleum observation, Marcos made personal appeals to these groups, stressing the nature of the economic emergency and the urgency of reform. In the 1989c price increase, Aquino served as the deference player before veto players, securing their support for the increases. The next proposition list what happens when the conditions change to a competitive institutional interplay.

> Proposition 7: When the formal-informal interplay shifts to "competitive," using the deference of the COG fails.

In the 1991 and 1994 petroleum price observations, Aquino and Ramos used their institutional powers to sway veto players. The tripartite and multisectoral corporatist system was intended to filter out other social players, narrowing down their numbers and demands. However, the increased numbers, which included religious and industrial groups, had no remedy or agents to adjust their costs, save a full or partial reversal of prices. A deference strategy from the COG to social veto players in a competitive institutional arrangement does not work. This mechanism, however, does work in reverse, as social veto players can seek deference from the COG, who then request the ERB to review the increases.

> Proposition 8: The CN should not apply the external-internal strategy in a mainly regulatory context.

In the 1984 petroleum observation, Marcos framed the price increases as a national emergency, the social veto actors protested and demonstrated against the prices, but ultimately, they accepted the adjustment. In 1991,

CN Estanislao revealed before the ERB that the IMF wanted an increase in prices. This information was the first formal link between the IMF program and price increases, but in the other observations, the media has often made the link. Price increases fail to bring out supporters for oil company profits. While revealing the link with the IMF provides no benefits, there is no evidence that its linkage would ensure reform.

> Proposition 9: The CN should use the external-internal tactic to apply to distributive conflicts that have accommodative institutional arrangements.

External-internal tactics can link better international conditions to domestic players. The Uruguay Round and AFTA were significant international and regional developments that provided benefits to supporters of import reform and costs to opponents. In the 1991 and 1994 import observations, the GOP made the link that drew the international context together with the domestic industrial landscape. Linking the two arenas can be assisted by changing the narrative, a tactic that is far from novel. The GOP changed the domestic agenda from import reform to industrialization, which was sloganized by Ramos into the "Philippines 2000" campaign. This context tacitly labels opponents of reform as unpatriotic. As a demonstration that an external-internal and agenda changing tactic should be applied, we can view the outcome with its absence. The 1991 import observation promoted the New Economic Program as a Filipino program, one designed and implemented by the GOP, insulating itself from international and regional forces, was rolled out and eventually repealed.

CAUSAL MECHANISMS

This work focuses on providing a viable research program that examines the interplay between the Two-Level Model and IMF-recipient county negotiations. On the model, this work teases out two causal mechanisms that focus on the domestic CN, the deference, and external-internal sequencing patterns. Taking each tactic individually at *each* observation, this work shows the equifinality pattern produced by the tactic. For instance, the application of the tactic in the context of other variables produces a particular outcome. Taking each tactic individually at *all* the observations demonstrates how one cause—the tactic—can produce multifinality outcomes such as success, limited success, qualified success, reversed limited success, reverse failure, and no attempt. The causal mechanisms for each level are examined and assessed. The propositions listed above lay out the groundwork for these mechanisms. The external-internal mechanism has a more significant effect to move Level IIA

Table 12.2 External-Internal Typology at Level IIB

External-Internal Pattern Type	*External-Internal Pattern Type*	*International Level IIB Leverage*	*Domestic Institutions*	*Primary Level IIB Actors*	*Level IIB Veto Players*	*Evidence*	*Notes*
Deflected Source	Reversed Limited Success	Clear	Complimentary	IMF Principals and IMF	Manufacturing Industries and Cronies	1984 I	IMF link revealed. Credit in limited quantities. Social veto players viewed GOP as problem.
No Agreement Increase Cost	Qualified Success	Clear	Complimentary	Japan/PAP, World Bank, and IMF	Manufacturing industries and businesses	1989a I	Japan, PAP, and World Bank provided necessary leverage to shift Level IIB veto player preferences.
Non-seqeuenced, Isolated Costs	Reversed Failure	Clear	Complimentary	None	Manufacturing Industries, Legislature, and Bureaucracies	1989b I	Not linked with international trade negotiations. Simultaneous and quick consultations with broad-based veto players involved legislature and other bureaucracies.
Connection and Contact	Reversed Failure	Clear	Accommodating	IMF	Labor, Transportation, Manufacturing, and Religious groups	1994 P	IMF link made. It consulted with public veto actors and legislature. Failed to assuage broad and powerful costs of veto players.
Limited Costs Affect	Limited Success	Clear	Substitutive	IMF Principals and IMF	Labor and Transportation	1991 P	BANTA costs have marginal effect on broad-based coalition. Extended drawn-out process.

Certain Probability	Success	Clear	Substitutive	Uruguay Round and AFTA	Manufacturing Industries	1991 I and 1994 I	Linked international free trade negotiations with import reforms. Increased costs and benefits for support and veto players.
Increased Costs of No Agreement	Reversed limited success	Clear	Substitutive	IMF, United States, and World Bank	Import Substituting Industries and Cronies	1984 I	United States and World Bank refused to provide credit unless an IMF program was in place.
Hidden Connection	Qualified Success	Unclear	Complimentary	IMF	Labor and Transportation	1984 P	No link made with IMF. Reforms implemented in stages. Tripartite weakened "excluded" labor veto players.
Limited Benefits, Significant Costs	Reversed Limited Success	Unclear	Accommodating	World Bank and IMF	Manufacturing Groups	1986 I	World Bank and IMF programs direct costs on social actors, but provide little benefits for trade groups as no market access to other countries or coalition building in trade negotiations.
Hidden Connection and Crisis	Success	Unclear	Complimentary	IMF	Labor, transportation	1989b P and 1989c P	Military coup attempt and national emergency calls absorbed costs of broad-based and violent veto actors.

objections to reform, including shaping the preferences of the COG (see Table 12.1). It is limited in shaping the preferences of Level IIB social veto players in a mostly regulatory context such as petroleum prices (see Table 12.2). It becomes more effective in the context of shaping landscape of domestic Level IIB veto players in the distributive context such as import reform. This causal mechanism has been able to demonstrate compelling reasons for reform, but its weakness has been its capacity to explain all the reform limitations.

The deference causal mechanism has also been shown to illuminate affects at Levels IIA and IIB. At Level IIA, the tactic shows a critical starting point for CNs, but it also shows that the process has its limitations as the COG is subject to deference tactics from international and domestic Level IIBs. It short, the deference tactic at the beginning of the reform process cannot deal adequately with the deference at the end of the decision-making procedure. In this regard, the tactic can explain nuances to the outcomes. Its limitations are also derived from the propositions listed above.

The external-internal and deference causal mechanisms are highly probable explanations for the outcomes, but other causal explanations could have applications. To be sure, I test the two CN driven mechanisms with the IMF initiated technocratic alliance. The CN initiated tactics have more explanatory power than the alliance tactic, which has limitations on most of the observations. It explains the transfer of the Washington Consensus ideas via the acceptance of the reform at Level IIA and the absence of individuals to transfer the ideas from the IMF to the COG. Also, the alliance mechanism ignores the meeting of the minds already in play, where the IMF staff and CNs already have similar or identical preferences, with differences at the implementation phase. Furthermore, the technocratic alliance strategy glosses over the formal-informal institutional arrangements in play, whether or not they be complementary, accommodating, or substitutive.

This work also treats the outcomes as deviant cases as differences in the processes and procedures can lead to similar outcomes but based on different causes and different outcomes based on similar causes. The typologies of the deference and sequencing tactics are shown in Tables 12.1–12.4 with the pattern type in the first column and the result in the second. The tactics are the antecedents. The main and minor conditions laid out are the condition variables that determine the size or strength of the impact on the outcome.[3] International leverage and formal-informal institution interplay provide the main conditions that produce the thirty-five outcomes (1986 P and 1989a P observations do not contain Level IIB attempts).

Sequencing matters at Level IIA. Not only does securing support one step at a time matter, but an external-internal causal path must precede a deference path. External-internal sequencing provides the CN with significant leverage at Level IIA (see Table 12.1). International support can provide resources or linkage with issues to permit the CN to argue against cabinet opposition (1989b P, 1991 P, and 1994 P). The evidence suggests that an external-internal sequence favors reform over an internal-external pattern that uses a collective path to consensus as CNs apply a synergy tactic (the IMF and its access to financing) to a reverberating one (build coalitions). The effectiveness of the External-Internal pattern varies with the formal and informal institutional IMF principal interaction. In the 1986 petroleum and 1986 import observations, the GOP used the internal-external sequence, where it successfully resisted IMF insistence to reform petroleum prices and succumbed to IMF pressure to limited import reform. A complementary arrangement can ignore IMF pressure, while a substitutive arrangement can eventually bow to pressure. Deference, in either complementary or accommodating contexts, can secure support of the COG, assuming that external actors send a clear signal (see Table 12.3). If the causal path from the IMF and its principals to the domestic COG is clear, then external-internal and deference tactics can matter, whereas an obfuscated causal path weakens the CN's leverage. This pattern is reflected in the 1984 Import and 1986 Import observations (see Principal Uncertainty, Table 12.1). If these strategies fail at Level IIA, then the issue is moot at Level IIB.

At Level IIB, external-internal sequencing appears to be more effective than a simultaneous tactic (see Table 12.2). Reform in the 1989b I observation was presented as Philippine owned, isolated from any international context. After negotiations and the implementation of EO 413, the GOP was inundated with social veto player, legislative, and bureaucratic calls for reversal. In contrast, the 1991 Import observation was external-internal sequenced as a product of international and regional trade reforms, permitting both supporters and opponents of reform to petition their preferences.

Deference has a greater impact in regulatory arrangements than distributive. At Level IIB petroleum observations, deference secured labor with a wide net via the corporatists tripartite arrangement and the COG used suasion, subsidy differentiation, and some degree of coercion for social veto actors to absorb costs. For the CN, deference matters at the beginning of the Level IIB coalition building process. It becomes a liability later in the process as veto players can also use the tactic to reverse, partially or wholly, reforms as shown in Table 12.4 under Back-Ended Deference, Janus-Faced COG, and Two Game COG. In contrast, external-internal has little effect on increasing the costs of affected petroleum price increase. This tactic under a regulatory arrangement may not matter as shown in Table 12.2 External-Internal Pattern Type—Connection and Contact. This tactic has limited influence on veto players and does little to flesh out proponents. Any applied costs are reversible. In contrast, the external-internal tactic has a more

Table 12.3 Deference Typology at Level IIA

Deference Pattern Type	*Outcome Result*	*International Level IIB Leverage*	*Domestic Institutions*	*Primary International Actors*	*Domestic Deference Players*	*Evidence*	*Notes*
Structural Change leverage	Success	Clear	Complimentary	Uruguay Round, AFTA	Trade Ministry	1989b I, 1991 I, and 1994 I	International and regional trade environment provide context for trade positions
Broad Principal Conditionality Leverage	Success	Clear	Complimentary	IMF Principals, PAP, and World Bank	Trade Ministry and Security Cluster	1989a I, 1989b P, and 1989c P	Principals, World Bank import requirement reforms, and PAP funding link with IMF program
Principal Conditionality Leverage	Success	Clear	Complimentary	IMF Principals	Cluster E	1991 P	IMF Principals petroleum reform requirement link with IMF program
Traditional IMF Leverage	Success	Clear	Complimentary	IMF	Cluster E	1994 P	IMF requires fiscal reforms
Country Timing Leverage	No Attempt	Clear	Complimentary	PAP	Cluster E	1989a P	No leverage to sway Cluster E
Direct Veto Player Contact	Qualified Success	Unclear	Substitutive	IMF	Marcos	1984 P	IMF requires petroleum reform

Hands Tied Deference	Qualified Success	Clear	Complimentary	IMF, United States	Populist Side of Cabinet	1986 P	No international leverage; price reduction makes increase implausible. Consensus rules in cabinet with broad-based principals
COGs Dialogue	Success	Unclear	Substitutive	IMF Principals and World Bank	Populist Side of Cabinet	1986 I	Aquino defers to de Larosière
Convergence of Interests	Qualified Success	Unclear	Substitutive	IMF Principals and World Bank	Trade Minister	1984 I	IMF Principals turn on Marcos

Table 12.4 Deference Typology at Level IIB

Deference Pattern Type	*Outcome Result*	*International Level IIB Leverage*	*Domestic Institutions*	*Level IIA Deference Players*	*Level IIB Veto Players*	*Evidence*	*Notes*
Broad-Based Deferenc Type	Limited Success	Clear	Complimentary	COG (Aquino)	Congress, business, labor, the media, and NGOs	1989c P	Deference to the multisectoral meetings helped to minimize the size of the groups that reacted to the price-hike, muting the effects.
Divide and Absorb	Success	Clear	Complimentary	COG (Aquino)	Labor and Transportation	1989b P	Tripartite used to secure broad labor veto players. COG appeals to "excluded" veto players. Opposition coalition weakened. Remaining veto players pursue remedies from bureaucracies.
Maximum Deference Value	Success	Clear	Complimentary	Trade Ministry	Congress, bureaucracies, manufacturing, other social veto players	1991 I and 1994 I	Trade ministry consulted with other bureaucracies, congress, manufacturing, and social veto players.
Timed Divide and Absorb	Qualified Success	Clear	Accommodating	COG (Marcos)	Labor and Transportation	1984 P	Use Tripartite and reform process quickly and reform in stages. COG appeals to "excluded" veto actors. Opposition coalition weakened. Remaining veto players pursued remedies from bureaucracies.

Solitary Deference Player	Qualified Success	Clear	Complimentary	Trade Ministry	Manufacturing and other social veto players	1989a I	Trade Ministry installed safeguards to protect domestic industries.
Continued Negotiation Deference	Reversed Limited Success	Clear	Substitutive	COG (Aquino)	Trade Minister	1986 I	COG not used, but social veto actors accessed bureaucracies for side-payments or reversed some reforms.
Janus-Faced COG	Reversed Limited Success	Clear	Substitutive	COG (Marcos)	Import Substituting Industries and Cronies	1984 I	Unexpected level of resistance. COG used as deference for and against reforms and other forms of access to dilute reforms.
Back-Ended Deference	Reversed Failure	Unclear	Accommodative	Trade Minister	Import and Export Substituting Industries	1989b I	Unexpected level of resistance. Veto players petitioned bureaucracies, legislature, and COG.
Two Game COG	Reverse Limited Success	Unclear	Accommodative	COG (Aquino)	Labor and Transportation	1991 P	Unexpected level of resistance. Tripartie and other forms of consultation applied, but "excluded" some veto players. COG used as deference for and against reforms and other forms of access to dilute reforms.
Single Deference Reversal	Reversed Failure	Unclear	Accommodative	No deference players	Labor, Transportation, Church, Manufacturing Industries	1994 P	Tripartite used; no COG or legislature used for deference. Unexpected resistance veto players and strength petitioned COG for reversal.

significant effect on distributive conditions as it can bring out supporters of "external" benefits, those who benefit from synergy. Furthermore, this tactic benefits more from trade agreements (e.g., AFTA and Uruguay Round) rather than from IMF or World Bank programs. Reform supporters receive benefits not only from reduced costs of imports, but also from more open markets in other countries.

Researchers may also examine the Maximum Probability External-Internal tactic in Table 12.1. The same causal mechanism applied in two microeconomic issues, petroleum and imports, have the same outcome at Level IIA. This typology suggests that there may be variables here that are common and generalizable to the study of political economy. Six observations, both petroleum and import, at Level IIA reflect the same tactic and the same success outcome. Using a similar approach, both tactics applied at Level IIB on 1991 and 1994 on imports produce the same effect. See Table 12.4 under the Maximum Deference Value and Table 12.2 Certain Probability typologies. Researchers may be able to tease out variables that may be generalizable across cases.

The thirty-five typologies are far from parsimonious. The categorizations, however, take the Two-Level model and push it more toward a theory, defined by Shepsle as "the embellishment of a model in which features that are abstract in the model are made more concrete and specific in the theory."[4] Shepsle adds to his distinction between model and theory, "to move from an abstract formulation of group choice to a more concrete application involving a specific group (the U.S. Congress) at a specific time (1862 or 1932 or 1986) on a specific issue (raising national revenue) requires us to nail our colors to the mast on a whole variety of matters."[5] The typology presented in this book is not the end; it is merely a stepping stone. It provides researchers with opportunities to subsequently test these propositions against new evidence.[6] It also contributes to the type of studies Bennett refers to as "scientific realism," which contrast with the philosophies of science enunciated by Thomas Khun and his views of scientific progress via paradigms versus Imre Lakatos and his alternative the research program.[7] Scientific realism focuses on how combinations of mechanisms interact in specified and often recurrent scope conditions or contest to produce outcomes."[8]

IMPLICATIONS

On the theoretical level, I produce a verbal negotiation midrange theory where policy makers use strategies as leverage variables to produce desired outcomes. These two tactics—deference and external-internal sequencing—can shift veto player preferences to produce policy reforms, at either Level IIA or IIB. The theory also demonstrates the conditions under which these tactics have general limitations—unclear leverage signals and institutional arrangements. This typology provides significant theoretical consequences. It demonstrates that outcomes cannot be categorized as merely success or failures. Instead, I show that there are significant patterned variations in outcomes as well as different typologies of failures; as a result, I produce more refined conceptual patterns.

I likewise produce typologies that can push the boundaries of verbal negotiation theories. It encourages more research into other international bargaining situations or more IMF-recipient country negotiations to examine other variations of external-internal, deference, and simultaneous bargaining. I also generate more research questions. Under what conditions does cooperative simultaneous negotiations provide a productive route than sequencing? At what point should one switch from sequencing to a simultaneous tactic, or vice versa? Under what conditions can deference be removed from the end of the ratification process? How can the costs be increased in sequencing tactics? How can we better measure the levels of success and failures and the degree of IMF credibility? What other ways can substitutive and accommodative institutional arrangements be exploited to push for policy reform?

This work also shows the Two-Level framework, the connection that drives the external and domestic levels through the causal mechanisms, and how these conceptualizations are fused by the principal-agent institutional arrangement. Conceição-Heldt and Mello surveyed and categorized the literature based on the type of questions asked.[9] These questions ranged from the most satisfying level of analysis from which to explain foreign policy to the interaction and integration of domestic and international arenas.[10] The scale and spectrum of their review is impressive. They also suggest other areas for potential research, one of which is the combination of the Two-Level framework and the principal-agent relations, as presented by Oppermann.[11] More research in this area is needed. Furthermore, more research that examines the causal mechanisms that drive outcomes needs to be integrated with the framework.[12] A midrange theory that can capture IMF-recipient country interaction is critical, especially in an age where financial integration and global and regional trade is more pronounced. To be able to make predictions, we need a better understanding of the institutions as it provides a critical component for the analysis of action. The means toward stronger insights into these constraints is through the formal-informal typology and the institutional equilibrium model.

These institutional frameworks reveal the dynamics that constrain and provide incentives for individuals and their actions. By wedding the formal-informal typology with the principal-agent/institutional equilibrium model, we are able to examine the role of informal rules and the attempts by individuals to change rules, both formal and informal, to produce the desired equilibrium. It seems obvious that informal complementary rules promote more cooperation and an equilibrium outcome, but the rules may neither always provide efficiency nor equity to the rules. Rather, they establish the status quo where individuals and groups have expectations about the rules, its rewards, and punishments. To increase the capacity for both efficiency and equity, decision-makers must either design the institutions better or improve the capacity of the state to punish or enforce its rules. Design or change can also be accomplished by having creative or "imaginative" actors reshape the institutions. For instance, reshaping the rules of petroleum price regulation may have provided Philippine left-oriented transportation with greater expectations of inclusion into the process, transparency and equity; as a result, strikes and demonstrations in response to price increases may have been either muted or eliminated, reducing the possibility of military coups and ultimately promoting more stability. As it was, for the periods studied under this work, petroleum price increases were met with loud resistance, many times viciously loud and disruptive. The observations show that the GOP CNs and COG often engaged in rule changes to transmit their desired reforms, usually at the consultation levels. The GOP since the late 1990s has deregulated the petroleum industry and as such, prices are no longer determined by an energy board nor the GOP.

One should approach this essay's conclusions with caution. The study's research design may limit the model's ability to identify conditions beyond the IMF-Philippine relationship, which may contribute to the success or failure of deference or external-internal tactics. Furthermore, conditions have changed, even in the Philippines. Petroleum prices are deregulated and economic growth has been even in the Philippine case, the context has changed. As of this writing, the Philippines no longer uses the OPSF, it has had robust levels of economic growth and with a forecast of 6.8 percent GDP in the medium term.[13] These mechanisms may be used in cases in which one party is relatively evenly matched or shares a relatively close bond (e.g. Turkey). On the other hand, in cases involving other emerging market economies, the antecedent, conditional, and leverage variables may be sufficient to outweigh the kinds of positive effects found in the paper's microeconomic reform cases. To answer such an issue, researchers should examine other IMF-Asian country relationships. The IMF experiences in Indonesia and Thailand are ripe for further grounding of the hypotheses developed in this study.

The policy implications are also significant. I draw out the leverage variables for policy makers in the previous section. When sequencing under democratizing conditions, they face a dilemma where they must weigh their preferences for the outcome (reform) against the process, which is "manipulative, deceptive, coercive, and even plainly unethical."[14] The move toward transparency is the trend in developing countries and may rule out the capacity for sequencing.

We learn several things from this study. First, we reinforce the conclusion that a case study is a viable approach toward theory building.[15] Second, and in a related way, this approach can be designed with embedded observations to provide more analytical rigor to its conclusions. Third, since we have found different processes patterns, we have generated more research opportunities. Finally, in a practical way, we highlight that consistent international signals and the formal-informal arrangement of domestic institutions are the keys toward economic reform, suggesting that CNs and other policy makers should do what they can to get consistent signals from international Level IIB actors and help to redesign institutions to form more complementary formal-informal arrangements.

This work has both theoretical and practical purposes. While the former is significant, it is equal in comparison with the latter. Reform can be initiated through the institutional policy process only with great care. Policy makers must use all their acumen and skill to navigate their way through particular political arrangements. Through reform, they take the first and necessary, but not sufficient, step toward their greater end—human development.

NOTES

1. Lijpart, 692.
2. Stone, 59; and Shepsle 2017, 121–124.
3. Van Evera 11.
4. Shepsle 2010, 9.
5. Ibid.
6. Levy, 13.
7. Bennett 2013. For the discussion on paradigms, see Kuhn. For the discussion on research programs, see Lakatos; and Keohane 1986.
8. Bennett 2013, 470.
9. Conceição-Heldt and Mello.
10. Ibid., 5.
11. Ibid., 20 and Oppermann.
12. See Bennett. Notable examples of using causal mechanisms can be found in Schoppa; Mayer; and Paarlberg.
13. International Monetary Fund 2017.
14. Sebinus 1996, 344.
15. See Lijphart; and Levey.

Bibliography

Alburo, Florian A. 1991. Sustaining Trade Liberalization and Reforms in the Philippines. *Philippine Review of Economics and Business* 28(2): 154–170.

Alburo, Florian A., et al. 1986. *Economic Recovery and Long-Run Growth: Agenda for Reforms*, vol. I, Main Report. Manila, Philippines: Philippine Institute for Development Studies.

Alcuaz, Coco. 2011. Ex-DTI Chief Navarro Dies. *ABSCBN News*, 7 July 2011. http://news.abs-cbn.com/business/07/07/11/ex-dti-chief-navarro-dies. Accessed on 19 November 2017.

Alm, James, Susan Collins, Paul Krugman, and Eli Remolona. No Date. The Philippine Economy. Manuscript for the United Nations Development Program.

Almonte, Jose. 2016. National Security Advisor. Interview with Tony Abad. *Political Capital*. Published 28 March 2016. https://www.youtube.com/watch?v=ARZVCX-Pqrc. Accessed on 19 November 2017.

Alonzo, Ruperto P., Dante B. Canlas, Raul V. Fabella, Felipe M. Medalla, and Cayetano Paderanga, Jr. 1988. An Alternative Debt Strategy: Fire Our Negotiating Team. As reprinted in *Philippine Free Press*, 3 September.

Alonzo, Ruperto P., et al. 1990. New Economic Plan: Making it Work. Published in the *Manila Chronicle*, 16 and 17 June 1990, 12 and 14.

Antonio, Jr., Emilio T. 1992. *The Import Levy: Did It Work?* Pasig City, Philippines: Center for Research and Communication Staff Memos, 6.

Aquino, Corazon. 1986. Interview with Corazon Aquino. *Far Eastern Economic Review*, 28 August 1986, 30.

Baldwin, Robert E. 1975. *The Philippines: Foreign Trade Regimes and Economic Development*. New York, New York: National Bureau of Economic Research.

Banal III, Cornado R. 1986. IMF Mission Tackles PNB Case. *Business Day*, 12 March 1986, 3.

Bapat, Navin A. and Bo Ram Kwon. 2015. When Are Sanctions Effective? A Bargaining and Enforcement Framework. *International Organization* 69(1): 131–162.

Bautista, Romeo M. 1989. *Impediments to Trade Liberalization in the Philippines*. London, England: Grower for the Trade Policy Research Centre.

Bautista, Romeo, John Power, and Associates. 1979. *Industrial Promotion Policies in the Philippines*. Manila, Philippines: Philippine Institute for Development Studies.

Bello, Walden, David Kinley and Elaine Elinson, eds. 1982. *Development Debacle: The World Bank in the Philippines*. San Francisco, California: Institute for Food and Development Policy.

Bennett, Andrew. 2013. The Mother of All Isms: Causal Mechanism and Structured Pluralism in International Relations Theory. *European Journal of International Relations* 19(3): 459–481.

Bennett, Andrew and Jeffrey T. Checkel. 2015. Process Tracing: From Philosophical Roots to Best Practices. In *Process Tracing: From Metaphor to Analytic Tool*, edited by Andrew Bennett and Jeffrey T. Checkel. Cambridge, United Kingdom: Cambridge University Press, 3–37.

Bernardo, Romeo L. and Marie Christine G. Tang. 2008. *The Political Economy of Reform during the Ramos Administration (1992–1998)*. The International Bank for Reconstruction and Development/The World Bank.

Bird, Graham, 1996. The IMF and Developing Countries: A Review of the Evidence and Policy Options. *International Organization* (Summer), 50(3): 477–511.

Blitz, Amy. 2000. *The Contested State: American Foreign Policy and Regime Change in the Philippines*. New York: Roman and Littlefield Publishers, Inc.

Boudreau, Vincent G. 1996. Of Motorcades and Masses: Mobilization and Innovation in Philippine Protest. In *The Revolution Falters: The Left in Philippine Politics After 1986*, edited by Patricio N. Abinales. Ithaca, New York: Cornell Southeast Asia Program, 60–82.

Boughton, James M. 2001. *Silent Revolution: The International Monetary Fund 1979–1989*. Washington, D.C.: The International Monetary Fund.

Brillantes, Jr., Alex B. 1993. The Philippines in 1992: Ready for Take Off? *Asian Survey* (February), 2(33): 224–230.

Broad, Robin. 1988. *Unequal Alliance, 1979–1986: The World Bank, the International Monetary Fund, and the Philippines*. Manila, Philippines: Ateneo de Manila University Press.

Browne, Christopher. 1993. Can the RP Economy Take-Off with or without the IMF. Speech prepared for the Joint Foreign Chambers Luncheon Meeting, Makati, Manila, Philippines. 7 July.

Buchanan, James M. 1980. Rent Seeking and Profit Seeking. In *Toward a theory of Rent-Seeking Society*, edited by James Buchanan, Robert D. Tollison, and Gordon Tullock. College Station, Texas: Texas A & M Press, 3–15.

Bulletin Times. 1984a. 12 March, 13.

———. 1984b. 18 May, 1.

———. 1984c. 22 October, 1–4

———. 1984d. 23 October, 13–14.

———. 1986a. 10 March, 1.

———. 1986b. 12 March, 13.

Business Bulletin. 1988a. 8 September, 11.

———. 1988b. 16 October, 11.

———. 1988c. 18 October, 11.
———. 1988d. 16 November 11.
———. 1988e. 23 November, 11.
———. 1988f. 26 November, 11.
Business Day. 1984a. 3 January, 2.
———. 1984b. 13 February, 3.
———. 1984c. 14 February, 3.
———. 1984d. 6 March, 3.
———. 1984e. 14 March, 3.
———. 1984f. 29 March, 3.
———. 1984g. 30 March, 3.
———. 1984h. 22 August, 2.
———. 1984i. 5 September, 2.
———. 1984j. 7 September, 2.
———. 1984k. 14 September, 2.
———. 1984l. 1 September, 3.
———. 1984m. 3 September, 3.
———. 1984n. 12 September, 2.
———. 1984o. 1 October, 3.
———. 1984p. 2 October, 2.
———. 1984q. 12 October, 2.
———. 1984r. 30 October, 2.
———. 1984s. 2 November, 2.
Business International. 1989. 29 September, 1.
Business World. 1993. 11 January, 1.
———. 1994. 1 December, 1.
———. 1997. 29 April, 1.
Calvert, Randall. 1995. The Rational Choice Theory of Social Institutions: Cooperation, Coordination, and Communication. In *Modern Political Economy: Old Topics, New Directions*, edited by Jeffrey S. Banks and Eric A. Hanushek. New York, New York: Cambridge University Press, 216–269.
Caraway, Teri L., Stephanie J. Rickard, and Mark S. Anner. 2012. International Negotiations and Domestic Politics: The Case of IMF Labor Market Conditionality. *International Organization* (66)1, 27–61.
Casper, Brett A. 2015. IMF Programs and the Risk of a Coup d'état. *Journal of Conflict Resolution* (59)5: 1–33.
Central Bank of the Philippines. 1984a. Central Bank Memorandum to Authorized Agents of Banks, No. 15, 15 February.
———. 1984b. Circular No. 1029, Consolidated Rules and Regulations to Govern Import Transactions. 12 October.
———. 1984c. 1983 Annual Report.
———. 1985. 1984 Annual Report.
———. 1986a. Central Bank Circular 1100. 30 April.
———. 1986b. 1985 Annual Report.
———. 1987. 1986 Annual Report.

———. 1988. 1987 Annual Report.

———. 1992a. 1991 Annual Report.

———. 1992b. Central Bank Circular 1347, 27 July.

———. 1992c. Central Bank Circular 1353, Further Liberalizing Foreign Exchange Regulations, 24 August.

———. 1992d. Central Bank Circular 1356, 25 September.

Chambers, Paul. 2014. Superficial Consolidation: Security Sector Governance and the Executive Branch in the Philippines Today. In *Security Sector Reform in Southeast Asia: From Policy to Practice*, edited by Felix Heiduk. New York, New York: Palgrave Macmillan, 102–130.

Chand, Nayan. 1984. Vote of no Confidence. *Far Eastern Economic Review*, 125, 13 September, 16.

Chaudoin, Stephen, Helen V. Milner and Xun Pang. 2015. International Systems and Domestic Politics: Linking Complex Interactions with Empirical Models in International Relations. *International Organization* 69(2): 275–309.

Checkel Jeffrey, T. 2015. Mechanisms, Process, and the Study of International Institutions. Process Tracing: From Philosophical Roots to Best Practices. In *Process Tracing: From Metaphor to Analytic Tool*, edited by Andrew Bennett and Jeffrey T. Checkel. Cambridge, United Kingdom: Cambridge University Press, 74–97.

Clarke, Gerard. 1994. Kabisig: Aquino's Lasting Legacy? *The Philippine Quarterly of Culture and Society* 22(1): 37–45.

Collas-Monsod, Solita. 1989. *Debt or Development: Philippine Imperatives and the Conventional Strategy for Debt Management*, paper presented to the Symposium on the Present and Future of the Pacific Basin Economy: A Comparison of Asia and Latin America. Tokyo, Japan: Institute of Developing Economies.

———. 1991. *The IMF Stabilization Program: Implications for the Philippines*. Speech at Freedom from Debt Coalition Forum, Makati Business Club, 15 February.

Collier, David, Henry E. Brady, and Jason Seawright. 2010. Critiques, Responses, and Trade-Offs: Drawing Together the Debate. In *Rethinking Social Inquiry: Diverse Tools, Shared Standards*, 2nd edn, edited by Henry E. Brady and David Collier. Lanham, Maryland: Rowman and Littlefield Publishers, Inc. 125–159.

Conceição-Heldt, Eugénia and Patrick Mello. 2017. Two-Level Games in Foreign Policy Analysis. *Oxford Research Encyclopedia of Politics* (June).

Davide Commission Report: The Final Report of the Fact-Finding Pursuant to R.A. No. 6832. 1990. Hilario G. Davide, Jr. Chair. Makati, Metro Manila, Philippines: Bookmark, Inc.

de Dios, Emmanuel S., ed. 1984. *An Analysis of the Philippine Crisis: A Workshop Report*. Quezon City, Philippines: University of the Philippines Press.

de Dios, Emmanuel S. 1992. The Aquino-Estanislao Recession. In *Bonds and Bondage: A Reader on Philippine Debt*, edited by Emmanuel S. de Dios and Joel Rocamora. Manila, Philippines: Transnational Institute, Philippine Center for Policy Studies, and Freedom from Debt Coalition.

de Dios, Emmanuel S. and Paul Hutchcroft. 2003. Political Economy. In *The Philippine Economy: Development, Policies, and Challenges*, edited by Arsenio M. Balisacan and Hal Hill. Oxford, United Kingdom: Oxford University Press, 45–73.

Del Rosario, Jr., Ramon R. 1996. Philippine Economic Prospects and the IMF Exit Program. *Manila Bulletin* 1995 Yearbook.

Department of Energy. n.d. Price Build-up. Secured from IBON Databank.

Dimanlig, Aarne B. 1984. *The Philippine Energy Sector: Bright Spot in a Darkened Environment.* Pasig City, Philippines: Center for Research and Communications, Staff Memos, 27 (December).

Diokno, Benjamin E. 1992. *Philippine Macroeconomic Policies Affecting Households.* Working Paper Series No. 92–17. Manila, Philippines: Philippine Institute for Development Studies. September.

Dohner, Robert S. 1988. Aquino and the Economy: An Assessment of the First Three Years. *Pilipinas* 9 (Fall): 1–32.

Dohner, Robert S. and Ponciano Intal, Jr. 1989. The Marcos Legacy: Economic Policy and Foreign Debt in the Philippines. In *Developing Country Debt and Economic Performance,* vol 3, edited by Jeffrey D. Sachs and Susan M. Collins. Chicago, Illinois: University of Chicago Press.

Dohner, Robert S. and Stephan Haggard. 1994. *The Politically Feasibility of Adjustment in the Philippines.* Paris, France: Development Center of the Organization for Economic Co-Operation and Development.

Dolan, Ronald E. 1991. ed. *The Philippines: A Country Study.* Washington, D.C.: Federal Research Division, Library of Congress.

Doronila, Amando. 1990. Orbos Counters Backlash. *Manila Chronicle*, 12 December, 4.

———. 1992. *The State, Economic Transformation, and Political Change in the Philippines, 1946–1972.* Singapore: Oxford University Press.

Dreher, Alex and Martin Gassebner. 2012. Do IMF and World Bank Programs Induce Government Crises? An Empirical Analysis. *International Organization* 66(2): 329–358.

Eckstein, Harry. 1975. Case Study and Theory in Political Science. In *Strategies of Inquiry*, edited by Fred I. Greenstein and Nelson W. Polsby. Reading, Massachusetts: Addison-Wesley Publishing Company, 79–131.

Economic Recovery and Long-Run Growth: Agenda /or Reforms. 1986. Main Report, Vol. 1. Manila, Philippines: Philippine Institute for Development Studies. 1 May.

Elster, Jon. 1990. When Rationality Fails. In *The Limits of Rationality*, edited by Karen Schweers Cook and Margaret Levi. Chicago, Illinois: The University of Chicago Press, 19–51.

Energy Regulatory Board. n.d. Historical Petroleum Product Prices (Per Date of Adjustment).

Estanislao, Jesus P. 1992. Finance, the Aquino Administration Record and Legacy (1986–1992), President Aquino and Her Cabinet. UP. Public Lectures on the Aquino Administration and the Post-EDSA Government (1986–1992), vol I. Diliman, Quezon City, Philippines: University of Philippine Press.

Evans, B. Peter. 1993. Building an Integrative Approach to International and Domestic Politics: Reflections and Projections. In *Double-Edged Diplomacy: International Bargaining and Domestic Politics*, edited by Evans, Harold K. Jacobson, and Robert D. Putnam. Berkeley, California: University of California Press, 397–430.

Finch, C. David. 1989. *The IMF: The Record and the Prospect, Essays in International Finance*, No. 175. Princeton, New Jersey: Princeton University, International Finance Section. May.

Fisher, Roger and William Ury. 1981. *Getting to Yes: Negotiating Agreement Without Giving In*. Boston, Massachusetts: Houghton Mifflin.

Frieden, Jeffry A., David A. Lake, and Kenneth A. Schultz. 2016. *World Politics: Interest, Interactions, and Institutions*, 3rd. New York, New York: W. W. Norton and Company.

Friman, H. Richard. 1993. Side Payments versus Security Cards: Domestic Tactics in Economic Negotiations. *International Organization* (Summer) 47(3): 387–410.

Galang, Jose. 1988a. Philippines Seeks New Credits. *Far Eastern Economic Review*, 141, 22 September, 92.

———. 1988b. Manila Asks for More. *Far Eastern Economic Review*, 142, 1 December, 78.

Gamboa, Reyanaldo D. 1991. *Oil Price Deregulation: A Solution or another Problem*. Pasig City, Philippines: Center for Research and Communication Staff Memos, 30.

Garchitorena, Gilbert G. 1991. *Petroleum Pricing Policy and Its Objectives*. Pasig City, Philippines: Center for Research and Communication, Staff Memos, 11.

General Agreement on Tariffs and Trade. 1986. Committee on Balance-of-Payments Restrictions. 1986 Consultation with the Philippines: Basic Document for Consultation. BOP/269. 24 November. https://docs.wto.org/gattdocs/q/GG/BOP/269.PDF. Accessed on 3 March 2018.

George, Alexander L. and Andrew Bennett. 2005. *Case Studies and Theory Development in the Social Sciences*. Cambridge MA: MIT Press.

George, Alexander L. and Richard Smoke. 1974. *Deterrence in American Foreign Policy: Theory and Practice*. New York: Columbia University Press.

Gigot, Paul A. 1986a. Ongpin's Vision. *Wall Street Journal*, 5 March, 13.

———. 1986b. Manila's Economic Revolutionary. *Wall Street Journal*, 5 March 13.

Goldstein, Judith. 1998. International institutions and domestic politics: GATT, WTO, and the liberalization of international trade. In *The WTO as an International Organization,* edited by Annie O. Krueger. Chicago, Illinois: University of Chicago Press, 133–152.

Gould, Erica. 2003. Money Talks: Supplementary Financiers and International Monetary Fund Conditionality. *International Organization* (57): 551–586.

Gregor, A. James. 1984. *Crisis in the Philippines: A Threat to U.S. Interest*. Washington, D.C.: Ethics and Public Policy Center.

Grief, Avner. 2006. *Institutions and the Path to the Modern Economy: Lessons from Medieval Trade*. New York, New York: Cambridge University Press.

Gutner, Tamar. 2017. *International Organizations in World Politics*. Thousand Oaks, California: CQ Press.

Haas, Peter. 1992a. *Saving the Mediterranean*. New York, New York: Columbia University Press.

———. ed. 1992b. Knowledge, Power, and International Policy Coordination. *International Organization* (Winter) special issue, 46: 1–35.

Haggard, Stephan. 1989. The Philippines: Picking Up After Marcos. In *The Promise of Privatization,* edited by Raymond Vernon. New York, New York: Council on Foreign Relations.

———. 1990. The Political Economy of the Philippine Debt Crisis. In *Economic Crisis and Policy Choice: The Politics of Adjustment in the Third World*, edited by Joan M. Nelson. Princeton, New Jersey: Princeton University Press.

———. 1991. *Pathways from the Periphery: The Politics of Growth in the Newly Industrialized Countries*. Ithaca, New York: Cornell University Press.

Hall, Peter and Rosemary C. R. Taylor. 1996. Political Science and the Three New Institutionalisms. *Political Studies* 44(5): 936–957.

Hawes, Gary. 1987. *The Philippine State and the Marcos Regime: The Politics of Export*. Ithaca, New York: Cornell University Press.

Hawkins, Darren G., David A. Lake, Daniel L. Nielson, and Michael J. Tierney, eds. 2006. *Delegation and Agency in International Organizations*. New York, New York: Cambridge University Press.

Hay, Colin. 2008. Constructivist Institutionalism. In *The Oxford Handbook of Political Institutions,* edited by R. A. W. Rhodes, Sarah A. Binder, and Bert A. Rockman. Oxford: Oxford University Press, 56–74.

Heldt, Eugénia. 2013. Two Level Games and Trade Cooperation: What do we now Know? *International Politics* (50)4: 579–599.

Helmke, Gretchen and Steven Levitsky. 2004. Informal Institutions and Comparative Politics: A Research Agenda. *Perspectives on Politics* (2)4: 725–740.

Hince, Kevin and Jacqui Lynch. 1997. Labour Relations and Tripartism: The Relevance of the Philippines Model to the Micro-States of the South Pacific. In *Current Research in Industrial Relations: Proceedings of the 11th AIRAANZ Conference,* edited by T. Bramble, B. Harley, R. Hall and E. Whitehouse. Brisbane: AIRAANZ, 153–163.

Hutchcroft, Paul D. 1991. Oligarchs and Cronies in the Philippine State: The Politics of Patrimonial Plunder. *World Politics* (April) 43(3): 414–450.

———. 1993. Selective Squander: The Politics of Preferential Credit Allocation in the Philippines. In *The Politics of Finance in Developing Countries*, edited by Stephan Haggard, Chung H. Lee and Sylvia Maxfield. Ithaca, New York: Cornell University Press, 165–198.

———. 1996. *The Philippines at the Crossroads: Sustaining Economic and Political Reform*. New York, New York: Asian Society.

———. 1998. *Booty Capitalism: The Politics of Banking in the Philippines*. Ithaca, New York: Cornell University Press.

Iida, Keisuke. 1993. When and How Do Domestic Constraints Matter: Two Level Games with Uncertainty? *Journal of Conflict Resolution* (September) 37(3): 403–426.

Ikenberry, G. John. 1988. Market Solutions for State Problems: The Internatinal and Domestic Politics of American Oil Decontrol. In *The State and American Foreign Economic Policy*, edited by G. John Ikenberry, David A. Lake, and Michael Mastanduno. Ithaca, New York: Cornell University Press, 151–178.

IMF Survey. 1985. 7 January, 12–13.

———. 1986. 3 November, 351.

International Monetary Fund. 1984. Philippines-Staff Report for the 1984 Article IV Consultation. Internal Document. EBS/84/117, 29 May.
———. 1985. Philippines, First Review Under Stand-by Agreement. Internal Document. EBS/85/109, 1 May.
———. 1986. International Monetary Fund Press Releases. IMF Releases 4th Tranche of $270 Million Loan. Obtained from IBON Databank.
———. 1986a. Letter of Intent. EBS/86/222. 28 October. Obtained from IBON Data Bank.
———. 1989a. Addition to the IMF Memorandum of Economic Policy, IMF Staff Report for the 1989 Article IV consultation, Review, Waiver, and Modification of the Extended Fund Facility. ESB 89/229, 1 December.
———. 1989b. Philippines—Staff Report for the 1989 Article IV Consultation and Review, Waiver, and Modification of the Extended Arrangement. EBS/89/229, 15 December.
———. 1991. Philippine-Staff Report for the 1990 Article IV Consultation and Request for Stand-by Arrangement and Purchase Under the Compensatory and Contingency Financing Facility, Internal Document, 6 February 1991, EBS/91/17.
———. 1994. International Monetary Fund Press Release. *Asian Wall Street Journal* Press release no. 94/43, 24 June.
———. 2002. Evaluation of Prolonged Use of IMF Resources: Evaluation Report. Independent Evaluation Office.
———. 1994. Confidential aide memoir as published to the *Manila Chronicle*, 1 October, 16.
———. 2002. Evaluation of Prolonged Use of IMF Resources: Evaluation Report. Independent Evaluation Office. Chapter 10.
———. 2017. IMF Staff Completes 2017 Article IV Mission to Philippines. Found at https://www.imf.org/en/News/Articles/2017/08/08/pr17321-philippines-imf-completes-2017-article-iv mission
Ivanova, Anna, Wolfgang Mayer, Alex Mourmouras, and George Anayiotos. 2003. *What Determines the Implementation of IMF-Supported Programs?* IMF Working Paper. WP/08/3
Jenkins, Shirley. 1954. *American Economic Policy toward the Philippines*. Stanford, California: Stanford University Press.
Jensen, Michael and William Meckling. 1976. Theory of the Firm: Managerial Behavior, Agency Costs and Ownership Structure. *Journal of Financial Economics* 3 October: 305–360.
Joaquin, Nick. 1990. *Jaime Ongpin, The enigma: A profile of the Filipino as manager*. Metro Manila, Philippines. J. V. Onpgin Institute of Business and Government, 1st.
Jose, Vivencio R., eds. 1984. *Mortgaging the Future: The World Bank and the IMF in the Philippines*. Quezon City, Philippines: Foundation for Nationalist Studies.
Kahler, Miles. 1992. External Influence, Conditionality, and the Politics of Adjustment. In *The Politics of Economic Adjustment: International Constraints, Distributive Conflicts, and the State,* edited by Stephan Haggard and Robert R. Kaufman. Princeton, New Jersey: Princeton University Press, 89–136.

———. 1993. Bargaining with the IMF: Two-level Strategies and Developing Countries. In *Double-Edged Diplomacy: International Bargaining and Domestic Politics,* edited by Peter B. Evans, Harold K. Jacobson, and Robert D. Putnam. Berkeley, California: University of California Press, Berkeley, 363–394.

Kamm, Henry. 1981. "Marcos Frees 341: Lifts Martial Law." *New York Times*, January 18. Accessed on 30 October 2017. http://www.nytimes.com/1981/01/18/world/marcos-frees-341-lifts-martial-law.html

Karnow, Stanley. 1989a. Reagan and the Philippines: Setting Marcos Adrift. *New York Times*. 19 March. Accessed on 30 October 2017. http://www.nytimes.com/1989/03/19/magazine/reagan-and-the-philippines-setting-marcos-adrift.html?pagewanted=all

———. 1989b. *In Our Image: America's Empire in the Philippines*. New York, New York: Random House.

Kaufman, Robert R. 1990. Stabilization and Adjustment in Argentina, Brazil, and Mexico. In *Economic Crisis and Policy Choice: The Politics of Adjustment in the Third World*, edited by Joan Nelson. Princeton, New Jersey: Princeton University Press, 63–112.

Keohane, Robert O. 1984. *After Hegemony: Cooperation and Discord in the World Political Economy*. Princeton, New Jersey: Princeton University Press.

———. 1986. Theory of World Politics: Structural Realism and Beyond. In *Neorealism and its Critics*, edited by Robert O. Keohane. New York: Columbia University Press, 158–203.

Killick, Tony, and Moazzam Malik, and Marcus Manuel. 1992. What Can We Know about the Effects of IMF Programmes? *World Economy* 15(5): 575–598.

King, Gary, Robert O. Keohane, and Sydney Verba. 1994. *Designing Social InquiryScientific Inference in Qualitative Research*. Princeton, New Jersey: Princeton University Press.

Kiondo, Andrew. 1992. The Nature of Economic Reforms in Tanzania. In *Tanzania and the IMF: The Dynamics of Liberalization*, edited by Horace Campbell and Howard Stein. Boulder, Colorado: Westview Press, 21–42.

Krugman, Paul, James Alm, Susan Collins, and Eli Remolona. 1992. *Transforming the Philippine Economy*. Metro Manila, Philippines. National Economic Development Authority and United Nations Development Programme.

Kuhn, Thomas S. 2012. *The Structure of Scientific Revolutions*, 4th edn. Chicago, Illinois: The University of Chicago Press.

Kuruvilla, Surosh. 1995. Economic Development Strategies, Industrial Relations Policies and Workplace IR/HR Practices in Southeast Asia. In *The Comparative Political Economy of Industrial Relations,* edited by Kirsten S. Wever and Lowell Turner. Madison, Wisconsin: Industrial Relations Research Association, 115–150.

Lakatos, Imre. 1970. Falsification and the Methodology of Scientific Research Programmes. In *Criticism and the Growth of Knowledge*, edited by Imre Lakatos and Alan Musgrave. New York, New York: Cambridge University Press, 91–196.

Lallana, Emmanuel C. 1994. The Political Economy of Philippine Tariff Reform. *Philippine Political Science Journal* 18(37–38): 29–56.

Lamberte, Mario B. 2003. *Central Banking in the Philippines: Then, Now, and the Future*. Manila, Philippines: Philippine Institute for Development Studies.

Lamberte, Mario B. and Josef T. Yap. 1991. *The Impact of the Gulf Crisis on the Philippine Economy*. Working Paper Series No. 91–03. Manila, Philippines: Philippine Institute for Development Studies, October.

Lavelle, Kathryn C. 2011. *Legislating International Organization: The US Congress, the IMF, and the World Bank*. Oxford, United Kingdom: Oxford University Press.

Levi, Margaret, Karen S. Cook, Jodi A. O'Brien, and Howard Faye. 1990. Introduction: The Limits of Rationality. In *The Limits of Rationality*, edited by Karen Schweers Cook and Margaret Levi. Chicago, Illinois: The University of Chicago Press, 1–18.

Levy, Jack S. 2008. Case Studies: Types, Designs, and Logics of Inference. Presidential Address. *Conflict Management and Peace Science* 25: 1–18.

Lichauco, Alejandro A. 1984. The International Economic Order and the Philippines Experience. In *Mortgaging the Future: The World Bank and the IMF in the Philippines*, edited by Vivencio R. Jose. Quezon City, Philippines: Foundation For nationalist Studies.

Lijphart, Arend. 1971. Comparative Politics and the Comparative Method. *American Political Science Review* (65)3: 682–693.

Lindsey, Charles W. 1993. The Economy. In *Philippines: A Country Study*, edited by Ronald E. Dolan. Washington, D.C.: GPO for the Library of Congress, 117–187.

Linz, Juan L. 1975. Totalitarian and Authoritarian Regimes. In *Handbook of Political Science, Macropolitical Theory*, vol. 3, edited by Fred I. Greenstein and Nelson Polsby Reading, Massachusetts: Addison-Wesley Publishing Co., 175–411.

Lopez, Mario Antonio G. 2002. Corporate Governance in East Asia: Implications for Economic Efficiency. In *Asian Economic Recovery: Policy Options for Growth and Stability*, edited by Tan Kong Yam. Singapore: Singapore University Press, 125–56.

Lowi, Theodore, J. 1964. American Business, Public Policy, Case-Studies, and Political Theory. *World Politics* 16(4): 677–715.

Lowndes, Vivien. 2010. The Institutional Approach. In *Theory and Methods in Political Science*, 3rd, edited by David Marsh and Gerry Stoker. London: Palgrave Macmillian, 60–79.

Macapagal-Arroyo, Gloria. 1991. Recent Trade Reforms in the Philippines. *The Philippine Economic Journal*, 30(70).

MacIntyre, Andrew, T. J. Pempel, and John Ravenhill, eds. 2008. *Crisis as Catalyst: Asia's Dynamic Political Economy*. Ithaca, New York: Cornell Studies in Political Economy.

Makasiar, Gary, No Date. *Government Intervention in the Energy Industry*. Paper presented to the Philippine Economic Society.

Makati Business Club. 1984a. *Economic Papers*. July, 10.

———. 1984b. *Economic Papers*. August, 3.

———. 1984c. *Economic Papers*. September, 11–12.

———. 1986. *Economic Papers*. October, 10–11.

———. 1990. *Economic Papers*. September, 3.
Manila Bulletin. 1986a. 13 April, 17.
———. 1986b. 22 April, 1.
———. 1986c. 6 May, 1.
———. 1986d. 7 May, 17.
———. 1986e. 22 May, 1.
———. 1986f. 23 May, 21.
———. 1986g. 8 June, 25.
———. 1986h. 18 July, l.
———. 1986i. 18 July, 17.
———. 1986j. 21 July, 17.
———. 1986k. 24 July, 13.
———. 1986l. 31 July, 22.
———. 1986m. 5 August, 13.
———. 1986n. 3 September, 17.
———. 1986o. 13 September, 17.
———. 1986p. 22 October, 23.
———. 1988a. 7 September, 11.
———. 1988b. 12 October, 25.
———. 1989a. 1 February, 1.
———. 1989b. 6 March, 5–6.
———. 1989c. 7 March, 5–6.
———. 1989d. 15 March, 3–5
———. 1989e. 16 March, 3–5.
———. 1989f. 21 April, 22.
———. 1990a. 6 February, 21.
———. 1990b. 4 March, 25.
———. 1990c. 5 April, 25.
———. 1990d. 26 June, 25.
———. 1990e. 27 June, 25.
———. 1990f. 29 June, 25.
———. 1990g. 2 August, 25.
———. 1990h. 6 December, 1.
———. 1991a. 31 January, 25.
———. 1991b. 11 June, 25.
———. 1991c. 17 June, 25.
———. 1991d. 15 July, 25.
———. 1991e. 28 August, 25.
———. 1992. 22 December, 25.
———. 1993a. 28 January, 25.
———. 1993b. 3 March, 25.
———. 1993c. 23 April, 23.
———. 1993d. 27 October, 22.
———. 1994a. 7 January, B–1.
———. 1994b. 11 January, B–1.

———. 1994c. 13 January, B–1.
———. 1994d. 14 January, B–1.
———. 1994e. 15 January, B–1.
———. 1994f. 24 January, B–1.
———. 1994g. 26 January, B–1.
———. 1994h. 28 January, B–1.
———. 1994i. 5 February, B–5.
———. 1994j. 19 February, 23.
———. 1994k. 14 July, 25.
———. 1994l. 22 July, 25.
———. 1994m. 6 September, 24.
———. 1994n. 12 October, 25.
Manila Chronicle. 1986a. 16 September, 1.
———. 1986b. 18 September, 1.
———. 1986c. 1 October, 4.
———. 1987. 30 December 1987, 1 and 2.
———. 1988a. 14 November, 18.
———. 1988b. 22 December, 1 and 2.
———. 1988c. 23 December, 1 and 2.
———. 1988d. 24 December, 1 and 3.
———. 1988e. 25 December, 1 and 2.
———. 1988f. 26 December, 1 and 3.
———. 1990a. 7 February, 3 and 5.
———. 1990b. 8 February 1990, 3 and 5.
———. 1990c. 19 May, 14.
———. 1990d. 15 June, 11.
———. 1990e. 16 June, 12.
———. 1990f. 17 June, 14.
———. 1990g. 20 June, 11 and 13.
———. 1990h. 21 June, 1 and 7.
———. 1990i. 22 June, 15 and 24.
———. 1990j. 27 June, 7 and 15.
———. 1990k 2 July, 15.
———. 1990l. 1 October, 16.
———. 1990m. 18 October, 9.
———. 1990n. 4 December, 11.
———. 1990o. 9 December, 13 and 17.
———. 1990p. 15 December, 1.
———. 1992a. 24 July, 7.
———. 1992b. 28 July, 7.
———. 1994. 1 October, 16.
Manila Standard. 1990a. Congressional leaders push price controls. 26 September 1990a, 2.
———. 1990b. Government studies rollback of prices. 26 September, 2.
———. 1990c. Cory Agrees on Wage Hike. 29 September, 2.

———. 1990d. Palace pushes affluence taxes. 2 November, 2.
———. 1990e. Aquino yields to Cabinet outcry over budget slash. 3 November, 2.
———. 1990f. Macaraig ouster as chief energy sought. 29 November 1990, 2.
———. 1990g. Welga aftermath: Reds call of ceasefire. 25 September, 2.
Manila Today. 1985. 3 January, 12–13.
Mansfield, Edward D, Helen V. Milner, and Jon C. Pevehouse. 2007. Vetoing Cooperation: The Impact of Veto Players on Preferential Trading Arrangements. *British Journal of Political Science* 37(3): 403–432.
March James G. and Johan P. Olson. 1984. The New Institutionalism: Organization Factors in Political Life. *American Political Science Review* 78: 734–749.
Martin, Lisa L. 2006. Distribution, Information, and Delegation to International Organizations: The Case of IMF Conditionality. In *Agency in International Organizations,* edited by Darren G. Hawkins, David A. Lake, Daniel L. Nielson, and Michael J. Tierney. Cambridge, UK: Cambridge University Press, 140–164.
Mayer, Frederick. 1992. Managing Domestic Differences in International Negotiations: the Strategic Use of Internal Side-payment. *International Organization* (Autumn) 46(4): 793–818.
McCoy, Alfred. 1999. *Closer than Brothers: Manhood at the Philippine Military Academy*. New Haven, Connecticut: Yale University Press.
Menardo, Amelia A. 2004. Report to APEC High-level Conference on Structural Reform, 8–9 September, Tokyo, Japan.
Milner, Helen. 1992. International Theories of Cooperation among Nations: Strengths and Weaknesses. *World Politics* 44(April): 466–496.
Milner, Helen V. and Peter B. Rosendorff. 1997. Domestic Politics and International Trade Negotiations: Elections and Divided Government as Constraints on Trade Liberalization. *Journal of Conflict Resolution* 41(1): 117–143.
Montes, Manuel F. 1987. *Macroeconomic Adjustment in the Philippines, 1983–85*. Working Paper Series No. 87075. Manila, Philippines: Philippine Institute for Development Studies.
———. 1991. *The Political Economy of Fiscal Policy in the Philippines*. Manila, Philippines: Philippine Institute for Development Studies.
———. 1992. *Economic Reform as Self-Flagellation: Testing the Limits of State-led Trade Reform in the Philippines, 1990–91*. Working paper, 10 May.
Montojo, Kenneth. 1999. "The Political Economy of Philippine Petroleum Deregulation." *Crossroads: An Interdisciplinary Journal of Southeast Asian Studies* 13(1): 65–104.
Moravcsik, Andrew. 1993. Introduction: Integrating International and Domestic Theories of International Bargaining. In *Double-Edged Diplomacy: International Bargaining and Domestic Politics,* edited by Peter B. Evans, Harold K. Jacobson, and Robert D. Putnam. Berkeley, California: University of California Press: 3–42.
Mosely, Paul. 1991. The Philippines. In *Aid and Power: The World Bank and Policy-based Lending: Case Studies*, vol. 2, edited by Paul Mosely, Jane Harrigan and John Toye. London, United Kingdom: Routledge: Chapter 12.

Mydans, Seth. 1987. Aquino, Under Pressure, Removes Her Closest Advisor. *New York Times*. 18 September. Accessed on 30 October 2017. http://www.nytimes.com/1987/09/18/world/aquino-under-pressure-removes-her-closest-adviser.html

National Census and Statistics Office, division of National Economic and Development Authority. 1985.

National Economic Development Authority. 1985. 1984 Economic Performance. *Philippine Economic Indicators*.

———. 1986. 1985 Economic Performance. *Philippine Economic Indicators*.

———. 1989. Philippine Development Report 1988.

———. 1990d. *Updates on the Medium-Term Philippine Development Plan: 1990–1992*. Manila: National Economic and Development Authority.

National Statistical Coordination Board. 1987.

National Statistics Office. 1984. *Philippine Yearbook*. Manila: National Statistics Office.

Ndikumana, Léonce, James K. Boyce, and Ameth Saloum Ndiaye. 2015. Capital Flight: Measurement and Drivers. In *Capital Flight from Africa: Causes, Effects and Policy Issues,* edited by S. Ibi. Ajayi and Lèonce Ndikumana. Oxford, United Kingdom: Oxford University Press, 15–54.

Ng, Willie. 1994. OPSF–the cause of it all. *Manila Bulletin*, 5 February.

North, Douglass C. 1981. *Structure and Change in Economic History*. New York, New York: W. W. Norton and Co.

———. 1990. *Institutions, Institutional Change and Economic Performance*. New York, New York: Cambridge University Press.

Odell, John S. 2000. *Negotiating the World Economy*. Ithaca, New York: Cornell University Press.

Odell, John and Barry Eichengreen. 2000. Changing Domestic Institutions and Ratifying Regime Arrangements. In *Negotiating the World Economy*, edited by John S. Odell. Ithaca: New York, 150–182.

Office of Energy Affairs. 1989. Price Build-Up Effective December 1.

———. 1990. Effective Price Build-Up Effective 22 September.

Official Handbook of the Philippines and Catalogue of the Philippine Exhibit. 1903. pt. I. Manila, Philippines: Bureau of Public Printing.

Ongpin, Roberto. 1986. The Binondo Central Bank. *Business Day*, 25 April, 6.

———. 2011. Interview with Pia Honteveros. 20 September 2011. ANC Strictly Politics. Accessed on 10 August 2016. https://www.youtube.com/watch?v=81HG4Cc10ME

———. 2013. Interview with del Mundo, Fernando. Special Report: Roberto V. Ongpin: Snake Pit beckons, Inquirer.net. http://newsinfo.inquirer.net/363947/roberto-v-ongpin-snake-pit-beckons. Accessed on 10 August 2016.

Oppermann, Kai. 2008. Salience and Sanctions: A Principal-Agent Analysis of Domestic Win-Sets in Two-Level Games—The Case of British European Policy Under The Blair Government. *Cambridge Review of International Affairs* (June) 23(2), 179–197.

Orejana, Rowena M. 1990. Inequity in oil pricing. *Manila Chronicle*, 11 December, 11.

Owen, Norman G. 1971. Philippine Economic Development and American Policy: A Reappraisal. In *Compadre Colonialism: Studies on the Philippines under American Rule*, edited by Owen. no. 3. Michigan: Michigan Papers on South and Southeast Asia, 49–64.

Oye, Kennth. 1979. The Domain of Choice: International Constraints and Carter Administration Foreign Policy. In *Eagle Entangled: U.S. Foreign Policy in a Complex World*, edited by Oye, Donald Rothchild, and Robert J. Lieber. New York, New York: Longman.

Paarlberg, Robert. 1997. Agricultural Policy Reform and the Uruguay Round: Synergistic Linkage in a Two-Level Game? *International Organization* (Summer) (51)3, 413–444.

Paderanga, Antonio R. W. and Cayetano W. Paderanga. 1988. The Oil Industry in the Philippines. *Philippine Economic Journal* (65)27, 89–110.

Payer, Cheryl. 1974. *The Debt Trap: The IMF and the Third World*. New York, New York: Monthly Review Press.

Philippine Daily Inquirer. 1985a. 25 December 1985, 7.

———. 1985b. 26 December 1985, 7.

———. 1986a. 2 March, 1.

———. 1986b. 6 May, 1.

———. 1986c. 10 May, 1.

———. 1992. 6 August, 25.

———. 1993a. 11 May, 25.

———. 1993b. 20 March, 25.

———. 1993c. 24 April, 25.

———. 1993d. 26 April, 25.

———. 1993e. 28 April, 25.

———. 1993f. 11 May, 25.

———. 1993g. 10 December, 1.

———. 1994a. 12 January, 1.

———. 1994b. 16 March, 1.

———. 1994c. 26 June, 1 and 6.

Philippine Free Press. 1988. 3 September, 1 and 5.

Philippine Star, 23 June 1991, 4.

Polak, Jacques J. 1997. The IMF Monetary Model at Forty. International Monetary Fund. WP/97/49.

Powell, Robert. 2015. Nuclear Brinkmanship, Limited War, and Military Power. *International Organization* 69(3): 589–626.

Presidential Management Staff Office of the President Manila, Philippines June 1992. The Aquino Management of the Presidency. Foundation for Policy Making. http://www.coryaquino.ph/coryaquino/assets/images/FoundationforPolicyMaking.pdf

Putnam, Robert D. 1988. Diplomacy and Domestic Politics: the Logic of Two-Level Games. *International Organization* (Summer) (42): 427–460.

Raffia, Howard. 1982. *The Art and Science of Negotiation*. Cambridge, Massachusetts: Harvard University Press.

Ragin, Charles C. 2014. *The Comparative Method: Moving Beyond Qualitative and Quantitative Strategies*. Oakland, California: University of California Press.

Ranis, Gustav. 1990. Contrast in the Political Economy of Development Policy Change. *Manufacturing Miracles: Paths of Industrialization in Latin America and East Asia*, edited by in Gary Gereffi and Donald L. Wyman. Princeton, New Jersey: Princeton University Press.

Reid, Robert H. and Eileen Guerrero. 1995. *Corazon Aquino and the Bushfire Revolution*. Baton Rouge, Louisiana: Louisiana State University Press.

Remolona, Eli M. Majar Mangahas, and Filologo Pante, Jr. 1985. *Studies on International Monetary and Financial Issues for the Developing Countries: Foreign Debt, Balance of Payments and the Economic Crisis of the Philippines in 1983–1984, Report to the Group of Twenty-Four*. New York, New York: UNDP/UNCTAD, Project INT/84/021. 16 July.

Republic of the Philippines. 1983. Executive Order No. 946, Amending Executive Order No. 918, Series of 1983.

———. 1983a. Ministry of Energy. Status of the Consumer Price Equalization Fund (As of June 30, 1983)—A Report to the President. Makati, Metro Manila, Philippines: Ministry of Energy.

———. 1984a. Memorandum of Economic Policy, as published in the *Bulletin Today*, 3 January.

———. 1984b. Presidential Decree No. 1956: Imposing an Ad Valorem Tax on Certain Manufactured Oils and Other Fuels, Bunker Fuel Oil and Diesel Fuel Oil; Revising the Rates of Specific Tax Thereon; Abolishing the Oil Industry Special Fund; and for Other Purposes.

———. 1986c. Board of Energy. Price Build-up Effective 24 March from IBON Databank.

———. 1986d. Executive Order No. 21, Amending Section 128 of the National Internal Revenue Code as Amended by Revision the Specific Tax Rates on Petroleum Products, 19 June.

———. 1987. Updated Philippine Letter of Intent, March 10 from IBON Databank.

———. 1988a. Administrative Order 91, Creating a Cabinet Committee on Matters Affecting Oil, 12 September.

———. 1988b. Republic Act No. 6647, An Act Further Restructuring the Import Duty Rates and Classification of Certain Articles Under Section 104 of the Tariff and Customs Code of 1978 (PD 1464), as Amended, 29 January.

———. 1989a. Republic Act No. 6727. 25 July. Found on the Republic of the Philippines Official Gazette website, http://www.gov.ph/1989/06/09/republic-act-no-6727/. Accessed on 11 July 2016.

———. 1989b. Memorandum on Economic Policy, 1989–92, reprinted in *Manila Bulletin*, 15–6 March, 3–5.

———. 1989c. National Economic and Development Agency, Philippine Development Report 1988. Manila: National Economic and Development Agency.

———. 1990a. National Emergency Memorandum Order No. 8, Modifying the Rates of Imported Duties on Certain Articles under Section 104 of Presidential Decree

No. 1464, Otherwise Known as the Tariff and Customs Code of 1978, as amended. 23 January.
———. 1990b. Supplementary Memorandum of Economic Policy during the Period 1989–1992, as reprinted in *Manila Chronicle*, 7–8 February 1990, 3, 5.
———. 1990c. Executive Order No. 422, s. 1990, Activating and Reorganizing the Energy Operations Board as the Administrative Machinery for the Efficient and Equitable Allocation and Distribution of Energy Under Certain Conditions., September 1990. Found on the Republic of the Philippines Official Gazette website, http://www.officialgazette.gov.ph/1990/09/11/executive-order-no-422-s-1990/. Accessed 11 July 2016.
———. 1990d. Republic Act 6965, An Act Revising the Form of Taxation on Petroleum Products from Ad Valorem to Specific, Amending for the purpose Section 145 of the National Internal Revenue Code as Amended by Republic Act Numbered Sixty-Seven Hundred Sixty Seven. 19 September.
———. 1990e. Executive Order No. 438, Imposing an Additional Duty of Five Percent (5percent) Ad Valorem on All Imported Articles Subject to Certain Exceptions and Conditions, 27 November.
———. 1991a. Executive Order No. 470, Modifying the Nomenclature and Rates of Import Duty of Certain Imported Articles Under Section 104 of the Tariff and Customs Code of 1978 (Presidential Decree No. 1464), as amended, 20 July.
———. 1991b. Executive Order No. 475, Reducing the Rate of Additional Duty Prescribed in Executive Order No. 443, Series of 1991, to Five percent (5percent) Ad Valorem, 15 August.
———. 1991c. Executive Order No. 478, Imposing Special Duties on Crude Oil and Oil Products Under Section 104 of the Tariff and Customs Code of the Philippines, as amended, 23 August.
———. 1992a. Executive Order No. 517, Lifting the Additional Duty of Five (5percent) Percent Ad Valorem Imposed Through Executive Order No. 443, As Amended, On all Imported Articles Subject to Certain Exceptions and Conditions, 30 April.
———. 1992b. Republic Act No. 7607, An Act Providing a Magna Carta of Small Farmers, 4 June.
———.1992c. Executive Order No. 1. Reducing the Rates of Import Duty on Electric Generating Sets Under Presidential Decree No. 1464, Otherwise known as the Tariff and Customs Code of 1978, As Amended, 30 June.
———. 1992d. Executive Order No. 2. Extending the Effectively of the Zero Rate of Import Duty on Cement and Cement Clinker Under Section 104 of Presidential Decree No. 1464, Otherwise known as the Tariff and Customs Code of 1978, As Provided For Under Executive Order No. 387, 1 July.
———. 1992e. Presidential Decree No. 1464, Otherwise known as the Tariff and Customs Code of 1978, As Provided for Under Executive Order No. 387, 1 July.
———. 1992f. Executive Order No. 8, Restructuring the Rates of Import Duties and Amending the Classification of Certain Articles under Section 104 of the Tariff and Customs Code of 1978, as Amended, 24 July.

———. 1993. Executive Order No. 115: Increasing the Special Duties on Crude Oil and Oil Products under Section 104 of the Tariff and Customs Code of the Philippines, as amended, 24 July.

———. 1994. Proposed Downstream Oil Industry Deregulation Action Plan. Draft Report, Task Force on Oil Industry Deregulation, 8 November.

———. 1995. *Medium-Term Philippine Development Plan, 1993–1998*. Pasig, Metro Manila, Philippines: National Economic and Development Authority.

Rhodes, R. A. W., Sarah A. Binder, and Bert A. Rockman. 2008. In *The Oxford Handbook of Political Institutions* edited by R. A. W. Rhodes, Sarah A. Binder, and Bert A. Rockman Oxford: Oxford University Press, xii–xvii.

Riedinger, Jeffrey. 1994. The Philippines in 1993: Halting Steps toward Liberalization. *Asian Survey* (February) 34(2), 139–146.

Romulo, Senator Alberto O. 1989. Privilege Speech. The Letter of Intent and the People. 13 February.

Rowley, Anthony. 1990. Passing the Hat. *Far Eastern Economic Review*, 15 November, 85.

Sacerdoti, Guy. 1984. Marching to the Beat of an Opposition Drum. *Far Eastern Economic Review*, 24 May, 84–86.

Sau, Ranjit. 1992. Financial Programing for Stabilization: Some Notes on IMF Model. *Economic and Political Weekly* (27)10/11: 531–534.

Schelling, Thomas C. 1966. *The Strategy of Conflict*. Cambridge, United Kingdom: Oxford University Press.

Schimmelfenning, Frank. 2003. *The EU, NATO and the Integration of Europe: Rules and Rhetoric*. Cambridge, United Kingdom: Cambridge University Press.

Schirmer, Daniel B. and Stephen Rosskamm Shalom, eds. 1987. *The Philippines Reader: A History of Colonialism, Neocolonialism, Dictatorship, and Resistance*. Boston, Massachusetts: South End Press.

Schoppa, Leonard J. 1993. Two level games and bargaining outcomes: Why Gaiatsu Succeeds in Japan in Some Cases but Not Others. *International Organization* (Summer), 47(3), 353–386.

Schotter, Andrew. 1981. *The Economic Theory of Social Institutions*. New York, New York: The Cambridge University Press.

Sciat, Gerardo P. Cesar. 2014. *Virata Life and Times: Through Four Decades of Philippine Economic History*. Quezon City, Philippines: The University of the Philippines Press.

Sebenius, James K. 1992. Challenging Conventional Explanations of International Cooperation: Negotiation Analysis and the Case of Epistemic Communities. *International Organization* (Winter) (46)1, 323–365.

———. 1996. Sequencing to Build Coalitions: With Whom Should I talk First? In *Wise Choices: Decisions, Games, and Negotiations*, edited by Richard J. Zeekhouser, Ralph L. Keeney, and James K. Sebenius. Boston, Massachusetts: Harvard Business School, 324–348.

———. 2012. *Level II Negotiations: Helping the Other side Meet Its "Behind the Table" Challenges*. Working Paper 13–004. Harvard Business School. 8 July. Accessible at http://hbswk.hbs.edu/item/level-ii-negotiations-helping-the-other-side-meet-its-behind-the-table-challenges

Sebenius, James K. and Michael K. Singh. 2012. Is A Nuclear Deal With Iran Possible? An Analytical Framework for the Iran Nuclear Negotiations. *International Security* 37(3): 52–91.

Shepard, Geoffrey and Aorian Album. 1991. The Philippines. In *Liberalizing Foreign Trade*, edited by Demetris Papageorgiou, Michael Michaely, and Armeane M. Choksi. Cambridge, Massachusetts: Basil Blackwell, Inc.

Shepsle, Kenneth A. 2008. Old Questions and New Answers About Institutions: The Riker Objection Revisited. In *The Oxford Handbook of Political Economy*, edited by Barry R. Weingast and Donald A. Wittman. Oxford, United Kingdom: Oxford University Press, 1031–1049.

———. 2010. *Analyzing Politics: Rationality, Behavior, and Institutions*, 2nd. New York, New York: W. W. Norton and Company.

———. 2017. *Rule Breaking and Political Imagination*. Chicago, Illinois: The University of Chicago Press.

Simon, Herbert A. 1979. Rational Decision Making in Business Organizations. *The American Economic Review* (69): 493–513.

———. 1956. Rational Choice and the Structure of the Environment. *Psychological Review* (63): 129–138.

Singer, J. David. 1961. The Level of Analysis Problems in International Relations. *World Politics* (14): 77–92.

Smith, Jennifer L. 1989. Debt and Democracy in the Philippines. *Multinational Monitor*. June 1989 v. 10. No. 6.

Stallings, Barbara. 1992. International Influence on Economic Policy: Debt, Stabilization, and Structural Reform. In *The Politics of Economic Adjustment: International: Constraints, Distributive Conflicts, and the State,* edited by Stephan Haggard and Robert R. Kaufman. Princeton, New Jersey: Princeton University Press.

Stauffer, Robert B. 1975. *The Philippine Congress: Causes of Structural Change*. Sage Research Papers on the Social Sciences, Comparative Legislative Series, No. 90–024. London, United Kingdom: Sage Publications.

Steinberg, David Joel. 1994. *The Philippines: A Singular and a Plural Place*. Boulder, Colorado: Westview Press.

Stone, Randall W. 2008. The Scope of IMF Conditionality. *International Organization*. 62(4): 589–620.

Tadem, Teresa S. Encarnacion. 2012. Virata: The Trials and Tribulations of a 'Chief Technocrat.' *Philippine Political Science Journal* 33(1): 23–37.

———. 2014. Philippine Technocracy and the Politics of Economic Decision-Making: A Comparison of the Martial Law and Post-Martial Law Periods. *Southeast Asian Studies* 3(2): 345–381.

———. 2016. The Rise and Fall of Virata's Network: Technocracy and the Politics of Economic Decision Making in the Philippines. *Southeast Asian Studies* 5(1): 35–71.

Tan, Norma. 1979. The Structure of Protection and Resources Flows in the Philippines. In *Industrial Promotion Policies in the Philippines*, edited by Romeo Bautista, John Power and Associates. Manila, Philippines: Philippine Institute for Development Studies.

Thacker, Strom C. 1999. The High Politics of IMF Lending. *World Politics* 52: 38–75.

Thomas, Alun and Uma Ramakrishnan. 2006. *The Incidence and Effectiveness of Prior Actions in IMF-supported Programs*. Working Paper. WP/06/213. Washington, D.C.: International Monetary Fund.

Thompson, Mark R. 1995. *The Anti-Marcos Struggle: Personalistic Rule and Democratic Transition in the Philippines*. New Haven, Connecticut: Yale University Press.

Thompson, Mark and Gregory W. Slaton. 1985. An Essay on Credit Arrangements between the IMF and the Republic of the Philippines: 1970–1983. *Philippine Review of Economics and Business* 22(1&2) (March–June), 59–80.

Thompson, W. Scott. 1992. *The Philippines in Crisis: Development and Security in the Aquino Era 1986–1992*. New York, New York: St. Martin's Press.

———. 2011. *Trustee of the Nation: The biography of Fidel V. Ramos*. Manila, Philippines: Anvil Publishing, Inc.

Tiglao, Rigoberto. 1988. IMF Talks Should be Different by Now. *Manila Chronicle*, 3 November.

———. 1989a. Manila tests its credit. *Far Eastern Economic Review*, 23 March, 74–75.

———. 1989b. IMF sets the pace, *Far Eastern Economic Review*, 8 June, 112.

———. 1990a. The Enemy is Us. *Far Eastern Economic Review*, 8 November, 52.

———. 1990b. Manila on the Brink. *Far Eastern Economic Review*, 8 November, 49.

———. 1990c. Frying Pan to Fire. *Far Eastern Economic Review*, 15 November, 84.

———. 1991a. A Question of Duty. *Far Eastern Economic Review*, 7 February, 57–58.

———. 1991b. Manna for Manila. *Far Eastern Economic Review*, 7 March, 53.

———. 1992. Free at Last. *Far Eastern Economic Review*, 27 August, 51.

———. 1994. Paralyzed by Politics. *Far Eastern Economic Review*, 12 May, 57.

———. 2016. Virata and the technocrats ran the martial law economy. *Manila Times*. March 3. http://www.manilatimes.net/virata-and-technocrats-run-the-martial-law-economy/248318/. Accessed on 15 November 2017.

Tiglao, Rigoberto and Rodney Tasker. 1992. Fidel's Challenge. *Far Eastern Economic Review*, 18 June, 17.

Timberman, David G. 1991a. The Philippines in 1990. *Asian Survey*31(2), A Survey of Asia in 1990: Part II (Feb., 1991), 153–163.

———. 1991b. *A Changeless Land: Continuity and Change in Philippine Politics*. Singapore: Institute of Southeast Asian Studies and Bookmark, Inc.

Times Journal. 1985. 15 October, l.

Truman, Edwin M. 2006. *Reforming the IMF for the 21st Century*. Washington, D.C.: Institute for International Economics.

US Foreign Assistance Reference Guide. 2005. US Department of State Publication and U.S. Agency for International Development. Accessed on the WWW on 10 November 2017 at http://pdf.usaid.gov/pdf_docs/Pnadc240.pdf

Van Evera, Stephen. 1997. *Guide to Methods for Students of Political Science*. Ithaca, New York: Columbia University Press.

Villegas, Bernardo M. 1983. *Strategies for Crisis: The Story of the Philippine National Oil Company*. Pasig City, Philippines: Center for Research and Communication.

Virata, Cesar. 1991. Transcript of Interview with Cesar Virata. 1991. World Bank History Project. Brookings Institution. November 29, Manila. Interviewed by Richard Webb and Devesh Kapur. Accessed on the WWW on 4 August 2016. http://oralhistory.worldbank.org/person/virata-cesar.

Virata, Cesar. and Associates. 1993. *Progress on Policy Reforms under Ramos, Philippines*. Manila, Philippines: US Agency for International Development.

Vreeland, James Raymond. 2003. *The IMF and Economic Development*. Cambridge, Massachusetts: Cambridge University Press.

———. 2007. *The International Monetary Fund: Politics of Conditional Lending*. New York, New York: Routledge Global Institutions.

Walker, Penelope. 1992. *Political Crisis and Debt Negotiations: The Case of the Philippines, 1983–1986*. Pew Case Studies in International Affairs. Case 133. Washington, D.C.: Institute for the Study of Diplomacy, School of Foreign Service, Georgetown University.

Wallander, Celeste. 1999. *Mortal Friends, Best Enemies: German-Russian Cooperation after the Cold War*. Ithaca. NY: Cornell University Press.

Walton, Richard E. and Robert B. McKersie. 1965. *A Behavioral Theory of Labor Negotiations: An Analysis of a Social Interaction System*. New York, New York: McGraw-Hill Book Company.

Waltz, Kenneth N. 1959. *Man, the State, and War: A Theoretical Analysis*. New York, New York: Columbia University Press.

———. 1979. *Theory of International Politics*. New York, New York: Random House.

West, Lois A. 1997. *Militant Labor in the Philippines*. Philadelphia, Pennsylvania: Temple University Press.

Williamson, John, ed. 1990. *Latin American Adjustment: How Much Has Happened?* Washington, D.C.: Institute for International Economics.

Woolcock, Stephen. 2005. European Union trade policy: Domestic institutions and systemic factors. In *The Politics of International Trade in the Twenty-First Century: Actors, Issues and Regional Dynamics,* edited by D. Kelly and W. Grant. Basingstoke, United Kingdom: Palgrave Macmillan, 234–251.

World Bank. 1985. *World Debt Tables: External Debt of Developing Countries, 1984-85 edition*. Washington, D.C., 140.

———. 1986a. *Philippines: A Framework for Economic Recovery*. Washington, D.C.: The World Bank. A World Bank Country Study, 1986.

———. 1986b. Memorandum, from the lbon Data Bank, 25 May.

———. 1989. *Adjustment Lending: An Evaluation of Ten Years Experience*. Washington, D.C.: World Bank.

———. 2017. Military Expenditures (percent of GDP), Philippines. Accessed on 17 November 2017 on https://data.worldbank.org/indicator/MS.MIL.XPND.GD.ZS?view=chart.

World Bank International Finance Corporation. 1986. Office Memorandum. Visit to the Philippines for Operational Discussions. 15 May.

Wurfel, David. 1991. *Filipino Politics: Development and Decay*. Quezon City, Philippines: Ateneo de Manila University Press.

Yap, Josef T. 1991. *The Philippines Recent Performance, Prospects for 1991–92, and Policy and Development Issues*. Working Paper Series No. 91–101. Manila, Philippines: Philippine Institute for Development Studies, January.

Yin, Robert K. 2008. *Case Study Research: Design and Methods*. Thousand Oaks, California: Sage Publications.

Young, Alasdair. 2002. *Extending European Cooperation: The European Union and the 'New' International Trade Agenda*. Manchester, United Kingdom: Manchester University Press.

Yuson, Alfred A. 2005. *The Life and Times of Jose B. Fernandez, Jr*. Quezon City, Philippines: Ateneo de Manila University.

Zuleta, Lito. 1990. Big Mac Shut Off from Circle of Power. *Manila Chronicle*, 9 December.

Subject Index

Author Index

About the Author

Kenneth Faulve-Montojo is a lecturer at Santa Clara University. He specializes in international political economy and comparative politics. He writes on Philippines, monetary, and economic policies.

www.ingramcontent.com/pod-product-compliance
Ingram Content Group UK Ltd.
Pitfield, Milton Keynes, MK11 3LW, UK
UKHW041423091026
15587UKWH00023B/271

9 781498 532693